Springer Tracts in Modern Physics
Volume 225

Springer Tracts in Modern Physics

Springer Tracts in Modern Physics provides comprehensive and critical reviews of topics of current interest in physics. The following fields are emphasized: elementary particle physics, solid-state physics, complex systems, and fundamental astrophysics.
Suitable reviews of other fields can also be accepted. The editors encourage prospective authors to correspond with them in advance of submitting an article. For reviews of topics belonging to the above mentioned fields, they should address the responsible editor, otherwise the managing editor.
See also springer.com

HP. Winter J. Burgdörfer

Slow Heavy-Particle Induced Electron Emission from Solid Surfaces

With 119 Figures

Springer

Professor Dr. Hannspeter Winter[†]
Institut für Allgemeine Physik
Vienna University of Technology
(† Nov. 8, 2006)

Professor Dr. Joachim Burgdörfer
Institute for Theoretical Physics
Vienna University of Technology
Wiedner Hauptstr. 8-10
1040 Vienna, Austria
Email: burg@concord.itp.tuwien.ac.at

Physics and Astronomy Classification Scheme (PACS):
34.50.Bw, 34.50.Dy, 78.70.-g, 79.20.Ap, 79.20.Rf

ISSN print edition: 0081-3869
ISSN electronic edition: 1615-0430
ISBN 978-3-642-08971-8
e-ISBN 978-3-540-70789-9

Springer is a part of Springer Science+Business Media
springer.com

Cover production: WMXDesign GmbH, Heidelberg

Preface

In the past few decades, great progress has been made in the field of dynamical interactions of photons and charged particles with surfaces. Developments in this field have largely been driven by technological advances. Foremost, the preparation and characterization of atomically flat surfaces has opened up opportunities to precisely and reproducibly investigate surface structure and dynamics in unprecedented detail.

An equally important advance represents the availability of new sources for projectiles interacting with and probing surfaces. Third-generation synchrotron sources provide high-brilliance photon flux over a broad range of energies extending from the ultraviolet to the hard X-ray regime. In the context of the primary topics of this book of even greater importance is the development of ion sources that provide slow, very highly charged ions with significant intensities covering charge states up to bare uranium. Currently pursued development projects such as HITRAP (highly charged trapped ions) at GSI Darmstadt hold the promise to deliver within the next few years cooled highly charged ions in the sub-eV kinetic energy regime.

These advances in experimental techniques have led to an increased focus on technological applications. Prominent examples include the controlled nanostructuring of surfaces for novel functionalities on the sub-micron scale and the search of surface materials suited for divertor surfaces in future thermonuclear fusion reactors, in particular to control and reduce physical and chemical sputtering. Concurrently, theory has made great strides. The understanding and solution of the many-body problem beyond the ground state, which is at the core of the description of dynamical surface processes, has considerably advanced. The development and application of time-dependent density functional theory and, more generally, of techniques of computational physics are key ingredients to the progress made over the last decade.

It is this backdrop that suggests the timeliness of a review of recent development of heavy-particle induced processes at surfaces with the emphasis on electron emission. This book is a collection of seven chapters written by experts in the various subfields. All contributors were asked to highlight the interconnections to other subfields. The chapters are not intended to serve as comprehensive reviews but rather as introductions to the most important developments of that subfield. This book is written for graduate students and researchers who wish to familiarize themselves with the topic of electron emission from surfaces.

Chapter one gives an introductory overview over theoretical concepts and methods. The scope of methods ranges from purely classical and semiclassical to fully quantum dynamical techniques. This chapter emphasizes the close connection to well-developed methods in the field of atomic collisions in the gas phase. The latter frequently serve as guide post for the much more complex problem of particle-surface collisions.

Heavy-particle induced electron emission shares many properties with the emission processes caused by other ionizing projectiles, in particular photons and electrons. Chapter 2 is intended to establish this connection and to highlight similarities with and differences to heavy-particle induced emission.

With the advent of slow highly charged ions (HCI), a novel electron emission channel, electron emission due the electronic potential energy of the projectile carried into the collision, has taken center stage. This process is conceptually different from kinetic emission which is driven by the kinetic energy of the impinging projectile and closely resembles binary collisions. Chapters 3 and 4 compare and contrast potential-energy and kinetic-energy driven electron emission.

The spin polarization of electrons in two-dimensional structures such as surfaces and quantum wells plays an increasingly important role in present and future technological applications ("spintronics"). Ion-induced electron emission from surfaces can provide surface-specific, selective information on short-range and long-range spin correlations, moreover, the spin-dependent response of many-electron systems to external perturbations, both spin polarized and unpolarized can be tested. Chapter 5 discusses recent advances in this field.

Electron emission from surfaces is traditionally considered a single particle-hole excitation with the particle, the electron, being in a positive-energy final state. Recently, the role of collective excitations as precursors and mediating processes has been appreciated and is the subject of Chap. 6 which discusses the role of plasmons.

Thin insulating films are known to be excellent electron emitters. They are applied in devices where the performance critically depends on the electron yield. The present understanding of slow ion-induced electron emission from such insulating films is discussed in Chap. 7.

The editors would like to thank all contributors to this volume who were willing to spend a considerable amount of time and effort on their chapters that are didactically written with the non-expert as potential reader in mind and, yet, appear well suited to convey the excitement and progress of their subfields. We express our hope that this volume, while reflecting the expertise of a large number of contributors, nevertheless approaches the coherence of presentation expected for a monograph.

Vienna
November 2006

Hannspeter Winter
Joachim Burgdörfer

Note Added to Preface

Just weeks after we wrote together the preface above, my co-editor and close colleague Hannspeter Winter passed away suddenly and unexpectedly on Nov. 8, 2006 as a result of a massive heart attack while pursuing his beloved spare-time activity, jogging in the Vienna Prater park.

Hannspeter was, to his last day, as vigorous, active, and "driving" as ever. As it happened, one of the last emails Hannspeter sent just hours before his untimely death went to the Springer publishing house concerning technical details for further processing of the present book.

While writing their chapters none of the contributors would have guessed what this volume would eventually turn into: a fitting scientific tribute to the lifetime research interest and work of the editor of this volume, Hannspeter Winter, a pioneer of electron emission studies from surfaces and of the physics of plasma-wall interactions.

Vienna
November 12, 2006

Joachim Burgdörfer

Contents

List of Contributors

Maite Alducin
Donostia International Physics
Center DIPC
20018 San Sebastian, Spain
wapalocm@sq.ehu.es

Friedrich Aumayr
Institut für Allgemeine Physik
Vienna University of Technology
1040 Vienna, Austria
aumayr@iap.tuwien.ac.at

Raúl Baragiola
Laboratory for Atomic
and Surface Physics
University of Virginia
Charlottesville, VA 29904, USA
raul@virginia.edu

Joachim Burgdörfer
Institute for Theoretical Physics
Vienna University of Technology
1040 Vienna, Austria
burg@concord.itp.tuwien.ac.at

Ricardo Díez Muiño
Unidad de Física de Materiales
Centro Mixto CSIC-UPV/EHU
20018 San Sebastian, Spain
rdm@sc.ehu.es

Pedro M. Echenique
Departamento de Física
de Materiales
Facultad de Ciencias Químicas
UPV/EHU
20018 San Sebastian, Spain
pedromiguel.echenique@ehu.es

J. Iñaki Juaristi
Departamento de Física
de Materiales
Facultad de Ciencias Químicas
UPV/EHU
20018 San Sebastian, Spain
wapjuoli@sc.ehu.es

Christoph Lemell
Institute for Theoretical Physics
Vienna University of Technology
1040 Vienna, Austria
lemell@concord.itp.tuwien.ac.at

Yury T. Matulevich
Samsung SDI
428-5 Gongse-dong, Giheung-gu
Yongin-si
Gyeonggi-do 446-577, South Korea
yury.matulevich@samsung.com

R. Carmina Monreal
Departamento de Física Teórica de la Materia Condensada
Universidad Autónoma de Madrid
28049 Madrid, Spain
r.c.monreal@uam.es

Max Rösler
Donostia International Physics Center DIPC
20018 San Sebastian, Spain
max.roesler@gmx.de

Wolfgang S.M. Werner
Institut für Allgemeine Physik
Vienna University of Technology
1040 Vienna, Austria
werner@iap.tuwien.ac.at

Hannspeter Winter
Institut für Allgemeine Physik
Vienna University of Technology
1040 Vienna, Austria
(† Nov. 8, 2006)

Helmut Winter
Institut für Physik
Humboldt-Universität zu Berlin
12489 Berlin-Adlershof, Germany
winter@physik.hu-berlin.de

Pedro A. Zeijlmans van Emmichoven
Department of Physics and Astronomy, Surfaces, Interfaces and Devices
Faculty of Science
Utrecht University
3584 CC Utrecht, The Nederlands
p.a.zeijlmans@phys.uu.nl

1

Theoretical Concepts and Methods for Electron Emission from Solid Surfaces

Joachim Burgdörfer and Christoph Lemell

1.1 Introduction

Theoretical descriptions of scattering of heavy particles (i.e. ions or atoms) at surfaces have remained a challenge for a multitude of reasons: such processes involve a large number of coupled degrees of freedom some of which can be treated classically, while others demand a quantum description. At the core is an interacting many-body system far from the ground state where ground-state theories such as density-functional theory (DFT) encounter difficulties and models that couple the electronic system near the ground state to the heavy particle motion (e.g. via a Car-Parrinello approach [1]) become inadequate.

For the experimental observable "electron emission", the topic of this book, additional complications result from the fact that the final state lies in the continuum part of the spectrum. Analytical or numerical determination of continuum states, their representation on either a numerical grid or basis states causes additional difficulties that are, even for much simpler systems such as ion-atom collisions, difficult to overcome. Therefore, a wide variety of simplifications and approximations are needed to render the problem of electron emission accessible to a theoretical description.

The present introductory overview highlights a few recently developed and applied methods and is not intended to represent an exhaustive review. More extensive discussions can be found in [2–5]. The selection of concepts and methods to be discussed is guided by the topics covered by the subsequent, mostly experimental, chapters.

One major dividing line between different models and approximations is provided by the projectile velocity. Slow (fast) collisions refer to projectile velocities v_p, small (large) compared to the characteristic electron velocity, $v_p \leq 1$ a.u. (or energies of 25 keV/amu). The Fermi velocity v_F plays frequently the role as characteristic electronic velocity in the target. Accordingly, slow collisions imply

$$v_p \leq v_F \ . \qquad (1.1)$$

J. Burgdörfer and C. Lemell: *Theoretical Concepts and Methods for Electron Emission from Solid Surfaces,* STMP **225**, 1–38 (2007)
DOI 10.1007/3-540-70789-1_1

Because of the intrinsic anisotropic nature of atom (ion)-surface collisions, the velocity components normal to the surface, $v_{p,\perp}$, and parallel to the surface, $v_{p,\parallel}$, must be separately considered when distinguishing between fast and slow collisions. Closely related to v_p is the distinction between "kinetic" (KE) and "potential" electron emission (PE). While the former implies that the kinetic energy of the projectile drives the emission process, the latter invokes the transfer of the potential energy of the projectile relative to its neutral ground state (e.g., metastable neutral atoms or ions with total ionization energy E_I) to be the agent causing ionization. Roughly speaking, KE is associated with fast collisions while PE occurs for slow collisions. Clearly, such a simplistic mapping cannot capture the subtleties of the interaction process and requires, as discussed below, revisions and amendments. Differences and similarities between kinetic and potential electron emission will be presented in Chaps. 3 and 4.

Another important control parameter is the strength of the interaction, in particular the strength of the interaction potential between the projectile and the electron, V. For $\eta = \langle V \rangle / W < 1$ (W: work function of the electron to be released) perturbation theory is applicable while in the opposite limit a non-perturbative treatment is called for. Frequently, for Coulomb interactions a dynamical rather than the static interaction strength parameter, the Sommerfeld parameter $\eta = Q/v_p$ (Q: charge of the projectile) controls the effective interaction strength. The latter plays a key role for highly charged ions ($Q \gg 1$). Linear response theory used in the context of the dielectric response and dynamical screening of Coulomb fields is a prototypical example of perturbation theory and relies on the assumption of weak perturbations, $\eta \ll 1$. Conversely, time-dependent density functional theory beyond the linear-response regime is a generic example of a non-perturbative treatment.

Theoretical models can be further characterized by the degree to which electronic processes are treated classically. While, on general grounds, there is no compelling reason to invoke the classical limit of quantum mechanics, it turns out that classical approximations are reasonably successful in describing the electronic dynamics in many aspects of surface collisions. The advantage of a classical calculation lies not only in the remarkable ease of its application compared to a full quantum treatment but also in the intuitive insights that can be gained. Clearly, its validity and applicability must be carefully assessed for each application. Often hybrid classical-quantum methods are found to be most suitable.

This chapter emphasizes the close connection to atom collision theory, from which many methods for theoretical description of atom-surface collisions are derived. Atomic units are used throughout the chapter unless indicated otherwise.

1.2 Interaction Potentials

The determination of effective, "dressed" interactions of charged particles and aggregates of charged particles near surfaces represents one important and non-trivial first step in the description of particle-solid interactions. It serves, on one hand, as input for the calculation of collision processes and, on the other hand, depends on the scattering dynamics itself. These potentials depend on both the atomic or ionic projectile as well as the target surface. Clearly, effective interactions strongly differ for metals (conductors) and insulators. Likewise, effective interaction potentials of neutral atoms strongly differ from those of highly charged ions with charge Q. A large variety of approximations are in use some of which are summarized below.

1.2.1 Image Potentials

At large distances from the surface and for small velocities of a charged particle, the classical approximation to the (static) image potential

$$V^{im}(R_z) = -\frac{Q^2}{4R_z} \tag{1.1}$$

is valid. In Eq. (1.1), R_z is the distance from the surface. Rather than taking the topmost atomic layer to be the position of the surface, for metals the so-called "jellium edge" located half a lattice constant outside the topmost atomic layer is often used as reference point when a jellium approximation, i.e. the smearing out of the ionic background, is invoked in the approximate description of metallic surfaces. It turns out, however, that the effective image plane is shifted with respect to the jellium edge. Equation (1.1) has therefore to be corrected for by an offset z_{im} [6]

$$V^{im}(z) = -\frac{Q^2}{4(R_z - z_{im})}, \qquad z_{im} \approx -0.2r_s + 1.25 \,. \tag{1.2}$$

r_s is the Wigner-Seitz radius of the electron gas, the radius of a sphere which on average contains one electron (target electron density $n = 3/4\pi r_s^3$). Equations (1.1) and (1.2) are valid in the regime of *linear* response of the surface to the external charge. For insulators, Eq. (1.2) can be generalized to [7]

$$V^{im}(R_z) = -\frac{Q^2}{4R_z}\frac{\varepsilon(\omega)-1}{\varepsilon(\omega)+1}, \tag{1.3}$$

where $\varepsilon(\omega)$ is the bulk dielectric function of the target while $(\varepsilon-1)/(\varepsilon+1)$ represents the optical limit $q \to 0$ of the surface dielectric function $\varepsilon(\mathbf{q},\omega)$. Ideal conductors and thus metallic surfaces correspond to the limit $\varepsilon(\omega) \gg 1$ for $\omega \ll 1$.

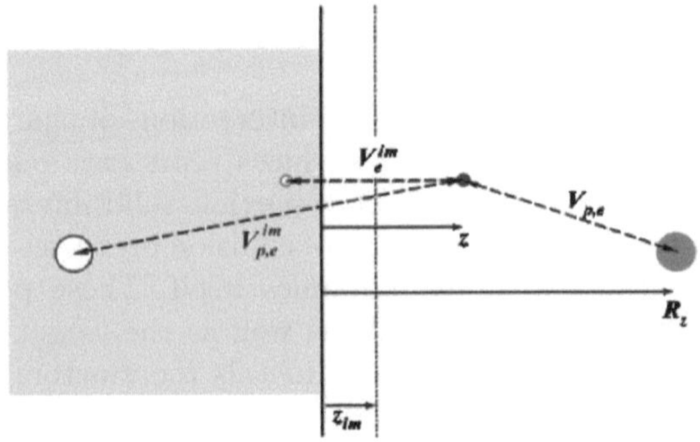

Fig. 1.1. Positions of the "active" electron (distance from surface z), the projectile (distance R_z), and the images (at $z - 2z_{im}$ and $R_z - 2z_{im}$ below the surface). Interaction potentials are indicated by *dashed arrows* (cf. text)

The potential in front of the surface for an electron at position $\mathbf{r} = (x, y, z)$ in presence of a projectile at position $\mathbf{R} = (0, 0, R_z)$ is determined by (see Fig. 1.1)

$$V_e(\mathbf{r}, \mathbf{R}) = V_e^{im}(\mathbf{r}) + V_{p,e}(\mathbf{r}, \mathbf{R}) + V_{p,e}^{im}(\mathbf{r}, \mathbf{R}) , \tag{1.4}$$

the sum of the interactions of the electron with the Coulomb field of the projectile $V_{p,e}$, given for hydrogenic projectiles by

$$V_{p,e}(\mathbf{r}, \mathbf{R}) = -\frac{Q}{|\mathbf{r} - \mathbf{R}|} , \tag{1.5}$$

the fields of its own and the projectiles image potentials, V_e^{im} and $V_{p,e}^{im}$, respectively. The latter two potentials are markedly different for metals and for insulators [8–10]. $V_{p,e}$ for non-hydrogenic ions can be approximated by single-particle core potentials [11]. Figure 1.2 displays the potential landscape for a multiply charged ion in front of a jellium surface with $r_s = 3$ a.u.

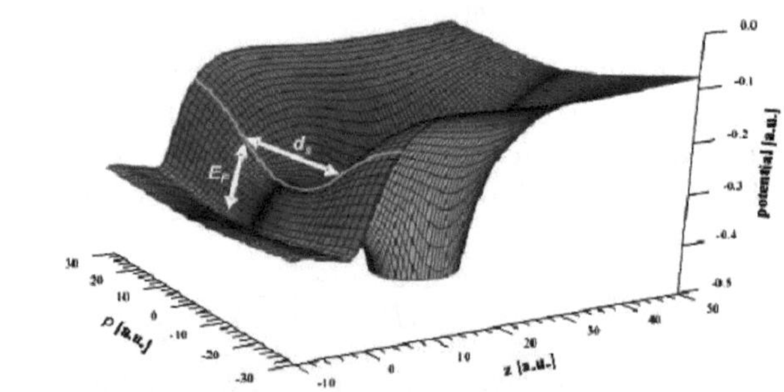

Fig. 1.2. Potential landscape for a slow ($v_p = 0.05$ a.u.) multiply charged ion ($Q = 3$) in front of a jellium surface ($r_s = 3$ a.u.). Electron capture proceeds predominantly over the barrier separating ion and target. The width and the height of the barrier at the Fermi energy E_F are indicated by *arrows*

At this point the question of the validity of this quasistatic approximation to the interaction potentials arises. Target electrons can be assumed to react instantaneously to the external perturbation as long as the projectile velocity is considerably slower than the Fermi velocity of the electron gas. This condition is met in most experiments on potential emission. The validity of Eqs. (1.1) and (1.3), furthermore, requires sufficiently large distances from the surface such that quantum corrections to screening can be neglected, i.e.,

$$z > \lambda_D = v_F/\omega_s \ , \tag{1.6}$$

where ω_s is the surface plasmon frequency [7].

1.2.2 Dielectric Response

For small distances or large velocities the response of the target electron gas described by the dielectric function ε beyond the asymptotic classical limit Eq. (1.1) has to be taken into account. Its inverse, ε^{-1}, describes the modification of the field in the presence of the medium,

$$\frac{1}{\varepsilon(\mathbf{q},\omega)} = \frac{\rho_{ext}(\mathbf{q},\omega) + \rho_{ind}(\mathbf{q},\omega)}{\rho_{ext}(\mathbf{q},\omega)} \ . \tag{1.7}$$

ρ_{ext} and ρ_{ind} are the external and induced charges, respectively. ε^{-1} is further related to the susceptibility χ of the system via [12]

$$\frac{1}{\varepsilon(\mathbf{q},\omega)} - 1 = \frac{4\pi}{q^2}\chi(\mathbf{q},\omega) \ . \tag{1.8}$$

Equation (1.8) represents the dynamic, i.e. frequency ω and wave number $\mathbf{q}$ dependent polarizability of the medium. This expression as functions of $\mathbf{q}$ and ω is valid for systems which are translationally invariant with respect to both position $\mathbf{r}$ and time t. The charge density fluctuation is space- and time-dependent with Fourier components $\rho(\mathbf{q},\omega)$. For isotropic systems, ε will depend only on the magnitude of $|\mathbf{q}| = q$ and ω. Equation (1.8) is directly applicable only for homogeneous systems such as bulk solids when the lattice structure can be neglected (e.g., jellium).

Due to the breaking of translational symmetry in the direction of the surface normal z, the response function for a jellium surface becomes nonlocal in z while retaining its dependence on the wave vector in the plane of the surface. The dielectric response theory for the bulk requires therefore modifications near surfaces. The extension of Eq. (1.8) to the case of broken translational invariance along the surface normal can be guessed as [13]

$$\rho_{ind}(\mathbf{q}_{\|},\omega,z,R_z) = \int dz' \, \chi(\mathbf{q}_{\|},\omega,z,z')\Phi_{ext}(\mathbf{q}_{\|},\omega,z',R_z) \ , \tag{1.9}$$

where R_z is the coordinate of the external point charge causing the response. Three-dimensional vectors are denoted by ($\mathbf{q} = (\mathbf{q}_{\|}, q_z)$ with $q^2 = q_{\|}^2 + q_z^2$), while 2D vectors in the surface plane are denoted by $\mathbf{q}_\parallel$. In the two-dimensional Fourier representation, the Coulomb potential of an external unit point charge is given by

$$\Phi_{ext}(\mathbf{q}_\parallel, \omega, z', R_z) = \frac{2\pi}{q_\parallel} e^{-q_\parallel |z' - R_z|} . \tag{1.10}$$

The response function χ is still local in the Fourier variables $(\mathbf{q}_\parallel, \omega)$, but nonlocal in the coordinate z in which the translational symmetry is broken. It is convenient to introduce a Dirac-bracket matrix notation [14] with respect to the nonlocal coordinates. Accordingly, ρ_{ind} of Eq. (1.9) will be identified as an off-diagonal matrix element of the operator $\underline{\underline{\rho}}_{ind}(\mathbf{q}_\parallel, \omega)$,

$$\rho_{ind}(\mathbf{q}_\parallel, \omega, z, R_z) = \langle z | \underline{\underline{\rho}}_{ind}(\mathbf{q}_\parallel, \omega) | R_z \rangle , \tag{1.11}$$

where the induced charge density at the coordinate z is generated by a source located at R_z. Analogously, the external potential can be expressed as the matrix element of the operator $\underline{\underline{\Phi}}_{ext}(\mathbf{q}_\parallel, \omega)$,

$$\Phi_{ext}(\mathbf{q}_\parallel, \omega, z, R_z) = \langle z | \underline{\underline{\Phi}}_{ext}(\mathbf{q}_\parallel, \omega) | R_z \rangle = \frac{2\pi}{q_\parallel} e^{-q_\parallel |z - R_z|} . \tag{1.12}$$

In this matrix notation, Eq. (1.9) simplifies to

$$\underline{\underline{\rho}}_{ind}(\mathbf{q}_\parallel, \omega) = \underline{\underline{\chi}}(\mathbf{q}_\parallel, \omega) \underline{\underline{\Phi}}_{ext}(\mathbf{q}_\parallel, \omega) . \tag{1.13}$$

Within the framework of the "self-consistent field" (SCF) method or "random-phase approximation" (RPA) we can approximate Eq. (1.13) by

$$\underline{\underline{\rho}}_{ind}(\mathbf{q}_\parallel, \omega) = \underline{\underline{\chi}}_0(\mathbf{q}_\parallel, \omega) \underline{\underline{\Phi}}_{SCF}(\mathbf{q}_\parallel, \omega) , \tag{1.14}$$

where χ_0 is the independent-particle susceptibility, and the self-consistent potential is given by

$$\underline{\underline{\Phi}}_{SCF}(\mathbf{q}_\parallel, \omega) = \underline{\underline{\Phi}}_{ind}(\mathbf{q}_\parallel, \omega) + \underline{\underline{\Phi}}_{ext}(\mathbf{q}_\parallel, \omega) . \tag{1.15}$$

Combining Eqs. (1.12–1.15) we find an operator integral equation for the response operator $\underline{\underline{\chi}}$

$$\underline{\underline{\chi}}(\mathbf{q}_\parallel, \omega) = \left[1 - \underline{\underline{\chi}}_0(\mathbf{q}_\parallel, \omega) \underline{\underline{K}}(\mathbf{q}_\parallel, \omega) \right]^{-1} \underline{\underline{\chi}}_0(\mathbf{q}_\parallel, \omega) \tag{1.16}$$

and the corresponding equation for the induced potential:

$$\underline{\underline{\Phi}}_{ind}(\mathbf{q}_\parallel, \omega) = \underline{\underline{K}}(\mathbf{q}_\parallel, \omega) \underline{\underline{\chi}}(\mathbf{q}_\parallel, \omega) \underline{\underline{\Phi}}_{ext}(\mathbf{q}_\parallel, \omega) . \tag{1.17}$$

In the RPA the kernel of the integral equation (1.16) is given by

$$\langle z|\underline{\underline{K}}_{RPA}(\mathbf{q}_{\|},\omega)|z'\rangle = \frac{2\pi}{q_{\|}}e^{-q_{\|}|z-z'|}\,. \tag{1.18}$$

Improvements beyond the RPA can be accomplished, for example, by including exchange-correlation corrections of density functional theory into the kernel,

$$\langle z|\underline{\underline{K}}_{LDA}(\mathbf{q}_{\|},\omega)|z'\rangle = \frac{2\pi}{q_{\|}}e^{-q_{\|}|z-z'|} + \Phi'_{xc}[n_0(z)]\delta(z-z')\,, \tag{1.19}$$

where

$$\Phi'_{xc}[n_0(z)]\delta(z-z') = \left.\frac{\delta\Phi_{xc}[n]}{\delta n}\right|_{n=n_0} xs \tag{1.20}$$

is the functional derivative taken at the unperturbed, but z-dependent, density $n_0(z)$. The local density approximation (LDA) is invoked in Eq. (1.19). The use of kernels from DFT [Eq. (1.19)] in the calculation of χ Eq. (1.16) is often referred to as time-dependent density functional theory (TDDFT). Equation (1.16) represents, however only its linear-response (LR) limit. A more detailed discussion is given in Sect. 1.5.

The starting point for the calculations of the many-body response function $\chi(\mathbf{q}_{\|},\omega,z,z')$ is the noninteracting particle density-density correlation function $\chi_0(\mathbf{q}_{\|},\omega,z,z')$. An explicit expression for $\chi_0(\mathbf{q}_{\|},\omega,z,z')$ can be easily determined within the independent-particle model for the semi-infinite jellium. For later reference we point out that lattice effects and thus coupling to phonons are neglected. Using the Fermi function at zero temperature for the occupation numbers, $f(p)=\theta(k_F-p)$, $\chi_0(\mathbf{q}_{\|},\omega,z,z')$ can be written in terms of the one-electron wave functions and eigenenergies as

$$\chi_0(\mathbf{q}_{\|},\omega,z,z') = \sum_{p_z,p'_z,\mathbf{p}_{\|}} (f(\mathbf{p}_z)-f(\mathbf{p}'_z)) \left(\frac{\phi^*_{p_z}(z)\phi_{p_z}(z')\phi_{p'_z}(z)\phi^*_{p'_z}(z')}{\omega+\varepsilon_{p_z}-\varepsilon_{p'_z}-\mathbf{q}_{\|}\mathbf{p}_{\|}-q^2_{\|}/2+i\gamma}\right) \tag{1.21}$$

with

$$f(\mathbf{p}_z) = f(\mathbf{p}_{\|},p_z);\; f(\mathbf{p}_z) = f(\mathbf{p}_{\|}+\mathbf{q}_{\|},p'_z)\,. \tag{1.22}$$

Within the framework of DFT, $\chi_0(\mathbf{q}_{\|},\omega,z,z')$ is constructed from Kohn-Sham orbitals $\phi_{p_z}(z)$ with energies ε_{p_z} in the degree of freedom along the surface normal (see Sect. 1.5.1).

An example of the real part of the total potential, $\Phi=\Phi_{ext}+\Phi_{ind}$ near the surface calculated within the specular reflection model (SRM) [15] is shown in Fig. 1.3 for a 0.3 MeV/u proton moving parallel to a surface. The induced charge density tries to screen the external charge. Since the response of the electron gas takes a finite time ($\sim\omega_p^{-1}$) the screening cloud trails the swift ion. This results in charge density fluctuations behind the projectile ("wake"). The potential shown in Fig. 1.3 was used in [15] to model the so-called "rainbow scattering" of electrons from a swift ion in the selvedge of the electron gas.

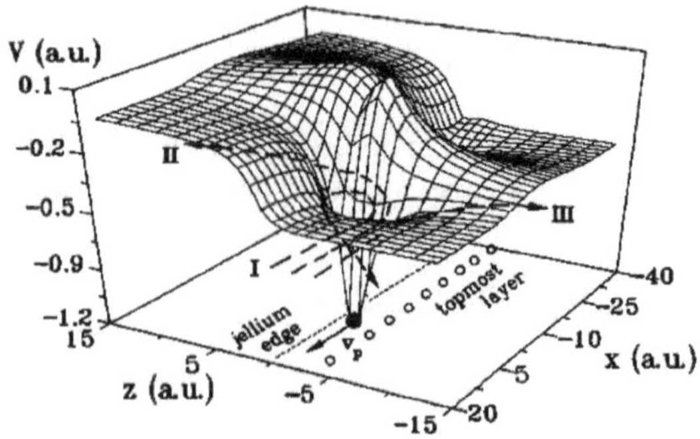

Fig. 1.3. Total electronic potential for a fast proton ($v \approx 3.5$ a.u.) moving parallel to an SnTe(001) surface ($\omega_p^{-1} \approx 1.8$ a.u.). The trailing response of the electron gas leads to the formation of the so-called wake. Trajectories of electrons scattered in the potential are plotted with *dashed lines* (see Sects. 1.3.1 and 1.3.2; from [15])

The imaginary part of the complex induced potential gives the position (**R**) dependent friction force or stopping power in terms of the matrix element

$$S(R_z, v_p) = \frac{1}{v_p} \int \frac{d^2\mathbf{q}_\parallel}{(2\pi)^2} \mathbf{q}_\parallel \mathbf{v}_p, \Im \left[\langle R_z | \underline{\underline{\Phi}}_{ind}(\mathbf{q}_\parallel, \omega = \mathbf{q}_\parallel \mathbf{v}_p) | R_z \rangle \right] . \quad (1.23)$$

An example of the stopping power for a charged particle moving parallel to and outside of a jellium surface as a function of the distance from the surface calculated using the DFT-kernel [Eq. (1.19)] is shown in Fig. 1.4. for $Q = 1$ and $v_p = 0.29$ a.u. For large distances R_z, $S(R_z, v_p)$ converges to an asymptotic R_z^{-3} dependence due to plasmon and interband excitations. At smaller distances, significant differences between the TDDFT-LR and the SRM calculations arise. The present TDDFT calculation includes the decay of long-wave plasmons due to interband transitions which are missing in standard jellium approximations.

Since a significant contribution to the friction force results from electronic target excitations to the continuum, it has been surmised [16] that, up to a constant of proportionality, the coefficient of KE, γ_{KE}, is proportional to the stopping,

$$\gamma_{KE} \propto S(R_z, v_p) . \quad (1.24)$$

The validity of the dielectric response description relies on the assumption of linearity of the response or, equivalently, the applicability of first-order perturbation theory. For highly charged ions (HCI) with charge Q its validity will be (nl) distance-dependent. The distance at which non-linear (nl) effects become important can be estimated to be [17]

$$R_{nl} \approx 2r_s\sqrt{Q} . \quad (1.25)$$

For slow HCI, the neutralization sequence and thus the effective reduction of Q starts at distances larger than R_{nl}. Therefore, the applicability of linear

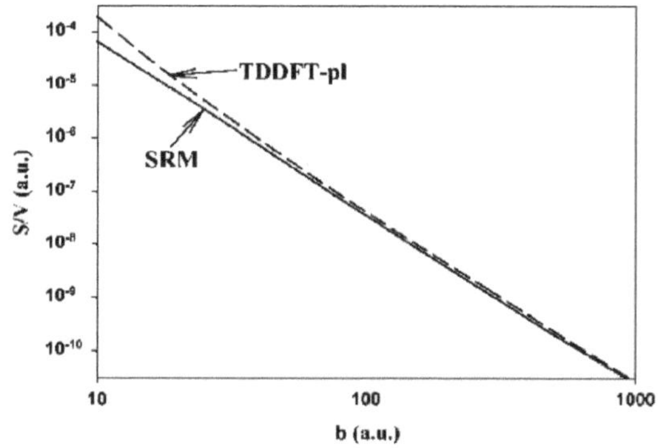

Fig. 1.4. Distance-dependent stopping power for a unit charge moving in front of an Al surface. *Solid line*: SRM, *dashed line*: TDDFT-LR using the kernel Eq. (1.19) and corrections due to interband transitions (from [14])

response extends to $Q \gg 1$. Going beyond linear response requires self-consistent solution of the many-body problem to all order in the strongly perturbing field of the projectile. Time-dependent density functional theory (Sect. 1.5.2) represents a promising avenue in this direction.

1.2.3 Atomic Potentials

The determination of the trajectories of ions (or atoms) near surfaces requires the knowledge of effective ionic (atomic) interaction potentials. While the long-ranged portion is well described by Coulomb-like image interactions for metal surfaces and Van der Waals interactions for insulators (see Sect. 1.2.2), the short-ranged portion is determined from the mutual penetration of the atomic charge clouds of the projectile atom and of the surface atoms. The problem is simplified in so far as, on an atomic length scale (≤ 1 a.u.), the surface can be viewed as a planar cluster of atoms. The effective potentials can therefore be deduced from atom-atom potentials known from elastic atomic scattering. Corrections for intermediate distances of the order of the lattice constant are included in embedded atom potentials [18]. Atom-atom potentials for a large class of collision partners and for a wide range of collision energies can be conveniently parameterized by approximate screened Coulomb potentials of the form [19]

$$V_{atom}(r) = \frac{Z_p Z_t}{r} f(r/a_s) , \qquad (1.26)$$

where $Z_{p,t}$ are the bare nuclear charges of the projectile and surface (target) atoms and f is the screening function. f approaches 1 in the limit $r \to 0$ and decays exponentially for large r. Different analytic forms of the interpolation function are in use. The following analytic form is most frequently employed

$$f(r/a_s) = \sum_i c_i \exp(-d_i r/a_s) \tag{1.27}$$

with $\sum_i c_i = 1$. Different parameterizations differ from each other by the choice of the screening length a_s and the vectors of coefficients c_i, d_i. The Molière potential [20], f_M, uses $c_i = 0.35, 0.55, 0.1$; $d_i = 0.3, 1.2, 6$ and the Firsov (or Thomas-Fermi) screening length

$$a_s = a_{TF} = 0.88534 \cdot (Z_p^{1/2} + Z_t^{1/2})^{-2/3} . \tag{1.28}$$

Other choices include the Bohr potential with only one exponential term in Eq. (1.27) with $c_1 = d_1 = 1$ and

$$a_s = a_B = (Z_p^{2/3} + Z_t^{2/3})^{-1/2} , \tag{1.29}$$

and the Ziegler-Biersack-Littmark (ZBL) potential [21] using the vectors $c_i = 0.1818, 0.5099, 0.2802, 0.02817$; $d_i = 3.2, 0.9423, 0.4028, 0.2016$ and the screening length

$$a_s = a_{ZBL} = 0.88534/(Z_p^{0.23} + Z_t^{0.23}) . \tag{1.30}$$

Improvements beyond these universal parameterizations can be derived from Hartree-Fock potentials for a given pair of projectile and target atoms [11, 22]. The basic limitations of these approximations is that quasi-static atomic potentials neglect the dynamic response of the system. The range of applicability will depend in a non-uniform manner on both the projectile velocity and the relevant distances close to the turning point. Screened Coulomb potentials [Eq. (1.26)] describe (quasi-) elastic scattering only for sufficiently large collision energies when the trajectory becomes insensitive to details of the Born-Oppenheimer potential curves but is predominantly determined by the repulsive core (typically a few 100 eV). At very low impact energies state-dependent Born-Oppenheimer potential curves (potential surfaces) are required as input for accurate trajectory calculations. Rainbow scattering has been used to test atomic scattering potentials in detail [23].

1.3 Classical Methods

1.3.1 Projectile Trajectories

Except for very low collision energies ("thermal" and "hyperthermal" energies ≤ 1 eV), classical dynamics is well justified and regularly applied to treat the motion of heavy particles near surfaces. The corresponding de Broglie wavelength $\lambda_{dB} = 1/\sqrt{2\mu E}$ (μ: reduced mass of the projectile-target system) is exceedingly small compared to typical atomic distances, $\lambda_{dB} \ll 1$ a.u. Note, however, that for grazing incidence with $E_{p,\perp} \ll E_{p,\parallel}$, λ_{dB} for the transverse motion need not to be small and quantum ("diffraction") effects may appear.

Classical trajectories of ions and atoms approaching and receding from surfaces provide the scenario for inelastic electronic processes to be discussed in the following sections. The trajectory "controls," to a large degree, the relative importance, the abundance, or the suppression of different classes of electronic transitions. One important control parameter for the study of surface interactions is the fraction of surface penetrating and reflecting trajectories. Ions on a penetrating trajectory experience both above-surface and below-surface (bulk) interactions. Since the electron density and target-atom density in the bulk is far greater, surface-specific processes can be easily overshadowed in experiments in which penetrating trajectories contribute significantly. Reflecting trajectories are therefore a *conditio sine qua non* for sensitive probes of distant ion-surface interactions.

Reflection in grazing-incidence surface collisions results, primarily, from collective scattering at many rather than an individual surface atom. This collective scattering is often referred to as "surface channeling". The effective atomic potential governing surface channeling is given by "string" (or "axial") or "planar" averaged potentials [24]. The planar surface potential is, e.g., given by

$$V_{planar}(R_z) = 2\pi n_a \int_0^\infty d\rho\, \rho V_{atom}\left(\sqrt{(R_z + a/2)^2 + \rho^2}\right) \tag{1.31}$$

with n_A the atomic density at the surface. For the potentials of the type (1.26) the integral (1.31) can be easily performed leading to

$$V_{planar}(R_z) = 2\pi n_a a_s Z_p Z_t \sum_i \frac{c_i}{d_i} \exp(-d_i(R_z + a/2)/a_s)\ . \tag{1.32}$$

A similar expression can be derived for string potentials. As expected for the specular reflection geometry, the effective potential depends only on the coordinate R_z. The parallel and perpendicular degrees of freedom of the motion become separable. The trajectory is therefore a superposition of a free-particle constant-velocity translation parallel to the surface and a perpendicular motion in the potential Eq. (1.32). The point of closest approach $R_{z,min}$ is implicitly determined by

$$E_{p,\perp} = V_{planar}(R_{z,min})\ . \tag{1.33}$$

The atomic potential does not yet include the long-range Coulomb-like tail of the image potential for ions with charge Q. Accordingly, in the presence of an image potential, Eq. (1.33) is modified to

$$E_{p,\perp} + \frac{Q}{4R_{z,min}} = V_{planar}(R_{z,min})\ . \tag{1.34}$$

The point of closest approach moves thus closer to the topmost atomic layer. Such a shift has consequences for the probability of inelastic electronic processes including electron emission to occur.

1.3.2 Binary Encounter Approximation

While employing classical mechanics for the heavy-particle motion is straightforward and a well justified approximation, treating the electronic degrees of freedom is, conceptually, far from obvious. Yet, classical approximation are frequently applied and remarkably successful. In fact, first treatments of atomic collisions predated quantum mechanics and invoked, inevitably, classical dynamics. One year before Bohr presented his atomic model, J.J. Thomson determined in 1912 the ionization cross section of a classical electron at rest with binding energy (work function) W by binary impact of an ion as [25]

$$\sigma_{ion}(E_p) = \frac{4Q^2}{E_p W}\left(1 - \frac{W}{4E_p}\right) . \tag{1.35}$$

Equation (1.35) can be considered the predecessor of the binary-encounter approximation (BEA). While the Thomson cross section [Eq. (1.35)] predicts a high-energy behaviour of $\sigma \sim E_p^{-1}$, the first Born approximation in quantum mechanics yields $\sigma \sim E_p^{-1} \ln E_p$ due to the contribution of optically allowed dipole excitations absent in classical mechanics [26]. With the improved understanding of the electronic bound states within the framework of quantum mechanics, it became obvious that the assumption of an electron at rest must be replaced by the Compton profile, i.e., an ensemble of electrons with the momentum distribution of the initially bound electron, $|\tilde{\phi}_i(\mathbf{q},t)|^2$, consistent with the virial theorem. Gryziński (1959 [27]) and Vriens (1967 [28]) formulated the BEA for ion-atom collisions exploiting the fact that quantum and classical microcanonical momentum distributions agree for hydrogenic systems [29].

Applications to ion-surface collisions is straight-forward. Binary-encounter collisions are prototypical for KE, for binary transfer of kinetic energy of the projectile to the target. For a jellium surface, σ_{ion} follows from the integral over all target-electron momenta $k < k_F$. The mapping of the initial momentum distribution due to the binary-encounter momentum transfer is schematically depicted in Fig. 1.5. Analytic results can be found for limiting cases. Assuming a head-on collision of a projectile with the electron, the maximum momentum transferred to the backscattered electron is given by $\Delta k = 2v_p$. Taking an electron with Fermi velocity v_F directed antiparallel to $\mathbf{v}_p$ the threshold projectile velocity v_{th} for transfer of sufficient energy and for KE to take place ($\Delta E > W$) is [16]

$$v_{th} = \frac{v_F}{2}\left(\sqrt{1 + W/E_F} - 1\right) . \tag{1.36}$$

For $v_p < v_{th}$ the maximum momentum transferred to the electron is insufficient to escape the target. More recently, Winter and Winter [30] have deduced for the coefficient of ("secondary") electron emission, γ, which is proportional to the total ionization cross section a $\gamma_{KE} \propto (v_p - v_{th})^2$ proportionality of the KE yield from simple phase space arguments for a free electron metal.

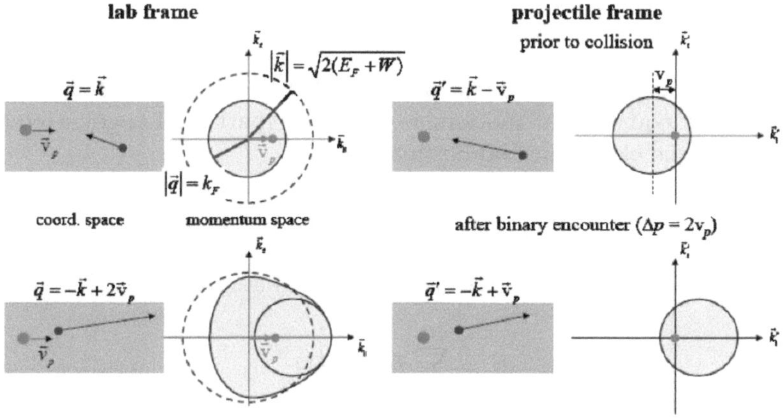

Fig. 1.5. Schematic diagram of a binary encounter in the laboratory (*left column*) and projectile frames (*right column*) before (*top row*) and after the collision (*bottom row*). The Fermi sphere is displaced in the projectile frame relative to the origin by v_p (*top right frame*). A momentum transfer of $\Delta p = 2v_p$ is imparted on backscattered electrons (*bottom right*). In the lab frame, parts of the shifted Fermi sphere have momenta large enough to escape the solid (*dashed circle*)

Although Eq. (1.36, predicts a well-defined threshold of KE, careful measurements in the threshold region have shown that KE can be observed also for $v_p < v_{th}$ [31] (see also Chap. 4). A more elaborate description of the electronic structure of the surface that takes the true momentum distribution of the target into account is therefore required (see Sect. 1.4.2).

1.3.3 Classical-Trajectory Monte-Carlo Methods

The BEA is, on a classical level, a perturbative approximation. It treats the binary collision system (= target electron-projectile) to all orders while neglecting the influence of the interactions within the solid surface (ionic cores, jellium potential, spectator electrons) during the collision. The BEA is thus the classical rendition of an impulsive approximation (IA). Going beyond the BEA within the framework of classical dynamics requires the solution of Hamilton's equation of motion for an ensemble of independent target electrons representing the phase space of the initial quantum state, the Wigner function $\rho_W(\mathbf{r}, \mathbf{p})$. Since ρ_W is not positive definite, alternative phase-space distributions, in particular the microcanonical ensemble $\rho_{MC} \propto \delta(E - H(\mathbf{r}, \mathbf{p}))$ are used. This is the essential idea of Classical trajectory Monte Carlo (CTMC) methods originally developed for atomic collisions [29]. Its main application is for (effective) one-electron problems. The many-body aspect, e.g., scattering of the ionizing electron with other electrons, can be accounted for indirectly by including stochastic forces into the equations of motion, i.e., by using a Langevin equation rather than Newton's equation of motion [32].

The time-dependent evolution of the electronic subsystem is calculated by the Monte-Carlo solution of the classical Liouville equation in terms of an ensemble of representative test particles that are propagated according to the stochastic equation of motion, i.e., a single-particle Langevin equation generated by the electronic Hamiltonian H_e

$$\frac{d\mathbf{r}}{dt} = \mathbf{p}\,, \quad \frac{d\mathbf{p}}{dt} = -\boldsymbol{\nabla} H_e(\mathbf{r}, \mathbf{R}, t) + \mathbf{F}_{stoc}\,. \tag{1.37}$$

The total Hamiltonian of the (electronic) system is given by

$$H = \sum_{j=1}^{N_e} H_e(\mathbf{r}_j, \mathbf{R})\,, \tag{1.38}$$

where N_e is the number of active electrons and $\mathbf{r}_j$ and $\mathbf{R}$ are the positions of the j-th electron and the projectile, respectively. The stochastic force $\mathbf{F}_{stoc}$ accounts for collisions with other electrons and ionic cores of the solid. Apart from inclusion of scattering and transport effects, the CTMC method is intrinsically a single-particle treatment. The deeper reason for this is that, classically, a many-electron system does not possess a stable ground state but spontaneously autoionizes.

The solution of the Liouville equation within CTMC proceeds along the following sequence of steps:

1. a large number of trajectories is started at $t \to -\infty$ sampling the initial phase-space distribution
2. Equation (1.37) is numerically integrated for each trajectory up to $t \to \infty$
3. the final phase space distribution is calculated from binning the phase-space points at the final time t_f.

This scheme is numerically demanding as the total number of electrons involved in the interaction, N_e (the sum of projectile electrons and target conduction band and core electrons within the interaction area), can become rather large and a large number of trajectories has to be followed to obtain reasonable statistics. Figure 1.3 shows an example of the CTMC simulation of the momentum distribution of electrons emitted in swift ion-surface interactions taking account of rainbow scattering of the convoy electrons near surfaces [33].

The potential landscape in front of an insulator surface (LiF) used in a simulation for charge transfer is shown in Fig. 1.6 [8]. The saddle region separating the ion and the surface is a complicated function not only of the distance but also of the lateral position of the projectile. The right-hand panels show the classical trajectory of a target electron as a highly charged ion approaches the surface. Initially, the electron orbits the ionic core located at (0,0,0). Later, the Coulomb potential of the projectile and its image charge deform the potential landscape and allow for an above-barrier transition from the surface to the ion. The initial conditions for the electron were chosen from

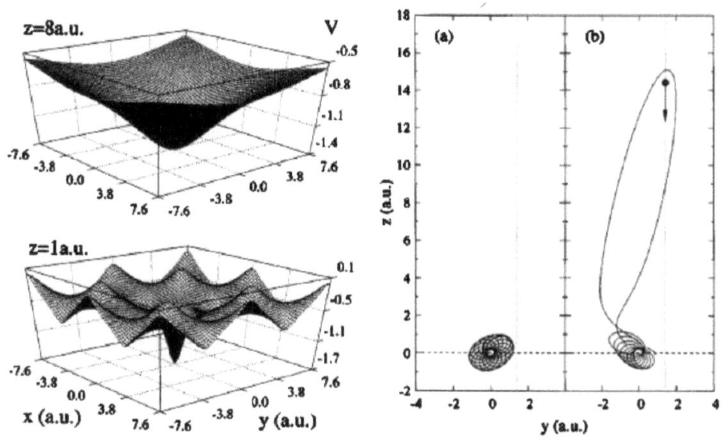

Fig. 1.6. *left panels*: Total electronic potential $V(\mathbf{r}, \mathbf{R})$ in planes parallel to the surface of the crystal. The projectile with charge $Q = 6$ is at position $\mathbf{R} = (0, 0, 12)$ a.u. *right panel*: Example of a classical trajectory of an electron as an HCI with $Q = 15$ approaches the surface along the dotted line. (**a**) The initial polarization of the classical orbit, (**b**) the more pronounced polarization followed by the passage over the saddle into the projectile at $R_z = 14.5$ a.u. (from Ref. [8])

a restricted microcanonical ensemble representing the 2p electrons of F^- ions. These calculations yield capture and loss rates which enter the rate equations for the populations of different n-shells in the classical over-barrier simulations discussed in the next section.

1.3.4 Classical Over-Barrier Model

A variant of a fully classical treatment is the classical over-the-barrier (COB) model. It has proven to be remarkably versatile and successful. Key is both a drastic simplification of the classical dynamics input compared to a full CTMC simulation as well as the possibility to easily, but approximately, incorporate multi-electron processes and quantum corrections.

The COB model was originally developed for one-electron capture into highly charged ions in ion-atom collisions by Ryufuku et al. [34] based on earlier work by Bohr and Lindhard [35] and later extended by Barany et al. [36] and Niehaus [37] to incorporate multi-electron transfer. Its extension to ion-surface collisions [8, 38, 39] provides a simple framework for the description of ion-surface interactions. The physical significance of the COB model is derived from the fact that only classically allowed over-the-barrier processes as opposed to tunneling are sufficiently fast to be effective within the characteristic interaction time of the ion with the surface.

Choosing for simplicity pure Coulomb potentials for all electronic interaction potentials entering Eq. (1.4) [in particular the self-image interaction V_e^{im} Eq. (1.1)] as the large-z limit for metals, the position of the saddle point

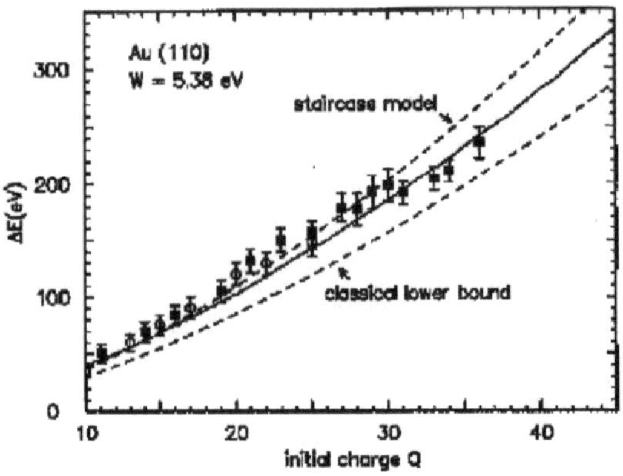

Fig. 1.7. Scaled energy gain $\Delta E/W$ due to image acceleration for different targets. Experimental data: $\circ$: Al; $\Delta, \square$: Au(110); $\bullet$: Au. Also shown are the COB model (see Eq. (1.42)) and the classical lower bound, from [42] and refs. therein

$z_s, dV_e(z_s, R_c)/dz = 0$ and the barrier height $V_e(z_s, R_c)$ can be determined analytically. The critical distance follows from [2, 8, 38, 39]

$$V_e(z_s, R_c) = E_F = -W \; , \tag{1.39}$$

where W is the work function of the surface, as

$$R_c(Q) \simeq \frac{\sqrt{2Q}}{W} \; . \tag{1.40}$$

The COB model predicts, furthermore, the critical quantum number n_c into which the resonant capture event takes place,

$$n_c = \frac{Q}{\sqrt{2W}} \left(\frac{1}{1 + \frac{Q-1/2}{\sqrt{8Q}}} \right)^{1/2} \; . \tag{1.41}$$

These predictions pertain to the early stages of the neutralization scenario. Equation (1.40) was first indirectly tested (Fig. 1.7) by measurements [40, 41] of the energy gain due to the image acceleration. Since the charge of the projectiles is reduced by one unit ($Q' = Q, Q-1...$) at each successive critical distance, $R_c(Q')$, Eq. (1.40), the energy gain along the neutralization sequence ("staircase") follows as [39]

$$\Delta E = \frac{W}{3\sqrt{2}} Q^{3/2} \tag{1.42}$$

in agreement with experimental data [40–42]. Similar estimates can be derived for insulators [8].

Direct tests of Eqs. (1.40) and (1.41) became available with the help of novel targets, metallic nanocapillaries, pioneered by the group of Yamazaki et al. [43]. Interaction with the internal walls of capillaries allows to select trajectories that avoid a close encounter with the surface thus allowing spectroscopic as well as charge-state analysis of the early stages of the neutralization scenario [44].

The COB predictions for charge transfer can be incorporated into a hybrid classical-quantum simulation which can also account for processes of interacting electrons for which fully classical simulations fail. Prime among are autoionization and Auger decay. Starting point is a set of coupled rate equations for the populations P_n of the n-th projectile shell [38]

$$\begin{aligned} v_\perp(R)\frac{d}{dR}P_n(R) &= I_n^{RC}(R) - I_n^{RI}(R)P_n(R) \\ &\quad +\frac{1}{2}\sum_{n'>n} A_{n,n'}P_{n'}^2(R) \\ &\quad -P_n^2(R)\sum_{n'<n} A_{n',n} , \end{aligned} \tag{1.43}$$

$$\begin{aligned} v_\perp(R)\frac{d}{dR}P_I(R) &= \frac{1}{2}\sum_{n,n'<n_I} A_{n,n'}P_{n'}^2(R) \\ &\quad +v_\perp(R)P_{n_I}(R)\left.\frac{dn_I(R)}{dR}\right|_{E_{n_I}=0} . \end{aligned} \tag{1.44}$$

I_n^{RC} and I_n^{RI} are the resonant capture and loss currents, respectively, $A_{n',n}$ describes the Auger decay rate of two electrons initially in shell n to a final state with one electron populating shell n' and the other electron in the continuum carrying the excess energy. Radiative decay processes have been neglected in [38] as they become important only for inner shells of ions with large core charges.

The additional Eq. (1.44) describes the differential increase of ionized electrons, $P_I(R)$, due to Auger processes and to the promotion of high-lying shells into the continuum. As the ion approaches the surface the energy level of an excited electron in the projectile is shifted upwards by the interaction with the image interaction. For a hydrogenic state the binding energy is given by

$$E_n = -\frac{Q_{eff,n}^2}{2n^2} + \frac{Q_{eff,n} - 1/2}{2R} . \tag{1.45}$$

For small distances R the inner shells get resonantly populated thus reducing $Q_{eff,n}$. At a given distance, a certain n-shell ($n = n_I$) enters the ionization continuum with rate $v_\perp dn_I(R)/dR$.

The capture rate I_n^{RC} entering Eq. (1.43) can be calculated within the COB model from the geometric cross section $\sigma \approx \pi d_s^2$ of the saddle area (see Fig. 1.2) and the current density along the surface normal. The loss rate I_n^{RI} follows from the ratio between σ and the surface of a sphere with Bohr radius $\langle r\rangle_n$ weighted with the orbital frequency of the Coulomb orbit

$\nu_n = 1/T_n = Q^2_{eff,n}/2\pi n^3$ with $Q^2_{eff,n}$ being the shielded core charge effective in shell n. For an intra-atomic Auger process and the fastest subset of possible channels ($ns^2 \to n's$) a simple scaling rule was found using the Cowan-code for isolated atoms in the gas phase [45],

$$A_{n',n} = \frac{5.06 \cdot 10^{-3}}{|\Delta n|^{3.46}} . \tag{1.46}$$

It should be noted that Eq. (1.46) holds for a wide range of combinations of n and n'. Extensions to other combinations of quantum numbers as well as to two-center Auger processes, Auger capture (AC), and Auger decay (AD) have been investigated [46]. In the case of AC an exponential reduction of the rate can be expected due to the exponentially decreasing probability of target electrons in the vacuum ($\propto \exp(-2k_t R)$). For AD rates a $1/R^3$ dependence at large distances was derived in [46]. Auger processes are sources of potential emission of electrons. For promotion electrons [second term in Eq. (1.44)] identification as PE is not as clear-cut and is valid only if the projectile motion is effectively decoupled from the weakly bound electronic shells. Accordingly, the effective potential V_p governing Newton's equation of motion

$$m\frac{d}{dR}v_\perp(R) = \frac{1}{v_\perp(R)}\frac{d}{dR}V_p(R) \tag{1.47}$$

is assumed to be weakly dependent on the population of diffuse outer shells.

An extended version of the COB model taking the spin of the electrons to be captured and quantum mechanical tunneling into account has recently been used to simulate the interaction of N^{6+} ions with a magnetized iron surface [47, 48]. Experimental data by Pfandzelter et al. [49] for energy and polarization of emitted electrons were compared to results of the COB model. As the experiment was performed with swift projectiles impinging on the surface under grazing incidence conditions, the total electron spectrum contains contributions from both potential and kinetic emission as well as secondary electrons produced in transport processes within the target. The experimental separation of kinetic and potential emission yields in medium-energy grazing incidence have been accomplished either by using single crystal target surfaces and measuring the dependence of the total yield on the azimuthal angle [50] or by measuring the energy loss of the scattered projectile and emitted electrons in coincidence [51]. A more detailed discussion can be found in Chap. 4.

For the analysis of the polarization of emitted electrons we concentrate on the K-Auger peak around 350 eV where contributions due to KE can be excluded. Figure 1.8 shows the measured (dots) and simulated (open circles) polarization of high energy electrons. Both results agree rather well. The measured polarization value close to the average polarization of the target conduction band as well as the simulation results indicate that direct inner-shell feeding of the L-shell by a tunneling process may only be a minor source for the electrons undergoing a KLL-Auger decay.

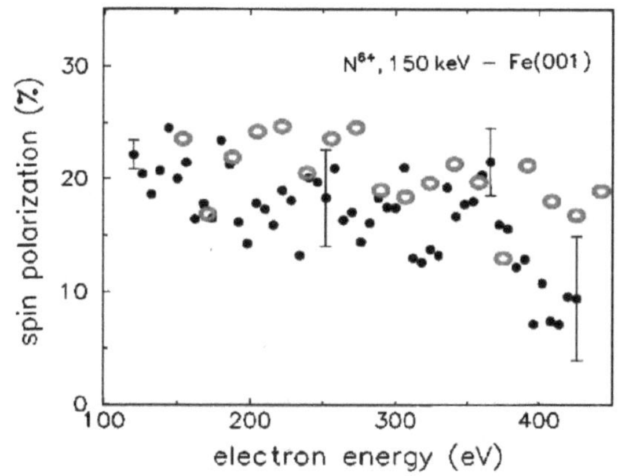

Fig. 1.8. Polarization of electrons emitted in N^{6+} magnetized iron interactions. Simulated data are shown by *open circles*, experimental data (dots) from [49]

An extension beyond rate equations for mean occupation numbers $P_n(R)$ has been introduced in terms of a classical transport theory (CTT) based on a multi-particle Liouville master equation [52]. It invokes four major ingredients:

- the explicit treatment of multi-electron processes by following the time evolution of the joint phase space density $\rho\left(t, \mathbf{R}, \{P^{(P)}\}, \{P^{(T)}\}\right)$ that depends on population strings of N-electron states in the projectile $\{P^{(P)}\}$ and $\{P^{(T)}\}$ target,
- the use of transition rates in the relaxation (or transport) kernel that are derived from quantum calculations (mostly, first-order perturbation theory) wherever available, e.g. two-center Auger capture and deexcitation rates [46], and
- the embedding of these processes into the framework of a classical phase space transport simulation for the ion.

The equation of motion of ρ is of the form of a Liouville master equation

$$\left(\frac{\partial}{\partial t} + \dot{\mathbf{R}}\nabla_R - \frac{1}{M}\frac{\partial V_p}{\partial R}\nabla_{\dot{\mathbf{R}}}\right)\rho = \mathcal{R}\rho\,, \tag{1.48}$$

where the "relaxation" (collision) operator includes single and double particle-hole (de-) excitation processes which represent resonant capture, resonant loss, hole hopping, ionization by promotion through the continuum, Auger capture, Auger deexcitation, and autoionization. These rates depend on both the local position of the ion, $\mathbf{R}$, and on the population strings $\{P^{(p)}\}$ and $\{P^{(T)}\}$.

Direct integration of the Liouville master equation Eq. (1.48) appears to be extremely difficult in view of the large number of degrees of freedom involved. Therefore, another ingredient,

- solution by test particle discretization and a Monte Carlo sampling for ensembles of stochastic realizations of trajectories comes into play. A large number of ionic trajectories with identical initial conditions for the phase space variables (R, v_p) is followed along an event – by – event sequence of stochastic electronic processes whose probability laws are governed by the rates of the underlying Liouville master equation. The resulting population strings $\{P_n^{(P)}(t)\}_\mu$ and $\{P^{(T)}(t)\}_\mu$ for a single stochastic trajectory μ are discontinuous functions of time. After sampling a large number of trajectories, one obtains smooth ensemble averages representing solutions of Eq. (1.48). Variants of the present approach have been previously employed for energetic electron transport through solids [32], ion transport through nanocapillaries [53], and, very recently, for the interaction of strong laser fields with large clusters [54]. An overview over methods for electron transport will be given in Chap. 2. One of the conclusions drawn from the simulations for ion-surface scattering is that for vertical incidence the hollow atom reaches the surface in a still multiply excited state. A significant fraction of relaxation and potential emission originates from (sub-) surface contributions. In contrast, near complete relaxation is possible in grazing-incidence collisions [55, 56].

1.4 Quantum Mechanical Methods

1.4.1 Time-Dependent Schrödinger Equation

The non-perturbative solution of the time-dependent Schrödinger equation is currently only feasible for effective one-electron problems. Two methods are frequently employed, the coupled-channel method [57, 58] using a basis expansion of the wavefunction $\psi(t)$ or wavefunction propagation on a space-time grid [59–61]. An example for the latter is the simulation of H^- detachment, which can be reduced to an effective one-electron problem. The electronic ground state of the surface, e.g. of LiF, provides an adequate representation of the channel potential. Quantum calculations for these problems have yielded valuable insight into the role of the Madelung potential for the detachment near LiF [59, 60] and the influence of the projected band gap of Cu [61]. Grid-based methods face the difficulty to extract final state occupation amplitudes through the projection onto asymptotic channel states, in particular of continuum states $|\mathbf{k}\rangle$, representing ionization. The latter originates from the presence of reflecting or absorbing boundaries of the grid. Basis expansions of the form

$$\psi(t) = \sum_j a_j(t)\phi_j(t) + \sum_{\mathbf{k}} b_{\mathbf{k}}(t)\phi_{\mathbf{k}}(t) \,, \tag{1.49}$$

where the index j extends over discrete bound states and $\mathbf{k}$ over discretized continuum states, suffer from truncation errors (only a finite number of bound and continuum states can be included). For simplicity we use the label j in Eq. (1.49) for any bound state including both core states and states of the band structure of a solid which will be treated as a quasi-continuum. The label

$|\mathbf{k}\rangle$ refers to unbound positive energy states in the ionization continuum. Employing Eq. (1.49) converts the Schrödinger equation into a system of coupled differential equations

$$\begin{aligned} i\dot{a}_j(t) &= \sum_{j'} \langle j|V_I(t)|j'\rangle a_{j'}(t) \\ &+ \sum_{\mathbf{k}} \langle j|V_I(t)|\mathbf{k}\rangle b_{\mathbf{k}}(t) \end{aligned} \tag{1.50}$$

$$\begin{aligned} i\dot{b}_{\mathbf{k}}(t) &= \sum_{j} \langle \mathbf{k}|V_I(t)|j\rangle a_j(t) \\ &+ \sum_{\mathbf{k}'} \langle \mathbf{k}|V_I(t)|\mathbf{k}'\rangle b_{\mathbf{k}'}(t) \ . \end{aligned} \tag{1.51}$$

In Eqs. (1.50) and (1.51) we use the intermediate or interaction representation, in which the evolution phases of the unperturbed system ($\sim \exp(-iH_0t)$) are included in the time dependence of the perturbation potential $V(t)$,

$$V_I(t) = \exp(iH_0t)V(t)\exp(-iH_0t) \ . \tag{1.52}$$

$V(t)$ includes the time-dependent perturbation by the ionic projectile and may include screening and image contributions (Sect. 1.2). A few cases in which Eqs. (1.50) and (1.51) were solved non-perturbatively, either numerically [60] or semi-analytically by employing a Wigner-Weisskopf approximation to the so-called memory kernel [58], have been reported.

1.4.2 Perturbation Approximation at High Energies

Born Approximation

A host of simpler approximations can be deduced from Eqs. (1.50) and (1.51). The Born approximation to electron emission follows by setting

$$\begin{aligned} a_j(t) &= \theta(j_F - j) \\ b_{\mathbf{k}}(t) &= 0 \ , \end{aligned} \tag{1.53}$$

where the index j extends over all states below the Fermi energy ($\varepsilon_j < \varepsilon_{j_F} = \varepsilon_F$). For each occupied orbital the ionization amplitude is

$$b_{\mathbf{k},j}(t) = -i \int_{-\infty}^{t} \langle \mathbf{k}|V_I(t')|j\rangle \, dt' \tag{1.54}$$

giving the total ionization probability

$$|b_{\mathbf{k},j}|^2 = \lim_{t\to\infty} \sum_{j\le j_F} |b_{\mathbf{k},j}(t)|^2 \ . \tag{1.55}$$

As the change in the initial occupation amplitude under the influence of $V_I(t)$ is neglected, Eqs. (1.54) and (1.55) is a first-order approximation. Equation (1.54) is expected to be valid when the change in the occupation amplitude is expected to be weak. This, in turn, implies that the speed of the projectile is sufficiently high ($v_p > v_F$) such that the mean action integral is small,

$$\int_{-\infty}^{\infty} dt\, \overline{\langle j|V_I(t)|j\rangle} \ll \pi \,, \tag{1.56}$$

where the average is taken over the relevant states j involved.

The final states in Eqs. (1.51) and (1.54) denoted by $|\mathbf{k}\rangle$ refer to the continuum state of the target, i.e., of the solid surface, to which the states $|j\rangle$ are initially bound, i.e., $|\mathbf{k}\rangle = |\psi_{\mathbf{k}}^{(T)}\rangle$. In fact, the derivation of Eq. (1.53) makes use of the orthogonality relation $\langle \mathbf{k}|j\rangle = 0$. In the presence of the ionic charge of the projectile which gives rise to projectile-centered bound and continuum states the final state is modified. Embedding the projectile continuum into the target continuum leads to strong perturbations of the continuum states which are most pronounced at electronic velocities (momenta) that vectorially match the velocity of the projectile ($\mathbf{k} = \mathbf{v}_p$) since the corresponding velocity in the frame of the projectile $\mathbf{k} - \mathbf{v}_p = \mathbf{k}'$ is small ($\mathbf{k}' \ll 1$) and the Coulomb-perturbation parameter $\eta = Q/\mathbf{k}'$ is large. This so-called "electron capture to continuum (ECC)" can be incorporated by replacing $|\psi_{\mathbf{k}}^{(T)}\rangle$ by the corresponding projectile-centered continuum state $|\psi_{\mathbf{k}}^{(P)}\rangle$ in Eq. (1.54). While the calculation of the transition amplitude is still of first order, the final-state interaction contains contributions to all orders in the effective projectile interaction. Calculations have been presented for both atomic Coulomb potentials [62, 63] as well as dressed potentials including dynamic screening and image potentials [64]. For the latter case, a comparison with experimental data for the so-called convoy peak in grazing incidence collisions confirmed the broadening of the ECC peak due to deviations from a pure Coulomb final-state interaction. The CTMC method (see Sect. 1.3.3) has been applied to convoy electron emission [15], where the distortion of the Coulombic final-state interaction is taken into account on a classical level treating, however, the collision non-perturbatively.

Distorted-Wave Approximations

Improvements beyond the first Born approximation can be achieved within the framework of distorted-wave Born (DWB) approximations. The underlying idea is to decompose the perturbation as follows

$$V_I(t) = U_I(t) + W_I(t) \,. \tag{1.57}$$

In Eq. (1.57) U_I represents the distortion potential and W_I the residual interaction, each of which taken to be in intermediate representation (I). The decomposition is, *a priori*, arbitrary and requires a judicious choice. In particular, the decomposition appropriate for the entrance channel $|\psi_j^{(T)}\rangle$ may be different from that for the exit channel $|\psi_{\mathbf{k}}\rangle$. Accordingly, different distortion potentials may be used in the entrance ($U^{(i)}$) and exit ($U^{(f)}$) channels subject to the constraint

$$U^{(i)} + W^{(i)} = U^{(f)} + W^{(f)} , \tag{1.58}$$

giving rise to different "post" and "prior" forms of the DWB approximations. The choice is dictated by the requirement that one should be able to include the distortion potentials U to all orders in the channel function. Moreover, one aims at including a major portion of V into U such that the residual interaction W is "weak". The strategy is now to employ a first-order Born approximation to the residual interactions only. Within the DWB approximations, the scattering amplitude is accordingly given by

$$b_{\mathbf{k},j}(t) = -i \int_{-\infty}^{t} \langle \chi_{\mathbf{k}}^{-}(t') | W_I(t') | \chi_j^{+}(t') \rangle \, dt' . \tag{1.59}$$

The superscript $+$ $(-)$ stands for outgoing (incoming) scattering boundary condition for the channel states.

A prototypical application of DWB theory is shown in Fig. 1.9 for the electron emission spectrum at different ejection angles resulting from glancing-angle scattering of protons at an Al(111) surface. The target electrons contributing to the emission [Eq. (1.54)] are decomposed into the atomic core electrons with $j \le j_{core}$ and the conduction electrons $j_{core} < j \le j_F$. For the core electrons, a DWB model well known from atomic collisions is employed, the so-called continuum distorted wave (CDW) with eikonal initial state (CDW-EIS) approximation [65]. In this case the entrance channel distortion is a Coulomb eikonal phase

$$\langle \mathbf{r} | \chi_j^{EIS} \rangle = \langle \mathbf{r} | j \rangle \exp\left[-i \frac{Q}{v_p} \ln(v_p x + \mathbf{v}_p \mathbf{x}) \right] , \tag{1.60}$$

where $\mathbf{x} = \mathbf{r} - \mathbf{R}$ is the coordinate of the electron to be ionized relative to the projectile. The CDW final state contains the full Coulomb scattering wave

$$\langle \mathbf{r} | \chi_{\mathbf{k}}^{CDW} \rangle = \langle \mathbf{r} | \mathbf{k} \rangle \Gamma(1 - iQ/v_p) \, {}_1F_1 \left[-i \frac{Q}{v_p}, 1, i(v_p x + \mathbf{v}_p \mathbf{x}) \right] , \tag{1.61}$$

whose asymptotic limit reduces to (1.60). The residual interaction is of the form

$$W_f = -\nabla_x \cdot \nabla_r \tag{1.62}$$

where the gradient ∇_x operates only on the distortion factor while ∇_r affects only the unperturbed wavefunction $\psi_j(\mathbf{r}, t)$. For the conduction band states, unperturbed jellium wavefunctions are taken as the initial and final states,

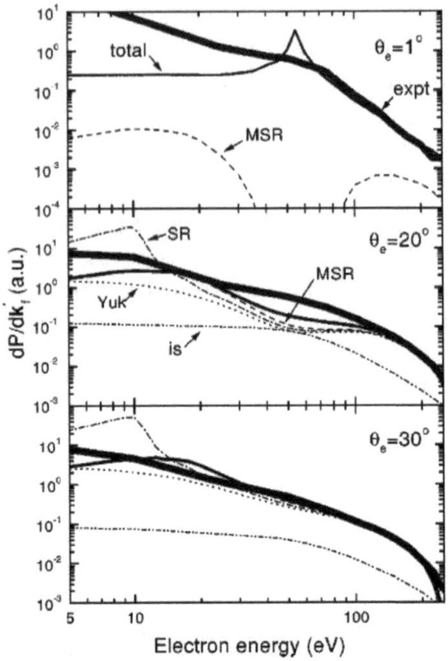

Fig. 1.9. Differential probability of electron emission for 100 keV protons impinging on an Al(111) surface with the incidence angle $\theta_{\text{in}} = 1°$. Three electron emission angles are considered: $\theta_{out} = 1°$, $20°$, and $30°$. Thin *solid line*, total probability of emission calculated by adding the binary valence and innershell contributions. *Dashed*, *dash-dotted*, and *dotted lines*, binary valence contributions evaluated by using the MSR, SR, and Yukawa potentials, respectively; *dash-dot-dotted line*, innershell contribution calculated with the CDW-EIS approximation. The *thick solid line* represents experimental data extracted from Ref. [67], normalized to the theoretical values (from [66])

$$|j\rangle = |\mathbf{k}_j\rangle \qquad (j_c < j < j_F, |\mathbf{k}| < k_F) \tag{1.63}$$

$$\langle\psi_{\mathbf{k}}^-| = \langle\mathbf{k}| \qquad (|\mathbf{k}| > k_F, E_f \geq 0)\,. \tag{1.64}$$

The effective residual interactions employed are different forms of the dynamical screening (wake or image) potential as discussed in Sect. 1.2. Results are shown for isotropic screening with a functional form of the Yukawa type and the specular reflection model (SRM) and a modified specular reflection model (MSR) [66]. At large ejection energies the result is independent of the approximation invoked while for low-energy emission the spectra differ. For forward electron emission core electron contributions overshadow the conduction band contribution rendering the spectrum "atomic-like". The discrepancy to the experiment is a signature of transport, i.e. multiple scattering (see Chap. 2). Multiple scattering is not included in the calculation presented in [66]. At larger emission angles, the valence-electron contribution becomes comparably more important and the agreement with the experiment is better signaling that multiple scattering contributions are less important. Clearly, large uncertainties in the low-energy part of the emission spectrum resulting from different models for the screened interaction within an effective one-electron problem (isotropic screening of a Yukawa-like potential, SRM) remain (see Chap. 6).

Impulse Approximation

The impulse approximation (IA) can be viewed as a modification of the DWB approximation. Alternatively, it can also be considered to be the quantum version of the binary-encounter approximation.

Starting point is the *exact* transition amplitude

$$b_{\mathbf{k},j} = -i \int_{-\infty}^{\infty} dt \, \langle \mathbf{k}(t) | V | \psi_j^+(t) \rangle \, ; \quad (j < j_F) \, , \tag{1.65}$$

where $|\psi_j^+(t)\rangle$ is the forward-propagated exact state of the electron j with initial condition

$$|\psi_j^+(t)\rangle \overset{t \to \infty}{\longrightarrow} |j\rangle \, , \tag{1.66}$$

and V is the channel perturbation not included in the unperturbed state $|j\rangle$. The idea is now to decompose the initial state $\phi_j(\mathbf{r},t)$ into its Fourier components

$$\phi_j(\mathbf{r},t) = \int \frac{d^3q}{(2\pi)^{3/2}} \tilde{\phi}_j(\mathbf{q},t) e^{i\mathbf{q}\mathbf{r}} \tag{1.67}$$

and consider the elastic scattering of a wavepacket of quasi-free particles composed of the Fourier components with momenta $\mathbf{q}$ at the channel perturbation, the projectile potential V,

$$\left(-\frac{1}{2}\nabla^2 + V\right) \psi_{\mathbf{q}}^+ = \frac{1}{2} q^2 \psi_{\mathbf{q}}^+ \, . \tag{1.68}$$

$\psi_{\mathbf{q}}^+$ are the exact continuum states in the presence of the channel potential, in the present case the projectile potential V. Accordingly, the exact scattering state is replaced in the IA by [63]

$$\psi_j^+ = \int d^3q \tilde{\phi}_j(\mathbf{q},t) \psi_{\mathbf{q}}^+ \, . \tag{1.69}$$

This replacement involves the "impulse" hypothesis: the neglect of the target potential, i.e., the potential that is responsible for binding of the electron to the solid surface in its initial state, during the scattering process. The properties of the target, specifically of the target initial state, enter only through its Compton profile $\tilde{\phi}_j(\mathbf{q},t)$. This is the direct analogue to the classical binary-encounter approximation. In general, the quantum impulse approximation involves the "off-shell" propagation of the scattered electron, i.e., the energy which is associated to each Fourier component, $q^2/2$, in Eq. (1.68) is different from the energy of the scattered electron. In other words, the free-particle energy-momentum relation, $E = q^2/2$, is, in general, not satisfied. The ionization amplitude in IA is thus given by

$$b_{\mathbf{k}_j} = -i \int dt d^3p d^3q \, \tilde{\phi}^*_{\mathbf{k}}(\mathbf{p}, t) \langle \mathbf{p} | V | \psi^+_{\mathbf{q}} \rangle \tilde{\phi}_j(\mathbf{q}, t) \,, \tag{1.70}$$

where $\tilde{\phi}_{\mathbf{k}}(\mathbf{p}, t)$ is the Fourier component $\mathbf{p}$ of the continuum final state $\mathbf{k}$. In the special case, that the latter can be replaced by a plane wave, $\tilde{\phi}_{\mathbf{k}}(\mathbf{p}) = \delta(\mathbf{k} - \mathbf{p})$, Eq. (1.70) reduces to

$$b_{\mathbf{k},j} = -i \int dt d^3q \, e^{ik^2t/2} \langle \mathbf{k} | V | \psi^+_{\mathbf{q}} \rangle \tilde{\phi}_j(\mathbf{q}, t) \,. \tag{1.71}$$

In Eq. (1.71) $\langle \mathbf{k} | V | \psi^+_{\mathbf{q}} \rangle$ can be identified to be the two-body T-matrix element $T_{fi}(\mathbf{k}, \mathbf{q})$ which, in general, is off the energy shell (i.e. $|\mathbf{k}| \neq |\mathbf{q}|$). If one approximates the amplitude $T_{fi}(\mathbf{k}, \mathbf{q})$ by its on-shell limit (i.e. $k^2/2 = q^2/2$), Eq. (1.71) can be related to the elastic scattering cross section of an electron at the projectile potential. The binary-encounter approximation is a special case of this on-shell limit and follows from the replacement [63]

$$\sigma_{el}(q^2/2, \cos\theta) = 16\pi^4 T(\hat{\mathbf{k}}q, \mathbf{q}) \tag{1.72}$$

leading to

$$\frac{d\sigma}{dE d\Omega} = \frac{k}{v} \int d^3q \, \sigma_{el}(q^2/2, \cos\theta) \tilde{\phi}_j(\mathbf{q} - \mathbf{v}_p) \cdot \delta(q^2/2 - q'^2/2) \tag{1.73}$$

with

$$(\mathbf{k} - \mathbf{v}_p)^2/2 = (\mathbf{q}' - \mathbf{v}_p)^2/2 - (\mathbf{q} - \mathbf{v})^2/2 \,. \tag{1.74}$$

Equations (1.71) and (1.73) are of particular interest for kinetic emission from the conduction band. One expects for a nearly free electron the Compton profile of the initial state j, $\mathbf{k}_j$, to be close to that of a free electron

$$\tilde{\phi}_j(\mathbf{q}) \approx \delta(\mathbf{k}_j - \mathbf{q}) \,. \tag{1.75}$$

Since $|\mathbf{k}_j| \leq k_F$, the differential ionization spectrum should have an upper-energy cut-off under the assumption of Eq. (1.65) at

$$\frac{q^2}{2} \leq \frac{(k_F + v_p)^2}{2} \,, \tag{1.76}$$

which is the theoretical underpinning of the threshold law for KE discussed in Sect. 1.3 Eq. (1.36). Conversely, if the Compton profile of the initial state features higher momentum components, i.e., $k_j > k_F$ at $\varepsilon_j = \varepsilon_F$, the emission spectrum extends to higher energies or, equivalently, emission below the threshold becomes possible.

Realistic momentum distributions near a metal surface are far from simple Fermi distributions derived from the free-electron model that does not feature momentum components above k_F. Instead, electron correlation and the corrugated surface potential induce the presence of higher momenta. Using such momentum distributions, it is possible to account for the observed sub-threshold KE [31] within the framework of IA or BEA (see Sect. 1.5.1).

1.4.3 Low-Energy Approximations

Quasi-Molecular Promotion: Fano-Lichten and Related Models

In 1965 Fano and Lichten [68] interpreted the ionization in slow inelastic Ar-Ar^+ collisions with the help of diabatic molecular-orbital diagrams of the interaction system. The collision velocity v_p was, by far, too low to result in a binary momentum transfer sufficient to overcome the ionization threshold, as discussed in the preceding section. Instead, Fano and Lichten realized that certain quasi-stationary energy levels of the transiently formed $(Ar–Ar)^+$ quasi-molecule are shifted to level energies above the ionization threshold creating inner-shell holes and autoionizing resonances. Key to this picture is that the electronic wavefunction can follow quasi-adiabatically the changes of the slowly moving atomic (ionic) cores. Only near narrow avoided crossings "jump" the electronic states from one adiabatic potential surface to another. Such connections between adiabatic potentials near avoided crossings are referred to as *diabatic* potential surfaces. The point to be noted is that the propagation of the collision system along diabatic surfaces allows for the continuous exchange of energy between the nuclear (heavy particle) and the electronic degree of freedom. Consequently, ionization by quasi-molecular promotion corresponds to *kinetic emission* even though no direct energy transfer in a binary collision between the electron and the heavy particle is operative.

A similar picture applies to electron emission spectra in ion-surface interactions if the projectile undergoes close collisions with target atoms [69]. Figure 1.10 shows the level diagram for the Ne-Ni system. Since the lattice constant of Ni is about $a \approx 3.5\,\text{Å}$ the level diagram for the atomic system can be considered a reasonable approximation, especially for the interesting range of distances below 1 Å. The conduction band of Ni is formed by its two 4s electrons. As the ion approaches the target atom, quasi-molecular levels are formed which converge for large interatomic distances to atomic levels. At certain distances, levels with the same symmetry can approach each other in a narrow region but cannot cross (Wigner-von Neumann non-crossing rule). Charge exchange between asymptotic projectile and target states takes place with a significant probability only in the region of these "avoided crossings". Transitions between potential curves can either produce a hole in a projectile state which gives rise to a Auger-type process or may shift the level above the vacuum level (see Fig. 1.10). Following a classical trajectory of the projectile near the surface where close atomic collisions occur and calculating the

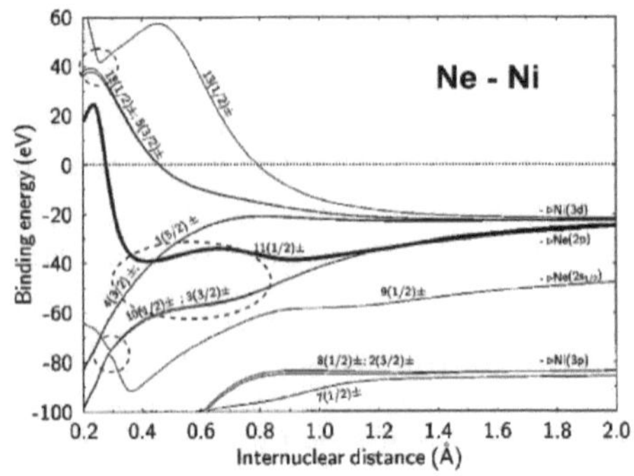

Fig. 1.10. Correlation diagram of the Ne-Ni collision system. Positions of avoided crossings are indicated by *dashed circles*. At internuclear distances $R = 0.3$, 0.45, and 0.8 Å molecular orbitals are shifted to above the ionization treshold (*dotted line*; from [71])

transition probabilities during single or multiple crossings, it is possible to estimate the contribution of level promotion and the filling of additional holes to the total electron yield [70]. As the pairwise interaction of levels is fairly localized, the system evolves along a potential curve until the next crossing occurs designating possible pathways of the system in Fig. 1.10 for a given projectile trajectory.

Landau-Zener-Stückelberg Approximation

The Landau-Zener-Stückelberg and Rosen-Zener approximations allow to compute the transition probabilities applicable to special cases for the shape of the adiabatic potential curves. Considering only a two-state model (a schematic picture of the region around the avoided crossing is shown in Fig. 1.11) the wavefunction can be written in the diabatic basis as

$$\psi(\mathbf{s},t) = A_1(t)\psi_1^d(\mathbf{s},R) + A_2(t)\psi_2^d(\mathbf{s},R) \ , \tag{1.77}$$

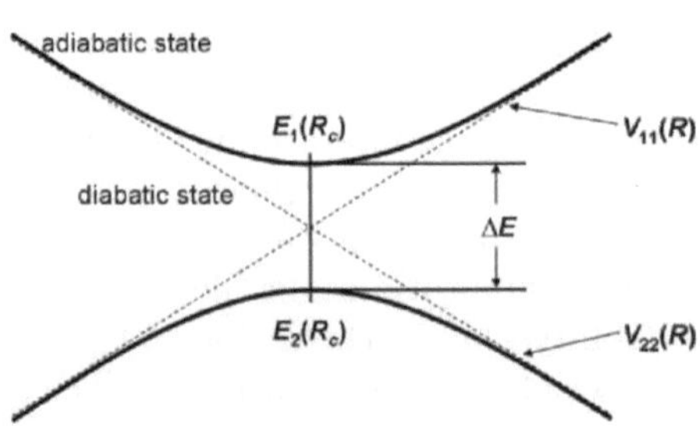

Fig. 1.11. Schematic diagram of an avoided crossing at $R = R_c$. Adiabatic levels are shown as *solid lines*, diabatic levels as *dashed lines*

where $\mathbf{s}$ is the electron-coordinate in the center-of-mass system and R the internuclear distance. Within a two-level approximation, the Schrödinger equation reduces to the system of coupled differential equations in the amplitudes,

$$i\dot{A}_1 = V_{11}(t)A_1(t) + V_{12}(t)A_2(t) \tag{1.78}$$
$$i\dot{A}_2 = V_{21}(t)A_1(t) + V_{22}(t)A_2(t) \tag{1.79}$$

with the elements of the coupling matrix $V_{ij}(t) = \langle\psi_i^d|H_e|\psi_j^d\rangle$. The initial boundary condition for the integration of Eqs. (1.78) and (1.79) along the (classical) trajectory are $A_1 = 1$ and $A_2 = 0$. At the crossing point $R = R_c$ the diabatic energies are equal ($V_{11} = V_{22}$) and the energy distance between the adiabatic levels is given by $\Delta E = 2|V_{12}(R_c)|$.

This set of equations can be simplified by removing the diagonal terms, i.e., by removing the evolution phase due to time (or R) dependent energy shifts of the diabatic curves. The new amplitudes C_i are related to the original amplitudes as

$$C_i(t) = \exp\left[i\int_0^t dt' V_{ii}(t')\,dt'\right] A_i(t) \tag{1.80}$$

leading to

$$i\dot{C}_1 = V_{12}(t)\exp(-i\delta(t))\,C_2(t) \tag{1.81}$$
$$i\dot{C}_2 = V_{21}(t)\exp(i\delta(t))\,C_1(t) \tag{1.82}$$

with

$$\delta(t) = \int_0^t [V_{22}(t') - V_{11}(t')]\,dt' \;. \tag{1.83}$$

To calculate the probability $P_{21} = |A_2(\tau > t_c)|^2$ for a transition from state 1 to state 2 at the crossing point $R = R_c$ passed at $t = t_c$, Landau [72], Zener [73], and Stückelberg [74] independently suggested to retain the linear term in the expansion of $[V_{22}(t') - V_{11}(t')]$ in R near the crossing point corresponding to the dashed line in Fig. 1.11 and approximate the off-diagonal element by a constant $V_{12} = \Delta E/2$, i.e., by its value at the crossing point. Note that the t and R dependences are used interchangeably ($R(t) = (\mathbf{v}_p)_R t$ where $(\mathbf{v}_p)_R$ is the radial component of the velocity). Under these assumptions and for a straight-line trajectory it is found that the probability for transitions between diabatic states is

$$P_{21}^D = 1 - \exp\left(-c \cdot \frac{\Delta E^2}{v_p}\right) \tag{1.84}$$

while those between adiabatic states is

$$P_{21}^A = \exp\left(-c \cdot \frac{\Delta E^2}{v_p}\right) \;. \tag{1.85}$$

For a "full" binary collision passing the region R_c twice (on the way in and on the way out) the joint probability for a transition from the potential curve 1 to the potential curve 2 is

$$P = 2P_{21}^D P_{21}^A = 2\exp\left(-c \cdot \frac{\Delta E^2}{v_p}\right)\left[1 - \exp\left(-c \cdot \frac{\Delta E^2}{v_p}\right)\right] . \tag{1.86}$$

At low velocities, the transition probability features a characteristic exponential suppression of inelastic transitions. This is a direct manifestation of the adiabatic limit (or of the uncertainty principle): As the collision takes infinitely long, all near-resonant Fourier components of the time-dependent perturbation that could drive the transition, $\omega \approx \Delta E$, have died out. An exponential decrease is also expected in the ionization spectrum as the emission of high-energy electrons where ΔE is large is suppressed. In the opposite limit of large velocities v_p we find $P_{21}^D \sim v_p^{-1}$ ($P_{21}^A = 1$), rather than the correct (first order perturbation) result $P_{21}^D \sim v_p^{-2} \ln v_p$. This reflects the restriction of the applicability of the Landau-Zener approximation to small velocities only.

Rosen-Zener Approximation

In Fig. 1.10 the large dotted circle emphasizes a region in which two levels with the same symmetry run in parallel over a large distance to finally repel each other at about $R = 0.4$ Å. No well-localized avoided crossing is recognizable. Yet, the two nearby levels interact allowing for population transfer modeled by the Rosen-Zener model also referred to as Demkov model [75]. It solves the coupled equations (1.78) and (1.79) assuming $[V_{22}(t) - V_{11}(t)] = \Delta E = const$ and $V_{12} = V_{21} = A/\cosh(Bt)$ in the transition region. These assumptions are somewhat complementary to those of the Landau-Zener model. The ansatz again allows for an analytic solution with the parameters A and B being functions of the velocity and the impact parameter. Integrating the transition probabilities over all allowed impact parameters ($b \leq R_c$; for larger b the transition does not occur), Bates derived a total cross section for the transition [76]

$$\sigma_{21}(v) = \frac{c_1 v^2}{\Delta E^2}\left(1 + \frac{c_2 \Delta E}{v}\right)\exp(-c_2 \Delta E/v) . \tag{1.87}$$

Before nearly perfect single-crystal surfaces could be used for measurements of the KE yield as in [31], promotion of quasi-molecular levels was a major obstacle to investigate the low-velocity threshold of kinetic emission. For very grazing angles of incidence the projectile trajectory has a turning point some a.u. in front of the topmost atomic layer. Since promotion to positive energies occurs for most target-projectile combinations only for distances smaller than 1.5 a.u., this channel for KE should not contribute. Surface imperfections (steps), however, destroy the channeling conditions along the smooth surface potential. Projectiles hitting surface steps get into close contact with target atoms reducing the internuclear distance. The term "surface assisted KE" was introduced to characterize this source of electron emission [69].

1.5 Many-body Effects: Density Functional Theory

1.5.1 Ground State Calculations

Describing stationary states as well as the evolution in time of a large system consisting of charged particle is, due to the large number of particles involved, out of reach. However, mean-field theories such as density-functional theory (DFT) promise a drastic simplification of the calculation. DFT is based on the notion that all physical observables can be written as a functional of the electron density. It has been shown that there exists a unique mapping between electronic potential and ground state density. Instead of the N-particle wavefunction, the density is constructed as the sum over single-particle pseudo-wavefunctions which are solutions of the set of Kohn-Sham (KS) equations [77],

$$H_0(\mathbf{r})\varphi_i(\mathbf{r}) = (T + V_{es}(\mathbf{r}) + V_{xc}[n(\mathbf{r})])\,\varphi_i(\mathbf{r}) = \varepsilon_i\varphi_i(\mathbf{r}) \tag{1.88}$$

$$n_0(\mathbf{r}) = \sum_i |\varphi_i(\mathbf{r})|^2\,, \tag{1.89}$$

where $n(\mathbf{r})$ is the electron density at position $\mathbf{r}$ and V_{es} and V_{xc} are the electrostatic and exchange-correlation potentials, respectively. The Kohn-Sham orbitals $\varphi_i(\mathbf{r})$ should not be interpreted as approximations to single-particle wavefunctions. Instead, their significance lies in the fact that they add up to the exact ground state density [see Eq. (1.89)] if the total energy of the system is minimized [78] and the exact exchange-correlation potential $V_{xc}[n(\mathbf{r})]$ were available. The latter, however, is in practice not the case. Different approximations to $V_{xc}[n(\mathbf{r})]$ have to be employed, in most cases optimized for the specific case investigated. For large systems, a *local* exchange-correlation potential which is given as a functional of the electron density $n(\mathbf{r})$ is frequently employed (local density approximation, LDA),

$$V_{xc}[n(\mathbf{r})] = \frac{0.611}{r_s(n)} - \frac{0.587}{(r_s(n) + 7.8)^2}(r_s(n) + 5.85)\,. \tag{1.90}$$

r_s is the Wigner-Seitz radius defined by $n = 3/4\pi r_s^3$.

One application of DFT for the groundstate of a metal surface for atom-surface collisions is the calculation of the Compton profile, i.e., the momentum density which enters the calculation of electron emission in the impulse approximation or, equivalently, the binary encounter approximation. Starting point are the KS-orbitals associated with band index n and wavevector $\mathbf{k}$ within the first Brioullin zone which can be written as

$$\psi_{n,\mathbf{k}}(\mathbf{r}) = \sum_{\mathbf{G}} c_{n,\mathbf{k}}(\mathbf{G})\, e^{i(\mathbf{k}+\mathbf{G})\mathbf{r}}\,, \tag{1.91}$$

where $\mathbf{G}$ is a reciprocal lattice vector. The KS orbitals $\psi_{n,\mathbf{k}}(\mathbf{r})$ can be combined to give the electron density

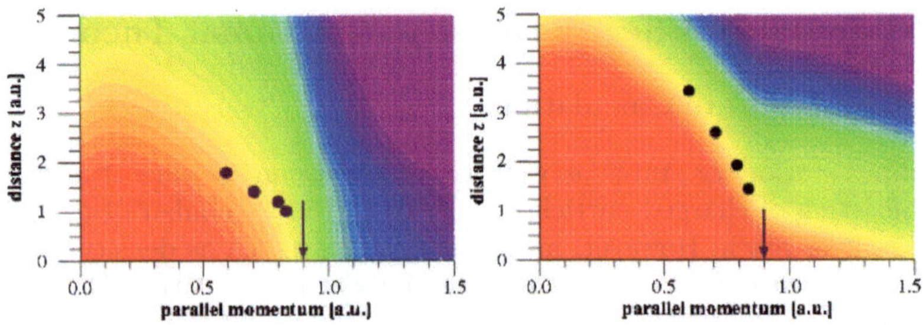

Fig. 1.12. Distribution of conduction-electron momenta parallel to the surface for Al(100) and Al(111) surfaces as a function of the distance from the topmost atomic layer. The intensity at k_F is indicated by the *black arrow*. Dots are experimental estimates [82] for a "local Fermi velocity" (from [79])

$$\rho(\mathbf{r}) = \sum_{n,\mathbf{k}} \mathrm{occ}(n,\mathbf{k})\, w_{\mathbf{k}}\, |\psi_{n,\mathbf{k}}(\mathbf{r})|^2 \tag{1.92}$$

with $\mathrm{occ}(n,\mathbf{k})$ and $w_{\mathbf{k}}$ being the occupation of the level with index $n, \mathbf{k}$ and a weighting factor for the symmetry of the point $\mathbf{k}$ in reciprocal space, respectively. Fourier transforms of the wavefunctions $\psi_{n,\mathbf{k}}(\mathbf{r})$ from spatial to momentum coordinates and summing over n and $\mathbf{k}$ give the desired momentum distributions. Figure 1.12 represents the two-dimensional Fourier transform [79] as a function of the momentum parallel to the surface, $|\mathbf{q}_{\|}|$,

$$\begin{aligned}\rho(\mathbf{q}_{\|}, z) &= \sum_{n,\mathbf{k}} \mathrm{occ}(n,\mathbf{k})\, w_{\mathbf{k}}\, \left|\psi_{n,\mathbf{k}}(\mathbf{q}_{\|}, z)\right|^2 \\ &\propto \sum_{n,\mathbf{k}} \mathrm{occ}(n,\mathbf{k})\, w_{\mathbf{k}} \sum_{\mathbf{G},\mathbf{G}'} c_{n,\mathbf{k}}(\mathbf{G}) c^*_{n,\mathbf{k}}(\mathbf{G}')\, e^{i(G_z - G'_z)z} \times \\ &\qquad \times \delta(\mathbf{q}_{\|} - (\mathbf{k}_{\|} + \mathbf{G}_{\|}))\delta(\mathbf{q}_{\|} - (\mathbf{k}_{\|} + \mathbf{G}'_{\|}))\,.\end{aligned} \tag{1.93}$$

As an example, we show calculations with the ABINIT code [80] using structure-optimized pseudopotentials in the generalized gradient approximation to the exchange-correlation potential taken from ref. [81]. The resulting momentum distributions (Fig. 1.12) display significant contributions from $\mathbf{q}$ above the Fermi wavenumber $|\mathbf{q}| > k_F$ to the Compton profile. They are the root cause of the so-called "sub-threshold" emission. Furthermore, the momentum distribution is found to depend on the orientation of the crystal structure (see Fig. 1.12).

Obviously, a realistic momentum distribution in a metal is far from a Fermi distribution derived from the free-electron model that lacks momentum components above k_F. Instead, electron correlation and the corrugated surface potential induce the presence of higher momenta. Using such a momentum distribution it is possible to explain the observed sub-threshold KE [82] within the framework of IA or BEA (see also Chap. 4).

1.5.2 Time-Dependent Density-Functional Theory

The term time-dependent density-functional theory (TDDFT) is used for two distinct methods: on the one hand, for linear response calculations accounting for exchange-correlation corrections from DFT [see Sect. 1.2.2, Eqs. (1.19) and (1.20)], on the other hand, for the time propagation of self-consistently calculated pseudo-wavefunctions. Here, we focus on the latter.

Having calculated the ground-state density of the system it is possible to propagate the pseudo-wavefunctions in time under the influence of an external perturbing potential, in the present case the time-dependent field of the scattered ion (atom). Theorems have been established [83] that are the time-dependent equivalents to those of Hohenberg and Kohn for the static ground state [78]. One of them proves the one-to-one mapping of time-dependent potentials and time-dependent densities. Therefore, the time-dependent density determines the external potential *uniquely* (up to an additive purely time-dependent function). On the other hand, the potential determines the time-dependent wavefunction, making the expectation value of any quantum mechanical operator a unique functional of the density. Consequently, one can construct the time-dependent version of Eqs. (1.88) and (1.89) as

$$H(\mathbf{r},t)\varphi_i(\mathbf{r},t) = \{T + V_{eff}[n(\mathbf{r},t)]\}\varphi_i(\mathbf{r},t) = i\frac{\partial}{\partial t}\varphi_i(\mathbf{r},t) \qquad (1.94)$$

$$n(\mathbf{r},t) = \sum_i |\varphi_i(\mathbf{r},t)|^2 \qquad (1.95)$$

with $V_{eff}(\mathbf{r},t)$ containing the time-dependent Hartree, exchange-correlation, and external potentials. Time-dependent wavefunctions of Eq. (1.94) are propagated from time t to time $t + \Delta t$ by

$$\varphi_i(\mathbf{r},t+\Delta t) = e^{-i\int_t^{t+\Delta t} H(\mathbf{r},t)\,dt} \cdot \varphi_i(\mathbf{r},t)\ . \qquad (1.96)$$

Using such a time-dependent simulation gives access to information on the early stages of the interaction. For example, the formation of a polarization layer at the surface which shields the inside of the solid against the external charge can be studied. Likewise, the first electron transfer above or through the potential barrier [84] can be observed, an example of which is shown in Fig. 1.13. The simulation of the complete neutralization sequence is still out of reach as two-electron processes (e.g. autoionization) are not yet included in this mean-field theory. Future "full" simulations will require to also calculate occupation probabilities for the electrons transferred from the surface to the incoming ion. This, however, is not a straight-forward matter of projecting KS-wavefunctions on projectile states. Only recently a functional has been developed that allows, in principle, for the extraction of time-dependent occupations for a two-electron model system [85].

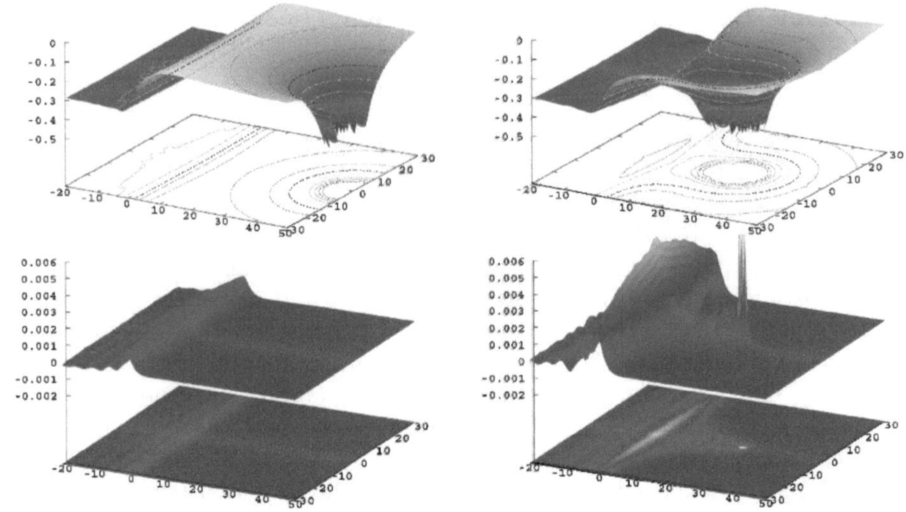

Fig. 1.13. Self-consistently calculated potential energy surfaces (*top pictures*) and induced density change (*bottom pictures*) for a triply charged projectile approaching a surface with $r_s = 3$ a.u. and $W \approx 0.33$ a.u. For large ion-surface distances R_z the projectile induces a polarization layer in the target (*left panels*). At smaller distances electron transfer to the ion becomes possible (*right panels*)

1.6 Open Questions and Future Directions

Ab-initio simulations of electron emission in ion-surface and atom-surface collisions remain a challenge for theory in the foreseeable future. Based on the progress made over the last decade, on can realistically expect that for a few effective one-electron processes fairly complete simulations of the electron dynamics within the framework of the time-dependent density-functional theory method will be forthcoming. Since for oblique incidence of the classical trajectory of the heavy particle the problem is fully three-dimensional without the opportunity to effectively reduce the number of degrees of freedom, even such "simple" problems remain a challenge. "Super-cell" methods invoking periodic boundary conditions to account for extended systems are intrinsically difficult to apply because of the long-range Coulomb field and the extended range of grazing-incidence trajectories. A prototypical case would be electron emission induced by protons scattered at a metal surface.

For more complex electronic processes involving electron correlation or "hard" electron-electron interactions beyond the mean-field level, an ab-initio treatment appears beyond current capabilities. Prototypical examples for the latter would be electron emission by Auger dacay, Auger capture, or Auger deexcitation. Here hybrid classical-quantum methods are likely to be most successful. Rates and matrix elements of effective one-electron processes may be calculated from TDDFT while those for (correlated) two-electron processes can be determined from more sophisticated multi-electron approaches including the multi-reference configuration interaction (MR-CI) method [86] and Bethe-Salpeter equations [87]. Those rates may then serve as input into multi-electron transport simulations based on either classical [52] or quantum [88] dynamics.

Future applications will likely address several technologically relevant issues. They include detection and detailed mapping out of surface ferromagnetism and anti-ferromagnetism by energy- and angle-resolved spectra. Another topic of interest would be electron emission accompanying nanostructuring by ion impact. Highly charged ions create dot-like defects ("blisters" or "craters" [89, 90]) involving the displacement of thousands and the emission of hundreds of surface atoms. Studying both theoretically and experimentally the correlated electron emission from a single dot would be of considerable interest, both for the understanding of blister formation and for diagnostic purposes. Another avenue of future investigations is the transmission through nano-capillaries. While ionic guiding effects have been observed [91–94] and simulated [53], little is known about the concomitant electron emission. Meanwhile also electron guiding through nanocapillaries has been detected [95].

Part of this work was performed in collaboration with S. Deutscher, C.O. Reinhold, K. Schiessl, B. Solleder, K. Tőkesi, X.-M. Tong, and L. Wirtz. Discussions with R. Abel, M. Alducin, A. Arnau, F. Aumayr, P. Echenique, R. Morgenstern, R. Schuch, N. Stolterfoht, Y. Yamazaki, HP. Winter, and H. Winter are gratefully acknowledged. Work supported by FWF special research program SFB016 "ADLIS" and by EU-projects HITRAP (HPRI-CT-2001-50036) and ITS-LEIF (HPRI-CT-2005-026015).

References

1. R. Car and M. Parrinello, Phys. Rev. Lett. **55**, 2471 (1985)
2. J. Burgdörfer: Atomic collisions with surfaces. In: *Review of Fundamental Processes and Applications of Atoms and Ions*, ed. C.D. Lin (World Scientific, Singapore 1993) pp. 517-614
3. H. Winter, Physics Reports **367**, 387 (2002)
4. A. Arnau et al., Surf. Sci. Rep. **27**, 113 (1997)
5. P.M. Echenique, F. Flores, and R.H. Ritchie, Sol. Stat. Phys. **43**, 229 (1990)
6. S. Ossicini, C. Bertoni, and P. Gies, Surf. Sci. **178**, 244 (1986)
7. R. Ritchie and A. Marusak, Surf. Sci. **4**, 234 (1966)
8. L. Hägg, C.O. Reinhold, and J. Burgdörfer, Phys. Rev. A **55**, 2097 (1997)
9. P.J. Jennings, R.O. Jones, and M. Weinert, Phys. Rev. B **37**, 6113 (1988); P.J. Jennings and R.O. Jones, Advances in Physics **37**, 341 (1988)
10. S. Deutscher, X. Yang, and J. Burgdörfer, Phys. Rev. A**55**, 466 (1997)
11. R.H. Garvey, C.H. Jackman, and A.E.S. Green, Phys. Rev. A **12**, 1144 (1975), and references therein
12. P Nozières and D. Pines: *The Theory of Quantum liquids*, Vol I, (Perseus Books, Reading, Massachusetts, 1998)

13. see e.g., A. Liebsch: *Electronic Excitations at Metal Surfaces* (Plenum, NY, 1997)
14. K. Tőkési, X.-M. Tong, C. Lemell and J. Burgdörfer, Phys. Rev. A **72**, 022901 (2005); Adv. in Quantum Chemistry **46**, 29 (2004)
15. C.O. Reinhold and J. Burgdörfer, Phys. Rev. A **55**, 450 (1997)
16. R.A. Baragiola, E.V. Alonso, and A. Oliva Florio, Phys. Rev. B **19**, 121 (1979)
17. P. Appell, Nucl. Instr. and Meth. in Phys. Res. B **23**, 242 (1987)
18. M.S. Daw and I. Baskes, Phys. Rev. B **29**, 6443 (1984); Phys. Rev. Lett. **50**, 1285 (1983)
19. I. Torrens: *Interaction Potentials*, (Academic Press, NY, 1972)
20. G. Molière, Z. Naturforsch. A **2**, 133 (1947)
21. J. Ziegler, J. Biersack, and U. Littmark: *The Stopping and Range of Ions in Solids*, (Pergamon, NY, 1985)
22. R. Shepard, I. Shavitt, R.M. Pitzer, D.C. Comeau, M. Pepper, H. Lischka, P.G. Szalay, R. Ahlrichs, F.B. Brown, and J.G. Zhao, Int. J. Quantum Chem. **22**, 149 (1988); H. Lischka et al., computer code COLUMBUS
23. A. Schüller, S. Wethekam, A. Mertens, K. Maas, H. Winter, and K. Gärtner, Nucl. Instr. and Meth. in Phys. Res. B **230**, 172 (2005)
24. J. Lindhard, Phys. Lett. **12**, 126 (1964)
25. J.J. Thomson, Philos. Mag. **23**, 449 (1912)
26. C. Reinhold and J. Burgdörfer, J. Phys. B **26**, 3101 (1993)
27. M. Gryziński, Phys. Rev. **115**, 374 (1959)
28. L. Vriens, Phys. Rev. **160**, 100 (1967)
29. R. Abrines and I.C. Percival, Proc. Phys. Soc. London **88**, 861 (1966)
30. H. Winter and HP. Winter, Europhys. Lett. **62**, 739 (2003)
31. HP. Winter, S. Lederer, H. Winter, C. Lemell, and J. Burgdörfer, Phys. Rev. B **72**, 161402 (2005)
32. J. Burgdörfer and J. Gibbons, Phys. Rev. A **42**, 1206 (1990)
33. C.O. Reinhold, J. Burgdörfer, K. Kimura, and M. Mannami, Phys. Rev. Lett. **73**, 2508 (1994)
34. H. Ryufuku, K. Sasaki, and T. Watanabe, Phys. Rev. A **21** 745 (1980)
35. N. Bohr and J. Lindhard, Dan. Vid. Sel. Mat. Phys. Medd. **78**, No. 7 (1954)
36. A. Bárány, G. Astner, H. Cederquist, H. Danared, S. Huldt, P. Hvelplund, A. Johnson, H. Knudsen, L. Liljeby, and K.G. Rensfelt, Nucl. Instr. and Meth. in Phys. Res. B **9**, 397 (1985)
37. A. Niehaus, J. Phys. B **19**, 2925 (1986)
38. J. Burgdörfer, P. Lerner, and F. W. Meyer, Phys. Rev. A **44**, 5674 (1991)
39. J. Burgdörfer and F. Meyer, Phys. Rev. A **47**, R20 (1993)
40. H. Winter, C. Auth, R. Schuch and E. Beebe, Phys. Rev. Lett. **71**, 1939 (1993)
41. F. Aumayr, H. Kurz, D. Schneider, M.A. Briere, J.W. McDonald, C.E. Cunningham and HP Winter, Phys. Rev. Lett. **71**, 1943 (1993)
42. C. Lemell, HP. Winter, F. Aumayr, J. Burgdörfer, and F.W. Meyer, Phys. Rev. A**53**, 880 (1996)
43. S. Ninomya, Y. Yamazaki, F. Koike, H. Masuda, T. Azuma, K. Komaki, K. Kuroki, and M. Sekiguchi, Phys. Rev. Lett. **78**, 4557 (1997)
44. Y. Morishita, R. Hutton, H.A. Torii, K. Komaki, T. Brage, K. Ando, K. Ishii, Y. Kanai, H. Masuda, M. Sekiguchi, F.B. Rosmej, and Y. Yamazaki, Phys. Rev. A **70**, 012902 (2004)
45. R.D. Cowan: *The Theory of Atomic Structure and Spectra*, (University of California Press, Berkeley, 1981)

46. J. Burgdörfer, C.O. Reinhold, and F.W. Meyer, Nucl. Instr. and Meth. in Phys. Res. B **205**, 690 (2003)
47. J. Burgdörfer, C. Lemell, K. Schiessl, B. Solleder, C. Reinhold, K. Tőkési, and L. Wirtz, in *Proceedings of ICPEAC XXIV* (World Scientific, Singapore, 2006); arXiv:physics/0605245
48. C. Lemell, M. Alducin, J. Burgdörfer, I. Juaristi, K. Schiessl, B. Solleder and K. Tőkési, Rad. Phys. Chem. **76**, 412 (2007)
49. R. Pfandzelter, T. Bernhard, and H. Winter, Phys. Rev. Lett. **86**, 4152 (2001)
50. I.G. Hughes, J. Burgdörfer, L. Folkerts, C.C. Havener, S.H. Overbury, M.T. Robinson, D.M. Zehner, P.A. Zeijlmans van Emmichoven, and F.W. Meyer, Phys. Rev. Lett. **71**, 291 (1993)
51. C. Lemell, J. Stöckl, J. Burgdörfer, G. Betz, HP. Winter, and F. Aumayr, Phys. Rev. Lett. **81**, 1965 (1998)
52. L. Wirtz, C. Reinhold, C. Lemell, and J. Burgdörfer, Phys. Rev. A **67**, 012903 (2003)
53. K. Schiessl, W. Palfinger, C. Lemell, and J. Burgdörfer, Nucl. Instr. and Meth. in Phys. Res. B **232**, 228 (2005); K. Schiessl, W. Palfinger, K. Tőkési, H. Nowotny, C. Lemell, and J. Burgdörfer, Phys. Rev. A **72**, 062902 (2005)
54. C. Deiss, N. Rohringer, J. Burgdr̈fer, E. Lamour, C. Prigent, J.-P. Rozet, and D. Vernhet, Phys. Rev. Lett. **96**, 013203 (2006)
55. S. Winecki, M. Stöckli, and C. Cocke, Phys. Rev. A **55**, 4310 (1997)
56. J. Burgdörfer, C.O. Reinhold, and F.W. Meyer, Nucl. Instr. and Meth. in Phys. Res. B **98**, 415 (1995)
57. G. Schiwietz, Phys. Rev. A **42**, 296 (1990); P. Grande, G. Schiwietz, G. Sigaud, E. Montenegro, Phys. Rev. A **54**, 2983 (1996); P. Grande and G. Schiwietz, Phys. Rev. A **58**, 3796 (1998)
58. J. Burgdörfer, E. Kupfer, and H. Gabriel, Phys. Rev. A **35**, 4963 (1987)
59. A. Borisov, J.P. Gauyacq, V. Sidis, A.K. Kazansky, Phys. Rev. B **63**, 045407 (2001)
60. A. Borisov and J. Gauyacq, Phys. Rev. B **62**, 4265 (2000)
61. H. Chakraborty, T. Niederhausen, and U. Thumm, Phys. Rev. A **70**, 052903 (2004)
62. U. Thumm, J. Phys. B **25**, 421 (1992)
63. J. Wang, C.O. Reinhold, and J. Burgdörfer, Phys. Rev. A **44**, 7243 (1991)
64. H. Winter, P. Strohmeier, and J. Burgdörfer, Phys. Rev. A **39**, 3895 (1989)
65. P. Fainstein, V. Ponce, R. Rivarola, J. Phys. B **22**, L559 (1989)
66. M.S. Gravielle and J.E. Miraglia, Phys. Rev. A **65**, 022901 (2002)
67. M.L. Martiarena, E.A. Sanchez, O. Grizzi, and V.H. Ponce, Phys. Rev. A **53**, 895 (1996)
68. U. Fano and W. Lichten, Phys. Rev. Lett. **14**, 627 (1965)
69. HP. Winter, F. Aumayr, J. Lörincik, and Z. Sroubek, Izvestia R. Acad. Nauk **66**, 548 (2002)
70. C. Lemell, J. Stöckl, J. Burgdörfer, G. Betz, HP. Winter, and F. Aumayr, In: *The Physics of Electronic and Atomic Collisions*, CP500, ed. by Y. Itikawa et al., (American Inst. of Physics, New York 2000) pp. 656–665
71. P. Kürpick, Phys. Rev. B **56** , 6446 (1997)
72. L. Landau, Phys. Z. d. Sowjetunion **2**, 46 (1932)
73. C. Zener, Proc. Roy. Soc. A **137**, 696 (1932)
74. E. Stückelberg, Helv. Phys. Acta **5**, 369 (1932)

75. N. Rosen and C. Zener, Phys. Rev. **40**, 502 (1932); Y.N. Demkov, Sov. Phys. JETP **18**, 138 (1964)
76. D.R. Bates, Discuss. Faraday Soc. **33**, 7 (1962)
77. W. Kohn and L.J. Sham, Phys. Rev. **140**, A1133 (1965)
78. P. Hohenberg and W. Kohn, Phys. Rev. B **136**, 864 (1964)
79. C. Lemell, A. Arnau, and J. Burgdörfer, Phys. Rev. B **75**, 014303 (2007)
80. X. Gonze et al., Comp. Mat. Science **25**, 478 (2002)
81. A.M. Rappe, K.M. Rabe, E. Kaxiras, and J.D. Joannopoulos, Phys. Rev. B **41**, 1227(R) (1990); Erratum Phys. Rev. B **44**, 13175 (1991)
82. HP. Winter, S. Lederer, H. Winter, C. Lemell, and J. Burgdörfer, Phys. Rev. B **72**, 161402 (2005)
83. E. Runge and E.K.U. Gross, Phys. Rev. Lett. **52**, 997 (1984)
84. C. Lemell, X.-M. Tong, K. Tökesi, and J. Burgdörfer, Nucl. Instr. and Meth. in Phys. Res. B **235**, 425 (2005)
85. N. Rohringer, S. Peter, and J. Burgdörfer, Phys. Rev. A **74**, 042512 (2006)
86. L. Wirtz, J. Burgdörfer, M. Dallos, T. Müller, and H. Lischka, Phys. Rev. A **68**, 032902 (2003)
87. A. Marini: A Many-body Approach to the Electronic and Optical Properties of Copper and Silver. In: *Correlation Spectroscopy of Surfaces, Thin Films, and nanostructures*, ed. by J. Berakdar, J. Kirschner (WILEY-VCH Verlag, Weinheim, 2004) pp 17–31
88. M. Seliger, C.O. Reinhold, T. Minami, and J. Burgdörfer, Phys. Rev. A **71**, 062901 (2005)
89. D. Schneider, M.A. Briere, M.W. Clark, J. McDonald, J. Biersack, and W. Siekhaus, Surf. Sci. **294**, 403 (1993)
90. I. Gebeshuber, S. Cernusca, F. Aumayr, and HP. Winter, Int. J. Mass. Spec. **229**, 27 (2005); F. Aumayr, private communication (2006)
91. N. Stolterfoht, J. Bremer, V. Hoffmann, R. Hellhammer, D. Fink, A. Petrov, and B. Sulik, Phys. Rev. Lett. **88**, 133201 (2002)
92. Gy. Vikor, R.T. Rajendra Kumar, Z.D. Pesíc, N. Stolterfoht, and R. Schuch, Nucl. Instr. and Meth. Phys. Res. B**233**, 218 (2005)
93. Y. Kanai, private communication (2005)
94. F. Aumayr, private communication (2005)
95. A.R. Milosavljevic, Z.D. Pesic, D. Sevic, S. Matefi-Tempfli, M. Matefi-Tempfli, L. Piraux, Gy. Vikor, and B.P. Marinkovic, Proceedings of SPIG 2006, 167 (Inst. of Physics, Belgrade, Serbia, 2006); A.R. Milosavljevic, Gy. Vikor, Z.D. Pesic, P. Kolarz, D. Sevic, and B.P. Marinkovic, Phys. Rev. A **75**, (2007, in print)

2

Photon and Electron Induced Electron Emission from Solid Surfaces

Wolfgang S.M. Werner

The emission of electrons from solid surfaces under photon and electron impact is discussed. The focus of this contribution is on the transport of electrons from their point of generation inside the solid to the surface. The physical quantities characterizing the electron–solid interaction are introduced and their sources in the literature are given. The theory for multiple scattering is outlined in some detail and spectrum processing procedures based on it are explained. The theoretical concepts are illustrated by means of experimental examples. The considered phenomena include elastic and inelastic electron reflection from solid surfaces, Auger– and photoelectron emission and secondary electron emission.

2.1 Introduction

Emission of electrons from solid surfaces can be brought about by a huge variety of different physical processes, such as bombardment of the surface with e.g., electrons [1–3], photons [4] or heavy charged particles [5–7], by heating a specimen giving rise to thermal emission [8], or by subjecting it to a high electric field leading to field–emission [9]. When the excitation process involves a sufficient energy transfer to the solid to ionize atoms in the specimen, electrons with an energy characteristic for the considered energy level of the solid, such as Auger– or photoelectrons can be created, that can be used to characterize the composition, the chemical, electronic, magnetic or crystallographic structure of a surface. Otherwise, the transferred energy is dissipated in a cascade of excitation processes involving the subsystem of free or weakly bound solid state electrons that eventually gives rise to emission of slow electrons which are observed in the low kinetic energy part of an electron spectrum as a broad and rather featureless peak typically below $\lesssim$10 eV.

Particle induced electron emission is commonly interpreted in the framework of a very general physical model, the so–called three step model, in which

W.S.M. Werner: *Photon and Electron Induced Electron Emission from Solid Surfaces*,
STMP **225**, 39–77 (2007)
DOI 10.1007/3-540-70789-1_2

the excitation of the emitted electrons, their transport from the point of liberation inside the solid into vacuum and their escape over the surface–vacuum barrier are treated as three separate steps. The electron excitation mechanism obviously depends on the details of the employed process, but similar mechanisms are relevant in different processes. For example, when a slow ion that approaches a surface is neutralized by electron capture from the surface, the sudden change in the potential gives rise to charge density oscillations ("plasmons") in the vicinity of the surface that lead to a slowing down of the incoming particle and decay after a few oscillations. In consequence, the incoming particle suffers a characteristic energy loss, which is difficult to observe experimentally. In further consequence, a slow electron with an energy characteristic of the frequency of the oscillation is emitted, a phenomenon which has been observed in several experiments and is known as potential emission of plasmons, treated in detail in Chap. 6 of this book. Likewise, when an atomic core level inside a solid is ionized during a photoelectric transition, the solid state electrons screen the core hole –that represents a strong and sudden perturbation– and a characteristic oscillating field is set up inside the solid, which decelerates the emitted photoelectron and eventually decays thereby transferring its energy to a solid state electron. The latter can be observed in the peak of slow electrons in a spectrum. In the case of photoemission the characteristic loss is easy to observe in the form of a plasmon loss structure accompanying the characteristic photopeak, the so–called *intrinsic* plasmons [4]. However, it is difficult to distinguish these from the excitations that occur when the electron traverses the solid on its way to the surface, the so–called *extrinsic* plasmons. Note that the excitation of extrinsic plasmons in photon induced experiments is similar to the mechanism giving rise to particle induced kinetic emission. In the case of photon induced electron emission, the liberation of characteristic photoelectrons is the most important feature of the process and the decay of the elementary excitations initiated, which also produces a broad peak of slow electrons in the spectrum, has not been studied as extensively as for particle induced electron emission.

The above makes it clear that the response of the solid to the incoming particles, be it photons, electrons, ions of any charge state and the like, always represents an important aspect of the electron excitation mechanism, even if the excitation mechanism per se is essentially different. This statement pertains in a much stronger form to the fate of the liberated electrons since they interact with the solid in *exactly* the same way, at least within the concept of the three–step model. In other words, the second step of the three step model, the electron transport to the surface, is common to all phenomena related to electron emission induced by bombarding a surface with electrons, photons or particles. Since electrons have a finite rest mass, energy and momentum conservation make it unlikely for them to be absorbed during an interaction with the solid. This has the beneficial effect that electron emission from solids is a commonly observed phenomenon (justifying the writing of the present book) but also has the side effect that electrons experience strong multiple scattering. This implies that the energy and angular distribution as well as the spin–statistics at the instant of generation are generally distinctly different from the respective distributions recorded in an experiment. Therefore, understanding the details of the electron transport process is essential for interpretation of experimental electron spectra. It is the purpose of the present chapter to review the fundamental physical quantities and theoretical basis for electron transport in solids and to highlight experimental results illustrating the phenomena discussed and concepts employed.

As noted above, one of the main questions in this connection is how a given energy, angular and spin distribution at the source is changed by the interaction and transformed into the measured spectrum. Experimentally this can be studied by directing a monochromatic, ideally collimated beam of electrons of a single spin state onto a surface and measuring the spectrum after the beam traverses the solid. Since electron transport is a linear phenomenon and the resulting spectrum is closely related to the "Green's function" of the transport problem, this allows one to model the spectrum for any arbitrary source distribution by superposition. In the range of electron energies of interest in the present book, the best realization of such an experiment is reflection electron energy loss spectroscopy. Therefore, the present discussion on electron transport phenomena is not restricted to electron emission experiments but also treats electron backscattering as an additional essential aspect of the field.

The discussion in this chapter is focussed on the transport of medium energy electrons, i.e. electrons with kinetic energies in the range of the binding energies between the most weakly and most tightly bound core electrons in the range between 50 eV and 100 keV. However, links to slow electron energy transport will be given whenever possible. Ion induced electron emission is treated in Chap. 7. Furthermore, to remain within the scope of this book, the electronic spin will be entirely disregarded and diffraction effects (coherent scattering) will be neglected throughout this chapter. This does not imply that the results are invalid for single crystals since imperfections in crystals lead to decoherence in the transport in which case the concepts outlined here can again be applied to the crystalline state [10].

The structure of this chapter is as follows: in the theory section the physical quantities governing the electron transport are introduced and discussed and illustrated with examples. The theory of electron transport is outlined in the next section using a Boltzmann–type kinetic equation as a starting point. On the basis of the solution of the kinetic equation, spectrum deconvolution procedures are treated in the subsequent section. Examples are given in the following sections on electron reflection, Auger and photoelectron emission and the emission of secondary electrons.

2.2 Theory of Electron Transport

2.2.1 Characteristics of the Electron–Solid Interaction

During the interaction with a solid, the momentum of an incoming electron is dissipated away over the large number of degrees of freedom of the solid. The degrees of freedom of the probing electron are subject to fluctuations caused by the interaction. Energy and momentum conservation require any momentum transfer to be accompanied by an energy transfer. However, for medium energy electrons large directional changes are often accompanied by a negligible energy transfer and vice versa [11]. In this energy regime, it is therefore useful to distinguish between *elastic* and *inelastic* processes. Here an elastic collision is understood as an interaction with the ionic subsystem of the solid in which the internal energy of the collision partner remains unchanged by the interaction. Inelastic processes on the other hand are dominated by the electronic excitations of the solid.

The different types of interaction processes relevant in the medium energy range are schematically shown in Fig. 2.1. The elastic interaction can be conceived as a scattering process between the incoming electron and a static screened Coulomb potential (e.g. a Thomas–Fermi–Dirac potential [12]) that is assumed to be unperturbed by the collision [6, 7]. The differential elastic scattering cross section can be obtained on the basis of the partial wave expansion method [13, 14]. The average distance between successive elastic collisions, the so–called elastic mean free path (EMFP), is related to the integral of the cross section over the unit sphere as:

$$\lambda_e^{-1} = N_a \int_{4\pi} \frac{d\sigma_e}{d\Omega}(\Omega) d\Omega \tag{2.1}$$

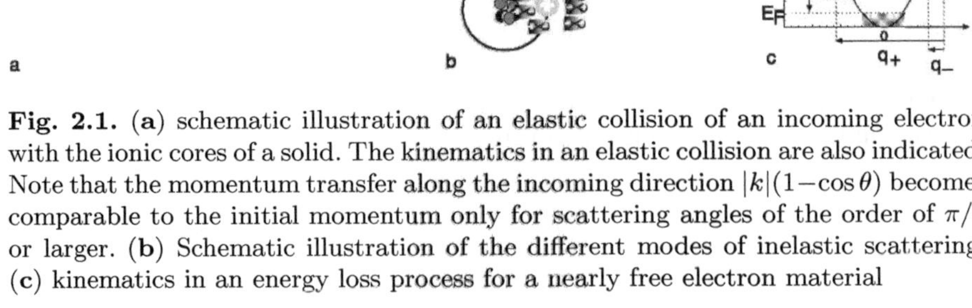

Fig. 2.1. (**a**) schematic illustration of an elastic collision of an incoming electron with the ionic cores of a solid. The kinematics in an elastic collision are also indicated. Note that the momentum transfer along the incoming direction $|k|(1-\cos\theta)$ becomes comparable to the initial momentum only for scattering angles of the order of $\pi/2$ or larger. (**b**) Schematic illustration of the different modes of inelastic scattering; (**c**) kinematics in an energy loss process for a nearly free electron material

where N_a is the atomic density. The kinematics in an elastic process are also indicated in Fig. 2.1a. The momentum transfer along the original direction is seen to be proportional to $(1 - \cos\theta_s)$ where θ_s is the polar scattering angle. The transport mean free path (TRMFP) measures the momentum transfer along the initial direction:

$$\lambda_{tr}^{-1} = N_a \int_{4\pi} (1 - \cos\theta_s) \frac{d\sigma_e}{d\Omega}(\Omega) d\Omega \tag{2.2}$$

Since for an elastic process one has $|k| = |k'|$, it follows that the momentum transferred along the initial direction becomes comparable to the initial momentum only when $\theta_s \geq \pi/2$. Therefore the transport mean free path is the typical distance an electron travels until it "forgets" its original direction. In other words, the transport mean free path is the characteristic distance for momentum relaxation.

In a solid, A large number of different inelastic excitation channels exist, that include plasmons, inter- and intraband transitions, electron hole pair generation, inner shell ionization etc. [15–17]. A physical model for the inelastic electron–solid interaction is provided by linear response theory that conceives an energy loss process as a decelaration of the probing particle in a polarization field set up during the interaction. Three different modes of the inelastic interaction can be distinguished (see Fig. 2.1b): Bulk (volume) excitations that occur inside an infinite medium when the polarization field is set up as a response to the probing particle (subscript "b" in the following). Surface excitations that occur as a consequence of the boundary conditions of Maxwell's equations at either side of an interface of two media with different electrical susceptibility (subscript "s" in the following). In the case that the above processes are experienced by a signal electron (i.e. an Auger or photoelectron) that was generated in the course of an ionization process these loss mechanisms are refered to as *extrinsic* losses, to distinguish them from an energy loss inherent to the ionization. The latter process, in which the decelarating field is set up by the response to the sudden appearance of the core hole, is refered to as an *intrinsic* loss (subscript "i" in the following).

The energy transferred from the probing electron to the elementary excitations of the solid is related to the frequency and wavenumber dependent dielectric function of the solid $\varepsilon(\omega, q)$ (see Chap. 1 for a fundamental discussion of linear response theory). In the bulk of the solid the distribution $W_b(T)$ of energy losses $T = \omega$ per unit pathlength and energy loss is given by the so-called differential inverse inelastic mean free path (DIIMFP) and can be expressed in terms of the dielectric loss function $Im\{-1/\varepsilon(\omega, q)\}$ via the relationship [18]:

$$W_b(T) = \frac{1}{\pi E} \int_{q_-}^{q_+} \frac{dq}{q} Im\left[\frac{-1}{\varepsilon(\omega, q)}\right] \tag{2.3}$$

Here E is the energy of the probing particle and q denotes the momentum transfer. Note that atomic units are used throughout this chapter, unless noted otherwise. The kinematics in an inelastic process are sketched in Fig. 2.1c for a nearly free electron material with a quadratic dispersion. The minimum and maximum momentum transfer for a given energy loss T follow from energy and momentum conservation and are seen to be given by

$$q_{\pm} = \sqrt{2E} \pm \sqrt{2(E-T)},$$

The hatched area corresponds to occupied states in the solid determining the maximum energy transfer allowed by Pauli's exclusion principle $T_{\max} = E - E_F$, E_F being the Fermi energy of the solid.

The average distance between successive inelastic collisions is given by the inelastic mean free path (IMFP):

$$\lambda_i(E)^{-1} = \int_0^{\infty} W_b(T,E)dT\ , \tag{2.4}$$

The mean energy loss in an individual collision is given by the first moment of the distribution of energy losses:

$$\langle T(E) \rangle = \int_0^{\infty} T w_b(T,E)dT \tag{2.5}$$

where $w_b(T,E)$ is the normalized DIIMFP $w_b(T,E) = W_b(T,E)\lambda_i$. As can be seen in Fig. 2.3 below, the normalized DIIMFP depends only weakly on the primary energy $w_b(T;E) \simeq w_b(T)$ and, as a consequence, the mean energy loss is also approximately energy independent $\langle T(E)\rangle \simeq \langle T \rangle$. The mean energy loss per unit path length, the stopping power, can therefore be written as:

$$\frac{dT}{ds}(E) = \frac{\langle T \rangle}{\lambda_i(E)} \tag{2.6}$$

Finally, the linear range R, is the path length the particle travels until its energy is entirely dissipated away to the solid:

$$R = \int_{E_{\min}}^{E_0} \left(\frac{dT}{ds}\right)^{-1} dT \tag{2.7}$$

where $E_{\min}$ is usually chosen as 50 eV, by convention.

Surface excitations are additional modes of the elementary excitations of the solid that arise near an interface as a consequence of the boundary conditions of Maxwell's equations. These modes decay rapidly with the distance from the interface and are orthogonal to the volume modes. In other words, surface and bulk excitations are coupled, making a distinction between them essentially artificial. For all practical purposes it is nonetheless useful to define volume excitations as the energy loss processes taking place in an infinite boundless medium and surface excitations as all changes in the excitation probability due to the presence of the boundary. As mentioned above, these modes are coupled. A simple physical model of this coupling is the partial depolarization of the bulk charge by the additional surface charge. Therefore, the strength of volume excitations is decreased in the vicinity of the surface at the expense of surface excitations, a phenomenon often referred to as "Begrenzungs"-effect in the literature, after the German word for boundary.

Several models for surface excitations have been put forward in the past decades [19–28]. The typical energy loss in a surface excitation, $\hbar\omega_s$ is somewhat smaller than for bulk excitations. In the Drude model the surface plasmon energy is given by: $\hbar\omega_s \sim \hbar\omega_b/\sqrt{2}$. The distribution of energy losses in a surface excitation depends in a complex manner on the direction of surface crossing, the depth below the surface and the energy loss. The depth scale where surface excitations contribute significantly to the overall energy loss is given by v/ω_s, where v is the electron speed. This characteristic depth is of the order of ~ 5Å for medium energy electrons. Since the typical penetration depth, governed by the IMFP, exceeds the thickness of the surface scattering zone by a factor of two or more, the infuence of surface excitations on the energy loss process is most conveniently given in terms of the normalized distribution of energy losses $w_s(T,\theta)$ (often refered to as differential surface excitation probability, DSEP) being equal to the inverse differential mean free path integrated over the surface scattering zone.

Tung and coworkers [25] give the following expression for the DSEP, $W_s(T)$:

$$W_s(\omega,\theta,E) = P_s^{+}(\omega,\theta,E) + P_s^{-}(\omega,\theta,E) \ , \tag{2.8}$$

where the quantity $P_s^{\pm}(\omega,\theta,E)$ is defined as

$$P_s^{\pm}(\omega,\theta,E) = \frac{1}{\pi E\cos\theta}\int\limits_{q_-}^{q_+}\frac{|q_s^{\pm}|dq}{q^3}\mathcal{I}m\left[\frac{(\varepsilon(\omega,q)-1)^2}{\varepsilon(\omega,q)(\varepsilon(\omega,q)+1)}\right] , \tag{2.9}$$

$\cos\theta$ denotes the off-normal surface crossing direction, and

$$q_s^{\pm} = \left[q^2 - \left(\frac{\omega+q^2}{\sqrt{2E}}\right)^2\right]^{1/2}\cos\theta \pm \left(\frac{\omega+q^2}{\sqrt{2E}}\right)\sin\theta \ . \tag{2.10}$$

The total surface excitation probability $\langle n_s(\theta)\rangle$ is obtained from expression (2.8) by integrating over the energy loss. This quantity is equal to the average number of surface excitations in a single surface crossing, giving the normalized differential surface excitation probability as $w_s(T,\theta) = W_s(T,\theta)/\langle n_s(\theta)\rangle$.

The dielectric function can be obtained from compilations of optical data (see e.g. Refs. [32–34]). An example is shown in Fig. 2.2a that compares the real and imaginary parts of ε for Cu in Palik's book [32] (dashed lines), values obtained with Density–Functional–Theory above the ground state [30, 31] and data obtained from analysis of Reflection Electron Energy Loss Spectra [35] (open and filles circles). Figure 2.2b shows the functions of ε in the integrands of Eqs. (2.3) and (2.9), refered to as the volume and surface "loss function" since these quantities determine the shape of the energy loss distribution in an individual inelastic process for zero momentum transfer. Note that the surface loss function shown in the upper panel of Fig. 2.2b has a negative excursion at about 28 eV, just at the position of the maximum in the volume loss function. This is a consequence of the coupling of the surface and bulk modes and is in agreement with the definition of surface excitations introduced above: The surface loss function is the difference between the "pure" surface loss term and the (negative) coupling term and becomes negative whenever the former exceeds the latter. Note that near the resonance ($\varepsilon \simeq -1$), the surface loss function can approximately be written as $\mathcal{I}m - 1/(1+\varepsilon)$, an expression often found in the literature. It is to be noted, however, that the shape of this simplified loss function is significantly different from the more accurate expression displayed in Fig. 2.2b.

The differential inverse inelastic mean free paths (DIIMFP) for Cu and SiO_2 derived from dielectric data with Eq. (2.3) are shown in Fig. 2.3 (a) and (b) respectively, on a semi–logarithmic scale for several primary energies. It can be seen that the electronic structure of these materials is distinctly different, the most prominent difference being the energy gap of SiO_2 of $\sim$9 eV, giving rise to a vanishing probability of energy losses below this value. Note also that the shape of the energy loss distribution in an individual collision depends only very weakly on the primary energy, at least for the most prominent part of the energy loss distributions $\lesssim$100 eV.

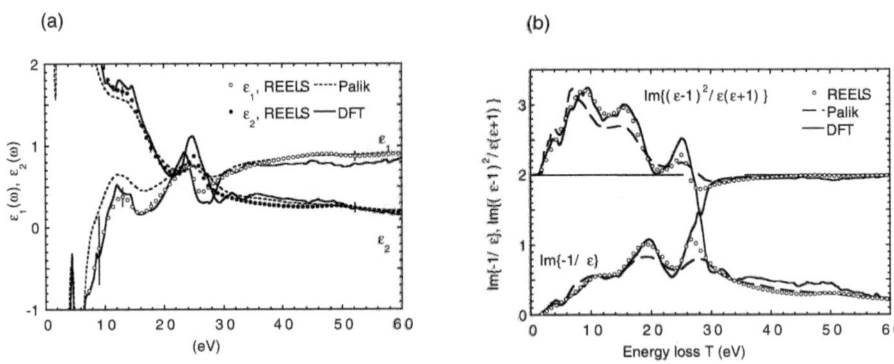

Fig. 2.2. (**a**) Dispersive, ε_1, and absorptive part, ε_2, of the dielectric function of Cu retrieved from REELS data presented in Fig. 2.9 (*open circles*) compared with Palik's data [29] (*dashed curves*) and with results from DFT calculations (*solid curves*) [30, 31]. *Open* and *filled circles*: REELS ε_1 and ε_2 respectively; Thick and thin *solid line*: DFT ε_1 and ε_2 respectively; Thick and thin *dashed line*: Palik's data; (**b**) Volume and surface loss function of Cu retrieved from REELS data shown in Fig. 2.9 below(open circles) compared with DFT results (*solid curves*) and Palik's optical data (*dashed curves*)

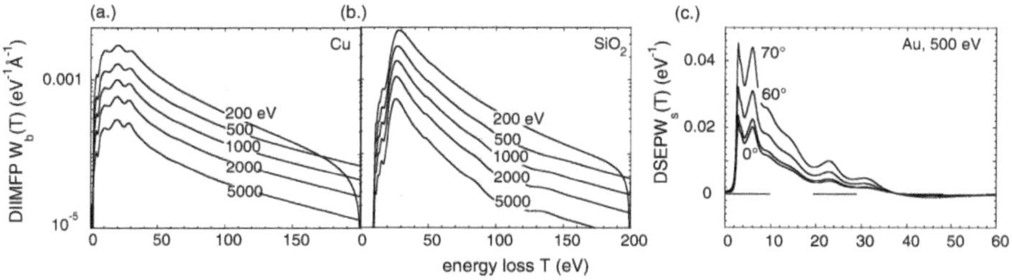

Fig. 2.3. Differential inverse inelastic mean free path (DIIMFP) for Cu (**a**) and for SiO_2 (**b**) for various energies on a semilogarithmic scale. Note that the shape of the DIIMFP is almost independent of the energy. (**c**) Differential surface excitation probability (DSEP) for 1000 eV electrons crossing a Cu surface for surface crossing directions of 0, 40, 60 and 70°

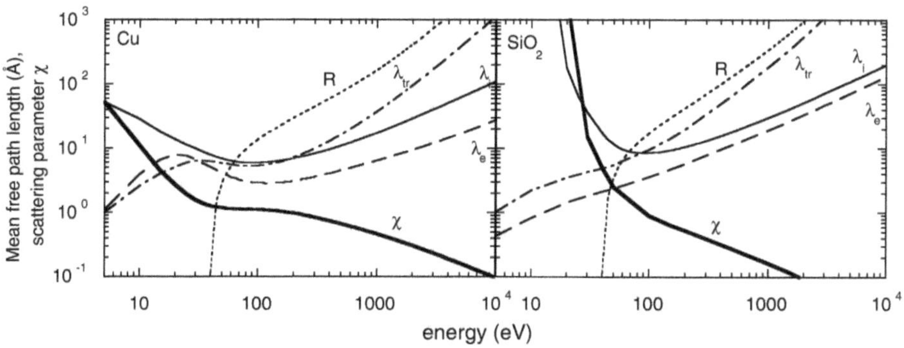

Fig. 2.4. Inelastic (λ_i), elastic (λ_e) and transport (λ_{tr}) mean free path and linear range (R) for Cu and SiO_2 as a function of the electron energy. The *thick solid curve* represents the scattering parameter $\chi = \lambda_i/\lambda_{tr}$

Figure 2.3c shows the differential surface excitation probability (DSEP) for 500 eV electron crossing a Au–surface for various off–normal surface crossing directions. Note, again, that the DSEP has a slight negative excursion at around 40 eV, being a consequence of the coupling of the surface and bulk modes, and that the shape of the energy loss distribution is very similar for any surface crossing direction.

The relevant pathlengths for electron transport, the inelastic mean free path (IMFP, λ_i), elastic mean free path (EMFP, λ_e) and transport mean free path (TrMFP, λ_{tr}) as well as the linear range (R) are summarized for Cu and SiO_2 in Fig. 2.4. The energy dependence of the elastic and inelastic mean free path as well as for the transport mean free path and the linear range is seen to be similar for both materials above $\gtrsim$200 eV. For lower energies the bandgap of SiO_2 causes a steep increase in the IMFP–curve that is absent for Cu.

The characteristic pathlengths discussed above are all measured along the particle's trajectory. In many cases the more interesting question is how an electron beam is attenuated when passing between two arbitrary points in space, i.e. the characteristic *distance*, λ_a for electron beam attenuation is the most relevant quantity. Within the framework of linear tranport theory, employing so–called the transport approximation, λ_a is found as [36, 37]

$$\lambda_a = \nu_0 \frac{\lambda_i \lambda_{tr}}{\lambda_i + \lambda_{tr}} \tag{2.11}$$

where ν_0 is the positive root of the characteristic equation

$$1 = \frac{\nu_0 \chi}{2(1+\chi)} \ln \frac{\nu_0 + 1}{\nu_0 - 1} , \tag{2.12}$$

and the quantity

$$\chi = \lambda_i / \lambda_{tr} \tag{2.13}$$

represents the so–called scattering parameter. For $\chi \lesssim 1$ one finds the approximate expression $\nu_0 \simeq 1 + 2\exp\{-2(\chi+1)/\chi\}$ [38]. The scattering parameter is shown as the thick solid curve in Fig. 2.4. For energies above ~50 eV, the value of the scattering parameter is seen to be less than unity, while it rises steeply for smaller energies. This behaviour is generally observed for any solid material and is very important since it defines two different regimes of the electron transport: for energies below $\lesssim$100 eV the transport is dominated by momentum relaxation, while for higher energies the attenuation is mainly determined by energy fluctuations. This can be seen by writing Eq. (2.11) in the two limiting cases of large and small scattering parameters as follows:

$$\begin{aligned} \lambda_a &\simeq \frac{1}{1+\chi} \lambda_i \approx \lambda_i, \quad \chi \ll 1 \\ \lambda_a &\simeq \frac{\chi}{1+\chi} \lambda_{tr} \approx \lambda_{tr}, \quad \chi \gg 1 \end{aligned} \tag{2.14}$$

Physically, these results can be understood as follows: when the transport mean free path is much larger than the inelastic mean free path, the particle suffers many inelastic collisions before it is deflected over a significant angle. Then, obviously, the inelastic mean free path determines the particle attenuation. Conversely, if the transport mean free path is much smaller than the inelastic mean free path, the particle is deflected many times before an energy loss takes place and the pathlength it travels between any two points in space significantly exceeds the distance between these points. Then the attenuation is governed by the transport mean free path. In conclusion, it can be stated, that for small values of the scattering parameter (e.g. in the medium energy range relevant for electron spectroscopy), the energy dissipation is dominated by energy fluctuations, while for large values of the scattering parameter (e.g. for low energies relevant for the emission of true secondaries) it is the relaxation of the particles momentum that governs the attenuation of the beam.

2.2.2 Multiple Scattering

The foundations for a theory describing multiple scattering in amorphous matter were established several decades ago. In 1940, Goudsmit and Saunderson studied the angular distribution of electrons transmitted through thin films [39]. They considered the case of small scattering angles, allowing them to make the approximation that all pathlengths are equal. Energy losses where alltogether neglected in this theory. A few years later, Landau studied the energy loss of particles passing through a foil [40]. He neglected deflections and considered a case where the total energy loss is small compared to the original energy of the particle. Then the variation of the interaction characteristics with the energy loss can be neglected, an approach commonly referred to as quasi–elastic approximation. At that time it was difficult to go beyond the approaches mentioned above since knowledge about the interaction characteristics was still rather sparse, a circumstance which has greatly improved since then. The commonly accepted model for electron transport in non–crystalline media, where diffraction effects can be neglected [10], combines these two early approaches in the framework of a Boltzmann type kinetic equation, as will be outlined below.

To study the electron transfer between the point of generation (either inside the solid in the case of emission experiments, or in vacuum in the case of reflection experiments), and detection in the analyzer it is usually assumed, for simplicity, that the energy $f_0(E')$, angular $g_0(\mathbf{\Omega}')$ and depth $c_0(z')$ distribution of the sources emitting the signal electrons are uncorrelated. The source function $S_0(E', \mathbf{\Omega}', z')$ can then be factorized into terms for the energy, angular and depth distribution:

$$S_0(E', \mathbf{\Omega}', z') = f_0(E') \times g_0(\mathbf{\Omega}') \times c_0(z') \tag{2.15}$$

Here and below the symbol $\mathbf{\Omega} = (\theta, \phi)$ is used to indicate the direction of motion of an electron, as well as a change in the direction of motion. It is obvious that except in the valence-band region in photoemission spectra, where the source energy and emission direction are known to be strongly correlated, this starting assumption is usually reasonable. The main question then is how the source spectrum $S_0(E', \mathbf{\Omega}', z')$ is changed by the electron transfer to give rise to the energy and angular distribution $Y(E, \mathbf{\Omega})$ detected in an experiment. The answer is provided by the Green's function $G(s, T, \mu)$ for the problem that is often referred to as the loss function in the present context. The generalized loss function $G(s, T, \mu)$ describes the distribution of energy losses T and net deflection angles $\theta = \arccos \mu$ after travelling a pathlength $s = |\mathbf{x} - \mathbf{x}'|$ between the two points $\mathbf{x}$ and $\mathbf{x}'$ in a homogeneous medium. Here and below the dependence on the azimuth will be suppressed since it is assumed that the scattering potentials are radially symmetric, giving rise to a cylindrical symmetry of scattering. Since the transport equation is linear, the solution satisfying the boundary conditions and source function for a specific problem can be found by superposition once this generalized loss function (that actually represents the Green's function of the problem) has been established. Following Landau [40], standard arguments can be used to derive the kinetic equation for the generalized loss function:

$$\frac{\partial G}{\partial s} = -\frac{1}{\lambda}\int_{-\infty}^{+\infty}\int_{4\pi}\Big\{G(s,T,\mu) - G(s,T_1,\mu_1)\Big\}w(T_2,\mu_2)d\Omega_2 dT_2 \tag{2.16}$$

In this equation, the energy loss and angular variables before (T_1, μ_1), during (T_2, μ_2) and after (T, μ) the collision are related via:

$$\begin{aligned} T_2 &= T - T_1 \\ \mu_2 &= \mu_1\mu + \nu_1\nu\cos(\phi - \phi_1) \end{aligned} \tag{2.17}$$

The symbols μ, ν are a shorthand notation for the cosine and sine functions respectively, and ϕ_1 and ϕ are the azimuths. The quantity $w(T,\mu) = \lambda W(T,\mu)$ is the normalized inverse differential mean free path for scattering (i.e. the distribution of scattering angles and energy losses in an individual collision) and λ is the total mean free path:

$$\frac{1}{\lambda} = \int_{-\infty}^{+\infty}\int_{4\pi} W(T,\mu)d\Omega dT \tag{2.18}$$

Note that in Equation (2.16) the energy loss range is formally extended down to $-\infty$ allowing for an energy gain in a collision (e.g. in electron–phonon scattering).

The second term on the right hand side of Equation (2.16) represents a convolution over the energy loss and deflection angles. Therefore, it is more convenient to solve this equation in Fourier-Legendre (FL) space [41, 42]. Introducing the collision number distribution $W_N(s)$, that describes the N–fold scattering probability as a function of the travelled pathlength:

$$W_N(s) = \mathcal{P}_N(s/\lambda) \equiv \frac{e^{-s/\lambda}}{N!}\left(\frac{s}{\lambda}\right)^N, \tag{2.19}$$

one finds the simple and general solution for the loss function [41]:

$$G(s,T,\mu) = \sum_{N=0}^{\infty} W_N(s)\Gamma_N(T,\mu). \tag{2.20}$$

The factor $\Gamma_N(T,\mu)$ in Equation (2.20) that represents fluctuations in the energy loss and deflection angles, is given by the $(N-1)$–fold selfconvolution of the differential mean free path, following the recursion relationship [43]:

$$\Gamma_N(T,\mu) = \int_{-\infty}^{+\infty}\int_{4\pi} \Gamma_{N-1}(T_1,\mu_1) w(T_2,\mu_2|\bar{E}_{N-1}) dT_2 d\mathbf{\Omega}_2 \tag{2.21}$$

Equation (2.20) represents a most general way to describe the particle transfer within the framework of a linearized Boltzmann equation. It expresses the loss function in terms of an expansion of the collision statistics and the associated fluctuations in the energy losses and deflections. As discussed above, for the medium energy range, one can separate the angular and energy part of the interaction by writing the normalized differential mean free path in the following form:

$$w(T,\mu) = \frac{\lambda}{\lambda_i} w_i(T) \frac{\delta(\mu)}{4\pi} + \frac{\lambda}{\lambda_e} w_e(\mu)\delta(T) \tag{2.22}$$

The quantity $w_e(\mu) = N_a \lambda_e (d\sigma_e/d\mu)$ represents the normalized elastic scattering cross section (see Equation (2.1)). The subscripts "i" and "e" indicate inelastic and elastic scattering.

Introducing the number of inelastic (n_i) and elastic (n_e) collisions after $N(= n_i + n_e)$ arbitrary collisions and inserting the above expression into Equation (2.20) gives:

$$G(s,T,\mu) = \sum_{n_i=0}^{\infty} W_{n_i}(s)\Gamma_{n_i}(T) \sum_{n_e=0}^{\infty} W_{n_e}(s)\Gamma_{n_e}(\mu) \tag{2.23}$$

Here the functions Γ_{n_i} and Γ_{n_e} now represent the self–convolutions of the inelastic and elastic terms in the differential mean free path. Likewise, the quantities $W_{n_e}(s)$ and $W_{n_i}(s)$ represent the stochastic process for elastic and inelastic scattering (see Equation (2.19)). The second factor, i.e. the elastic collision expansion, is identified as the distribution of pathlengths in an infinite medium, $Q(s,\mu)$ [39]:

$$Q(s,\mu) = \sum_{n_e=0}^{\infty} W_{n_e}(s)\Gamma_{n_e}(\mu) \tag{2.24}$$

and one obtains:

$$G(s,T,\mu) = \sum_{n_i=0}^{\infty} \Gamma_{n_i}(T) W_{n_i}(s) Q(s,\mu) \tag{2.25}$$

The first factor in the sum represents the energy loss fluctuations after a given number of inelastic collisions n_i, while the second and third factor together govern the collision statistics, i.e. the number of particles that suffer n_i collisions. The shape of the pathlength distribution determines whether the energy dissipation proceeds predominantly via energy fluctuations or by momentum relaxation: if the pathlength distribution is a sharply peaked function, the energy spectrum will exhibit the signature of the characteristic energy losses in the DIIMFP, since only a few collision orders contribute to a given energy loss. When the pathlength distribution is broad, many more collision orders contribute to a given energy loss and the characteristic losses will be smeared out. In consequence, the energy distribution resembles the distribution of pathlengths rather than the characteristic energy losses caused by energy transfers to the elementary excitations of the solid. It should be kept in mind that the value of the scattering parameter χ determines whether the pathlength distribution is a sharply peaked or a smooth and broad function. When $\chi \gg 1$, many deflections take place before a particle loses energy giving rise to a broad pathlength distribution and emphasizing the role of momentum relaxation in the energy dissipation process, while for small values of $\chi \ll 1$, energy fluctuations play the main role in the energy dissipation process.

Thus, if the condition Equation (2.22) is satisfied, the problem of solving for the particle transfer in a non-crystalline medium has been reduced to finding the pathlength distribution $Q(s, \mu)$ satisfying the boundary conditions of the considered problem. It should be noted that although an exact expression has been found for an infinite medium Eq. (2.24), it is by no means trivial to find the pathlength distribution for the particle transfer near a surface. In such cases one usually has to resort to numerical calculations (such as a Monte Carlo model, see e.g. Ref. [11, 44]) or employ the usual techniques of linear transport theory [38, 45, 46] (see Refs. [47–49] for applications to electron scattering).

When deflections are neglected ($\lambda_e \to \infty$), only the zero order term in Eq. (2.23) in the function $W_{n_e}(s)$ differs from zero and as a consequence $Q(s, \mu) = \delta(\mu)/4\pi$. This implies that the particles' direction remains unchanged when it travels an arbitrary distance in the solid, a situation coinciding with the rectilinear motion model. In this case the loss function becomes:

$$G(s, T) = \sum_{n_i=0}^{\infty} W_{n_i}(s) \Gamma_{n_i}(T) \tag{2.26}$$

This is just the collision expansion of Landau's original result [40, 43].

In the quasi–elastic case, when the interaction characteristics are approximately independent of the energy, the stochastic process for multiple scattering is given by the Poisson distribution, $W_N(s) = \mathcal{P}_N(s)$, Eq. (2.19). Beyond the quasi–elastic regime, when the number of collisions increases, the energy dissipation process is more and more determined by momentum relaxation rather than by energy fluctuations since the pathlength distribution becomes very broad [43]. Then Equation (2.20) may be further generalized to account for the energy dependence of the interaction characteristics by using [43]

$$W_N(s) = \mathcal{W}_N(s) \equiv \frac{\lambda_N}{\Lambda_N} \left(\frac{s}{\Lambda_N} \right)^N \frac{e(-s/\Lambda_N)}{N!} \tag{2.27}$$

Here Λ_N represents the average mean free path after N collisions:

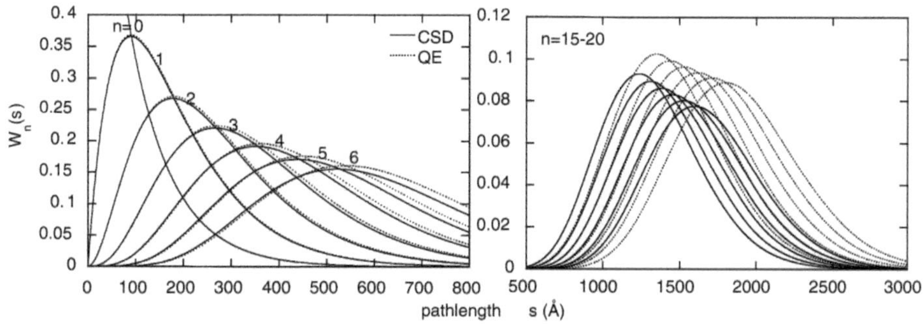

Fig. 2.5. Probability density $W_{n_i}(s)$ for n–fold inelastic scattering as a function of the travelled pathlength for 3 keV electrons traversing a Cu-target. *Solid curves*: Slowing down regime Eq. (2.27); *dotted curves*: Poisson distribution Eq. (2.19)

$$\Lambda_N = \frac{1}{N+1} R_N = \frac{1}{N+1} \sum_{k=0}^{N} \lambda_k \tag{2.28}$$

Where R_N is the linear range after N collisions (cf. Eqs. (2.5) and (2.7)) and the quantity λ_k is the total mean free path after k collisions. In Fig. 2.5, the quasi–elastic approximation for the stochastic process for inelastic scattering (Eq. (2.19), dotted curves) is compared with the result in the slowing down regime (Eq. (2.27), solid curves) for 3 keV electrons travelling in a Cu specimen. For the first few scattering orders ($n \lesssim 5$) the two approaches coincide approximately, when the number of inelastic processes increases ($n > 10$) the quasi–elastic approximation fails to describe the collision statistics quantitatively.

Landau's loss function is shown in Fig. 2.6. The left panel corresponds to Landau's original result that was derived in the quasi–elastic approximation. The thick solid line represents the relationship between the energy and the pathlength for a constant IMFP and stopping power. The middle panel represents the true slowing down case, by using Eq. (2.27) as the collision number distribution in the loss function Eq. (2.26). The panel on the right hand side displays the energy loss as a function of the pathlength modelled by a direct simulation approach in which each individual scattering process is explicitly simulated. Each energy-loss event taking place at a given energy of a particle for a given pathlength s travelled in the solid is marked in this figure by a dot. In other words, this "grayscale" representation of the slowing down process corresponds to the distribution $G(s, T)$ of energies lost after a certain pathlength irrespective of the direction of the particle, or, yet in other words, for pathlengths s measured *along* the trajectory. The thick solid line in figure (b) and (c) represents the energy loss in the continuous slowing down approximation (CSDA) that assumes a fixed relationship between the travelled pathlength and the energy loss

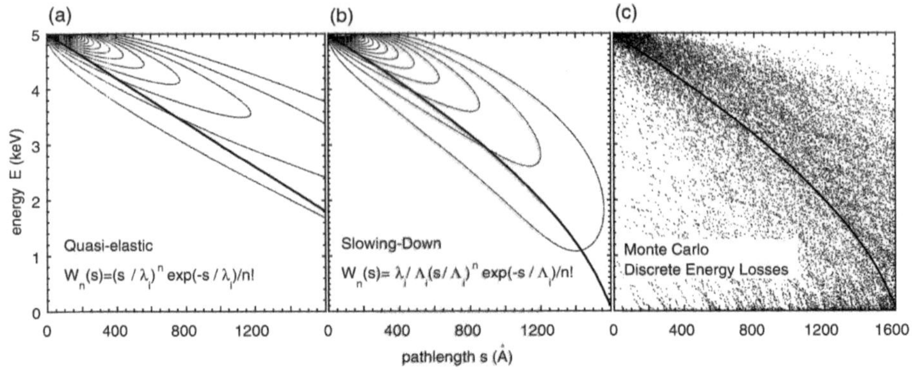

Fig. 2.6. Contour plot of Landau's loss function Eq. (2.26) for 5 keV electrons slowing down in Cu. The *thick solid line* represents the continuous slowing down approximation (CSDA) that neglects energy fluctuations. The separation between consecutive contours amounts to 2×10^{-4}eV^{-1}Å^{-1}. The value of the loss function at the lowest (outermost) contour amounts to 2×10^{-4}eV^{-1}Å^{-1}. (**a**) Landau's original result in the quasi–elastic approximation obtained by using the Poisson distribution Eq. (2.19) for the collision number distribution $W_{n_i}(s)$. The *thick solid line* represents the relationship between the mean energy and the travelled pathlength in the quasi–elastic approximation; (**b**) Loss function generalized to the true slowing case by using Eq. (2.27) for the collision number distribution; (**c**) Visualisation of the energy dissipation simulated with a Monte Carlo calculation (see text)

$$\bar{E}_{CSDA}(s) = E_0 - s\left\langle \frac{dT}{ds} \right\rangle \tag{2.29}$$

when E_0 is the primary energy. The above expression makes it clear that the CSDA completely ignores the fact that the energy loss is subject to fluctuations, making it i.a. impossible to model the peak of elastically backscattered electrons within the CSDA.

Having found the Green's function, the outgoing spectrum is obtained as a superposition of all trajectories travelling a given pathlength in the solid:

$$Y(E, \mu) = \int_0^\infty \int_0^\infty \int_{4\pi} G(s(z'), T, \mu'') S_0(z', E + T', \mu') dz' dT' d\Omega'' \tag{2.30}$$

Here and below it is assumed that we are dealing with a case of plane symmetry where the outgoing intensity is independent of the azimuth. Introducing the partial intensities C_{n_i} as the number of electrons that arrive in the detector after experiencing a certain number of inelastic collisions for the specific boundary conditions, the observed spectrum can be written in the convenient form:

$$Y(E, \mu) = \sum_{n_i=0}^{\infty} C_{n_i}(\mu) \Gamma_{n_i}(T) \otimes f_0(E + T) \tag{2.31}$$

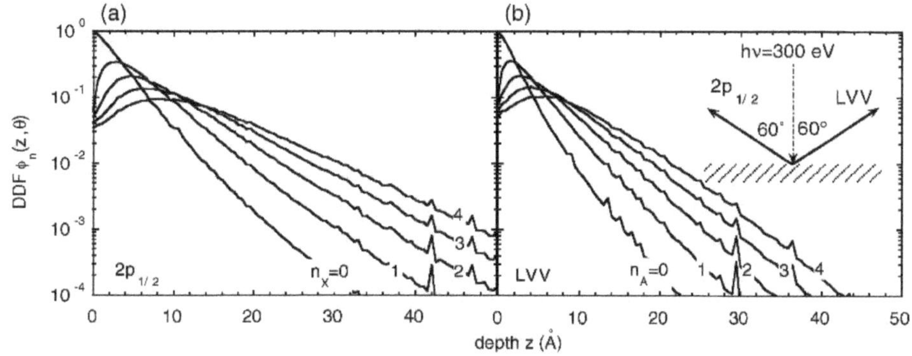

Fig. 2.7. (**a**) Partial escape distribution, or depth distribution function (DDF), $\phi_{n_X}(z, \theta_o)$ for Si $2p_{1/2}$ photoelectrons excited by 300 eV x-rays for a geometrical configuration shown in the inset. These data were simulated by means of a trajectory reversal Monte Carlo model [50, 51]. (**b**) Same as (**a**), $\phi_{n_A}(z, \theta_o)$, for Si-LVV Auger electrons

where $f_0(E)$ is the energy distribution at the source and the symbol $\otimes$ denotes a convolution. The partial intensities, are found from the above as:

$$C_{n_i}(\mu) = \int_0^\infty \int_{4\pi} W_{n_i}(s(z'))Q(s(z'), \mu'')c_0(z') \times g_0(\boldsymbol{\Omega}')d\boldsymbol{\Omega}'dz', \tag{2.32}$$

Introducing the depth distribution functions $\phi_{n_i}(z', \mu)$ that describe the probability for electrons generated at the depth z' to escape from the surface in the polar emission direction μ after experiencing n_i inelastic collisions:

$$\phi_{n_i}(z', \mu) = \int_0^\infty \int_{4\pi} W_{n_i}(s)Q(s, \mu'')g_0(\mu')\delta(s - s(z'))d\boldsymbol{\Omega}'ds, \tag{2.33}$$

the partial intensities may be expressed as an integral over the depth:

$$C_{n_i}(\mu) = \int_0^\infty \phi_{n_i}(z', \mu)c_0(z')dz' \tag{2.34}$$

The latter expression is particularly convenient in emission problems, relevant for AES, XPS, APECS and the like. Note that the depth distribution functions are defined as an average over the source angular distribution. Example of the emission depth distribution functions (DDF), or partial escape distributions, are given in Fig. 2.7 for Si 2p photoelectrons and Si-LVV Augerelectrons excited by 300 eV photons. These results were calculated by an efficient reverse trajectory Monte Carlo code [50] that benefits from the symmetry of the Boltzmann equation in that each electron trajectory is generated in reverse and traced back to its point of origin in the solid. For small solid angles of detection an efficiency enhancement of several orders of magnitude over conventional calculations is obtained in this way.

It is clear from Fig. 2.7 that the partial escape distributions do not follow a Poisson distribution, as predicted by the rectilinear motion model: first of all, the escape distributions for $n \geq 1$ are finite for $z = 0$, which is due to the fact that those electrons that are emitted at the very surface with an initial direction pointing towards the interior of the solid have a finite probability to be backscattered. Furthermore, the slope of the DDF is not constant with the depth. It turns out that for small escape depths $z \lesssim \lambda_{tr}/\mu_e$, where $\mu_e = \cos\theta_e$ is the cosine of the emission angle, the slope of the DDF is given by λ_i/μ_e, while for larger depths it becomes indepent of the emission angle and equal to λ_a as given by the upper stroke of Eq. (2.14). This is another example of the influence of momentum relaxation on the electron transport: for pathlengths beyond λ_{tr}, the electrons are multiple elastically scattered over large angles and any connection between their initial direction and the emission direction is lost.

In conclusion of this section, it is to be noted that the measured intensity also depends on a number of experimental factors like the analyzer transmission function, the detector efficiency etc. However, for spectrum analysis, the quantities of main importance are the *reduced* partial intensities γ_{n_i}:

$$\gamma_{n_i}(\mu) = C_{n_i}(\mu)/C_{n_i=0}(\mu), \tag{2.35}$$

i.e. the partial intensities divided by the area of the no–loss peak. In this case, all experimental factors cancel out.

2.2.3 Spectrum Analysis Procedures

As stated in the introduction, the purpose of developing a theory for the electronic transfer between the source and the detector is to gain information on the source distribution as well as the interaction characteristics. To see how this can be achieved, the spectrum Eq. (2.31) is written in Fourier space in the following form:

$$\widetilde{y}(\mathbf{\Omega}) = \sum_{n_i=0}^{\infty}\sum_{n_b=0}^{\infty}\sum_{n_s=0}^{\infty} \alpha_{n_i}(\mathbf{\Omega})\alpha_{n_b}(\mathbf{\Omega})\alpha_{n_s}(\mathbf{\Omega})\widetilde{w}_i^{(n_i)}\widetilde{w}_b^{(n_b)}\widetilde{w}_s^{(n_s)}\widetilde{f}_0 \tag{2.36}$$

Where a tilde ("$\sim$") is used to indicate a quantity in Fourier space. Here Eq. (2.31) was generalized to take into account the different types of excitations (surface, bulk and intrinsic). Note that here and below the symbol n_i is used to indicate the number of intrinsic inelastic collisions and should not be confused with the number of inelastic excitations in the previous paragraph. Furthermore, the important fact that the reduced partial intensities for the different types of excitations are uncorrelated has been utilized [52–54]:

$$\alpha_{n_i,n_b,n_s}(\mathbf{\Omega}) = \alpha_{n_i}(\mathbf{\Omega}) \times \alpha_{n_b}(\mathbf{\Omega}) \times \alpha_{n_s}(\mathbf{\Omega}) \tag{2.37}$$

The partial intensities for volume scattering can be calculated by means of linear transport theory, or, what is usually more convenient, by means of a numerical procedure, e.g. a Monte Carlo simulation [50]. The partial intensities for intrinsic and surface scattering approximately follow Poisson statistics [55–57]. Note that the information on the source angular and depth distribution is contained in the partial intensities.

The source angular distribution can be retrieved by inverse modelling of the measured angular distribution. In the case of Auger electron emission the source angular distribution is found to be isotropic, while in XPS it accurately follows the theoretically expected dipole shape [58, 59].

Information on the depth distribution of the emitting sources for compositional depth profiling is usually also achieved by inverse modelling of either the angular distribution [60], or the energy distribution of a given photoemission or Augerelectron peak [61, 62]. Since these procedures both involve inverse modelling it is generally difficult to assess the reliability of the results.

The remaining question then is how an experimental spectrum can be analysed to get information on the quantities $f_0(E)$, $w_i(T)$, $w_b(T)$ and $w_s(T)$. Information on the bulk and surface differential mean free path can be derived from reflection electron energy loss spectra (REELS). In this case, the source energy and angular distribution can be fully controlled in an experiment and no intrinsic excitations occur. Assuming that the partial intensities for surface and bulk scattering are known (e.g., by using a semiinfinite target) we are left with two unknowns in Eq. (2.36), $w_s(T)$ and $w_b(T)$. Obviously, a single spectrum cannot give the unique shape of these two quantities since one equation with two unknowns has no unique solution. Nonetheless it has been common practice in the past twenty years to retrieve the differential mean free path from a single REELS spectrum, but quantitative interpretation of the results is cumbersome [63]. When two REELS spectra $y_{L,1}(T)$ and $y_{L,2}(T)$ for which the relative contributions of surface and bulk excitations is sufficiently different, are measured, the pair of spectra can be deconvoluted to give the unique solution for the normalized DIIMFP and DSEP, using the formula [64, 65]:

$$\begin{aligned} w(T) &= \sum_{k=0}^{2}\sum_{l=0}^{2} a_{k,l}Y_{k,l}(T) \\ &- \int_{T'=0}^{T} \sum_{k=0}^{2}\sum_{l=0}^{2} b_{k,l}Y_{k,l}(T-T')w(T')dT' \end{aligned} \tag{2.38}$$

where the quantity $Y_{k,l}(T)$ is the (k,l)-th order cross convolution of the two REELS spectra:

$$Y_{k,l}(T) = \int_{0}^{\infty} y_1^{(k)}(T-T')y_2^{(l)}(T')dT' \tag{2.39}$$

and $y_1^{(k)}$ and $y_2^{(l)}$ are the $(k-1)$ and $(l-1)$-fold selfconvolution of the measured spectra. Since $w(T=0) \equiv 0$ the integration on the right hand side of Eq. (2.38) can always be carried out over the energy loss range for which the loss distribution is already known. In this way, the entire energy loss distribution is retrieved. The coefficients $a_{k,l}$ and $b_{k,l}$ are functions of the bulk partial intensities α_{n_b} [64] and are different for the surface and bulk single scattering loss distribution. The formula for retrieval of the DIIMFP and DSEP is identical and given by Eq. (2.38).

For analysis of the source energy distribution $f_0(E)$, e.g., in a photoemission spectrum, it is usually assumed that the bulk and surface differential mean free path as well as the partial intensities for surface and bulk scattering are known. Furthermore, for simplicity, the distribution of energy losses in an intrinsic excitation is assumed to be equal to the bulk differential mean free path. Then the remaining two unknowns are the average number of intrinsic excitations $\langle n_i \rangle$ in a single photoionization and the source energy distribution. Consecutive deconvolution of the spectrum, eliminating the different types of scattering one by one, can be achieved using the formula [11, 66]:

$$Y_{k+1}(E, \boldsymbol{\Omega}) = Y_k(E, \boldsymbol{\Omega}) - q_{k+1} \int Y_k(E+T, \boldsymbol{\Omega}) \Gamma_{k+1}(T) dT \ , \qquad (2.40)$$

where Γ_{k+1} is the k–fold selfconvolution of the differential mean free path of the considered type of inelastic scattering and the coefficients q_k are functions of the reduced partial intensities $\gamma_n = C_n/C_{n=0}$ given in [66]. Applying this procedure for $k = 1 \ldots K$ removes multiple scattered electrons up to the K–th scattering order from the spectrum. Repeating the whole procedure for all relevant types of scattering eventually gives the source energy distribution $f_0(E)$.

2.3 Examples

2.3.1 Backscattering of Electrons from Solid Surfaces

Figure 2.8 displays results for 500 eV electrons reflected quasi–elastically from a gold surface. In Fig. 2.8a, the angular distribution of the zero order partial intensities, i.e., the elastically reflected electrons, is shown for normal incidence and for a polar angle of acceptance of 4°. The dotted line is an analytic theory for the elastic backscattering due to Oswald, Kasper and Gaukler (OKG) [67] while open diamonds are the results of efficient trajectory reversal Monte Carlo simulations [41]. Both approaches are based on Eq. (2.23). The filled diamonds are results of conventional Monte Carlo calculations and the open circles between emission angles of 35° and 75° are experimental results [68]. Good agreement is seen between the different approaches, except for the OKG theory at normal emission. This is caused by some simplifying assumptions in the OKG-theory [41]. The shape of the angular distribution is closely related

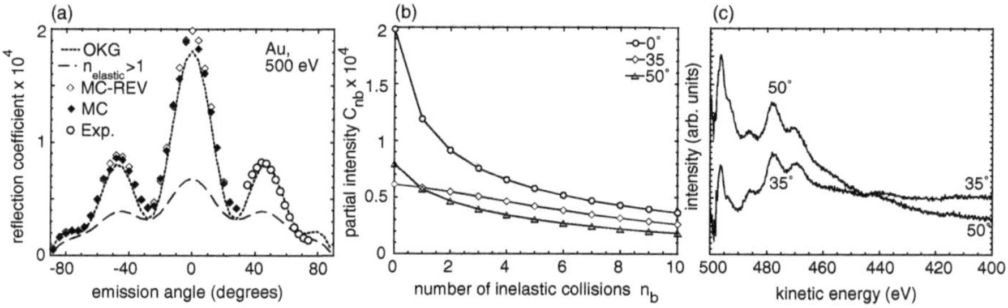

Fig. 2.8. Spectra of 500-eV electrons elastically backscattered from a Au surface for normal incidence. (**a**) Angular distribution of the peak intensity. *Dashed line*: Oswald–Kasper–Gaukler theory [67]; *Long dashed line*: contribution of electrons that experienced two or more elastic collisions; *Open diamonds*: results of trajectory reversal Monte Carlo calculations [41]; *Filled diamonds*: conventional Monte Carlo Calculations; *Open circles*: Experimental results [68]; (**b**) Partial intensities for volume scattering for (off-normal) emission angles of 0°, 35° and 50°; (**c**) Experimental reflection electron energy loss spectra for emission under 35° and 50°

to the shape of the (backward part) of the elastic scattering cross section. The maxima and minima are a result of the interference of the incoming electron, assumed as a plane wave, and the spherically symmetric outgoing wave. These features are sometimes refered to as generalized Ramsauer–Townsend oscillations since exactly the same phenomenon is responsible for the celebrated Ramsauer–Townsend effect [69].

The higher order partial intensities for a few emission directions are shown in Fig. 2.8b as a function of the bulk scattering order. The sequence of partial intensities for 35° is seen to be qualitatively different from the other two geometrical configurations. The explanation [57, 70] is that at 35°, about two third of the electrons making up the elastic peak are scattered twice or more, while for 50° about every second electron only experiences a single elastic collision. This can be seen by comparing the long dashed curve in Fig. 2.8a, that represents the higher elastic scattering orders, with the total elastic intensity. The reason is obviously the presence of the sharp and deep minimum in the cross section at ~27° and the maximum at ~45°: while for 45° there is a rather large probability for receiving the net deflection required to escape in a *single* elastic collision, multiple scattering is needed at emission directions around ~27°. This implies that in the latter case the pathlength travelled in the solid is larger and in consequence, the zero order partial intensity will be decreased, while the higher order partial intensities are enhanced.

The resulting spectra, shown in Fig. 2.8, clearly exhibit this behaviour in the energy loss distribution. It should be kept in mind that this apparent difference in the energy loss process is entirely caused by *elastic* scattering. Apart from the direction dependence of the surface excitation probability, the loss processes for the two geometries are *identical*. This is one example which clearly shows that the details of the energy dissipation are always governed by the combined influence of elastic and inelastic scattering.

REELS spectra simulated with Eq. (2.36) (with $\alpha_{n_i} = 1$ and $w_i(T) = \delta(T)$) are shown as solid curves in Fig. 2.9a for 1000 and 3000 eV. The corresponding experimental data are shown as open circles and differ from the simulations in absolute magnitude and relative intensity of the surface ($\lesssim$10 eV) and bulk loss features. Subjecting the experimental pair of spectra to the procedure summarized in Eq. (2.38) leads to the DIIMFP (open circles) and DSEP (filled circles) shown in Fig. 2.9b. The solid lines are obtained by fitting these data to Eq. (2.3) and Eq. (2.8) using an extended Drude-Lorentz model for the dielectric function to interpolate the data [35]:

$$\varepsilon_1(\omega, q) = \varepsilon_b - \sum_i \frac{A_i(\omega^2 - \omega_i(q)^2)}{(\omega^2 - \omega_i(q)^2)^2 + \omega^2\gamma_i^2}$$

$$\varepsilon_2(\omega, q) = \sum_i \frac{A_i\gamma_i\omega}{(\omega^2 - \omega_i(q)^2)^2 + \omega^2\gamma_i^2}, \tag{2.41}$$

where ε_b is the static dielectric constant, A_i represents the oscillator strength, γ_i the damping coefficient and ω_i is the energy of the i–th oscillator. A quadratic dispersion $\omega_i(q) = \omega_i + q^2/2$ was used for the transition described by the i–th oscillator. Note that a dielectric function of the form Eq. (2.41) implicitly satisfies the Kramers-Kronig dispersion relationships.

The resulting fit is quite good and in particular, the resulting curves for the real and imaginary part of the dielectric function, that are shown as open

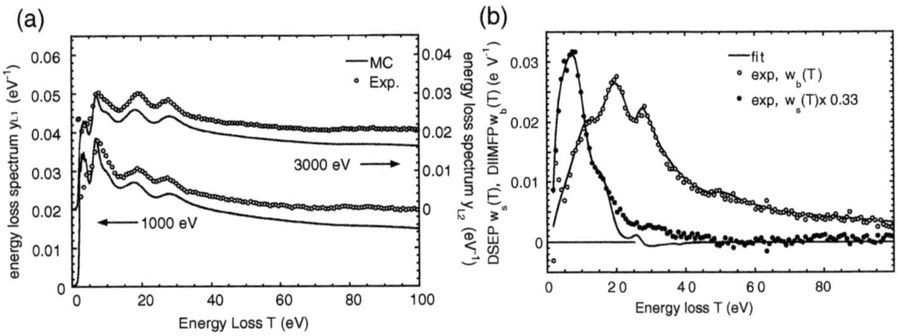

Fig. 2.9. (**a**) Experimental (*open circles*) and simulated (*solid curves*) REELS-spectra of Cu for 1000 and 3000 eV. (**b**) Normalized differential surface excitation probability (DSEP, $w_s(T)$) and normalized differential inverse inelastic mean free path (DIIMFP, $w_b(T)$) retrieved from the data in (**a**) by means of Eq. (2.38). The *solid line* is obtained by simultaneously fitting these data to Eqs. (2.3) and (2.8) for a set of Drude-Lorentz fit–parameters for the dielectric function, as described by Eq. (2.41). The resulting dielectric function is shown in Fig. 2.2a as *open* and *filled circles.* The data for the surface excitation probability were scaled by a factor of 0.33 in order to facilitate comparison

and filled circles in Fig. 2.2 are seen to agree excellently with DFT-calculations [30, 31], while significant differences with earlier optical data [29] are seen for $\omega \lesssim 30$ eV. However, it is noted that the experimental error in the dielectric data derived from REELS is quite large in this energy range, as indicated by the error bars in Fig. 2.2.

The results in Fig. 2.2 show that dielectric data can be extracted over a wide spectral range, from the visible to the soft x-ray regime, from two very simple measurements. These data can be utilized to calculate the mean free path, stopping power and related quantities for electrons, and various types of (multiple) charged particles in this way providing important data for quantitative spectrum interpretation in different fields of physics.

Surface excitations has been intensively studied in the recent past. A large number of REELS spectra have been measured as a function of the incident energy E, and the surface crossing direction θ [71–73]. An empirical relationship for the average number of surface excitations in an individual surface crossing was derived from this data set [71]:

$$\langle n_s(\theta, E)\rangle = \frac{1}{a\sqrt{E}\cos\theta + 1} \tag{2.42}$$

The value of the parameter a in this equation in units of the nearly free electron value $\sqrt{8a_0/\pi^2 e^2} = 0.17\ \text{eV}^{-1/2}$ can be estimated for an arbitrary material as

$$\frac{a}{a_{NFE}} = 0.039\hbar\Omega_p + 0.4,$$

where $\hbar\Omega_p = \hbar\sqrt{4\pi N N_v e^2/m_e}$ is the generalized plasmon energy in eV and N_v is the number of valence electrons.

A means to measure the IMFP of electrons in solids, that has attracted a lot of attention recently, is the use of elastic peak electron spectroscopy [76]. In this technique, the absolute value of the elastic peak of backscattered electrons, i.e. the zero order partial intensity, is related to the IMFP using the formula:

$$I_{elastic} = \alpha_{n_s=0}\alpha_{n_b=0} = \exp(-\langle n_s\rangle)\int_0^\infty Q(s)\exp(-s/\lambda_i)ds \tag{2.43}$$

where Poisson statistics is assumed to govern plural surface scattering and Eq. (2.32) is used for the bulk partal intensities. For a given pathlength distribution $Q(s)$, that can be calculated with a Monte Carlo simulation, a calibration curve relating the measured peak intensity to the IMFP can be established. Since absolute measurements of the elastic peak are difficult, a reference for which the IMFP is assumed to be known is often used. An example of this procedure is shown in Fig. 2.10 for the IMFP in Si using Cu as reference. Different values of the average number of surface excitations have been used to correct the peak intensities for surface scattering, leading to significantly different IMFP values. The theoretical correction factors published

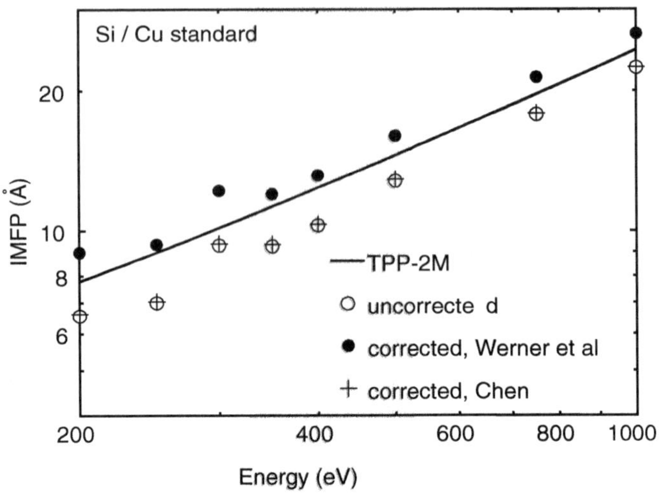

Fig. 2.10. Inelastic mean free path of Cu derived from Elastic Peak Electron Spectroscopy (EPES) measurements using Si as a reference material. *Open circles*: IMFP values obtained without correction of the elastic peak intensities for surface excitations; *Plus-signs* and *filled circles*: corrected for surface excitations using the values of the surface excitation parameter of Ref. [74] and Ref. [71] respectively; *solid curve*: IMFP values according to the TPP-2M predictive formula [75]

in Ref. [74] produce only a small correction, while the empirical surface excitation parameters $\langle n_s \rangle$ of Ref. [71] using Eq. (2.42) lead to reasonable agreement with IMFP values predicted by the TPP-2M formula [75].

The examples presented above show that in the quasi–elastic regime, the energy loss features in spectra of reflected electrons are governed by energy fluctuations (in surface and volume mode) and that the elastic interaction influences the collision statistics. When the energy loss increases, the situation changes significantly. This is illustrated in Fig. 2.11 that displays the energy spectrum of 5 keV electrons reflected from Au, Ag Cu and Al surfaces. The shape of the energy distribution is qualitatively different for the four studied materials: while the intensity decreases monotonically with increasing energy loss for Cu, Ag and Au, a maximum is observed for Al at around 4.3 keV. The reason is that the elastic scattering cross section increases with increasing atomic number, so for the light element, Al, the probability for a large net deflection reaches appreciable values only for a large pathlength. In consequence, the pathlength distribution exhibits a maximum at a large pathlength, which is seen as the maximum in the energy spectrum. In the quasi–elastic portion of the spectrum, shown in Fig. 2.11b, on the other hand, the energy spectrum is clearly dominated by energy fluctuations.

The theoretical approach to calculate the energy distribution [43] takes into account energy fluctuations and the energy dependence of the mean free paths, making the results much more realistic than the continuous slowing

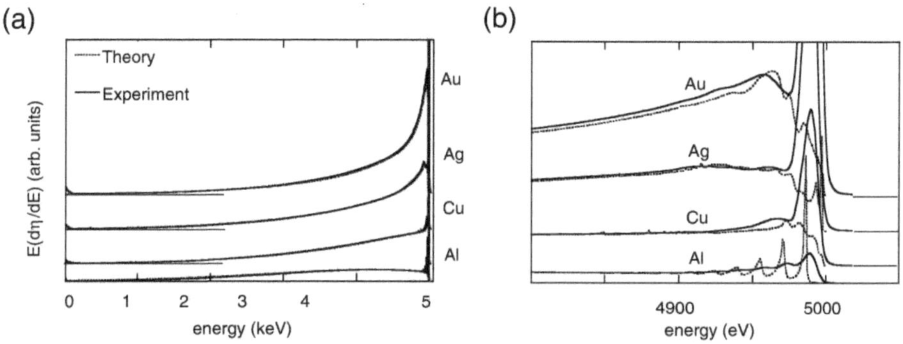

Fig. 2.11. Differential backscattering coefficient of 5 keV electrons reflected from Au, Ag, Cu and Al. *Dotted curves*: Monte Carlo calculations in the slowing down regime using Eq. (2.27) [43]; *Solid curves*: Experimental results by Goto et al. [77]. (**a**) Total slowing down energy range where momentum relaxation dominates the energy distribution. In this representation the differences between theory and experiment are hardly discernable. (**b**) Expanded view of the quasi–elastic energy range governed by energy fluctuations

down approximation where energy fluctuations are alltogether neglected (see Fig. 2.6). This is conveniently achieved merely by using Eq. (2.27) for the collision number distribution in the loss function.

2.3.2 Auger and Photoelectron Emission

The contributions of the various types of inelastic processes to an (unmomochromated) photoemission spectrum are illustrated in Fig. 2.12a and b for Si and SiO_2. The upper panel shows the raw data. In the case of Si, distinct plasmon losses can be distinguished ~15 eV below the Si2s and 2p peaks,

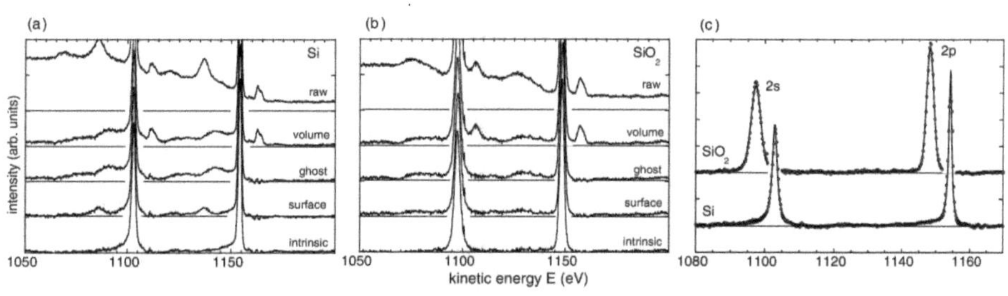

Fig. 2.12. Mg Kα excited Si 2s and 2p photoelectron spectra of (**a**) Si and SiO_2. The *upper panel* shows the raw data, subsequent panels show the spectra subjected to Eq. (2.40), consecutively eliminating features cuased by bulk scattering, the Mg K$\alpha_{3,4}$ and Kβ ghost x-ray lines, surface and intrinsic plasmons. The *bottom panel* represents the lineshape at the source $f_0(E)$. (**c**) Fit of the resulting source spectra to a Gaussian profile for SiO_2 and a Doniach-Sunjic lineshape for Si

while the inelastic background for SiO_2 is smoother exhibiting the characteristic loss peak seen in Fig. 2.3b. Furthermore, x-ray ghost line replicas of the peaks, excited by the Mg $K\alpha_{3,4}$ and $K\beta$ ghost x-ray lines, are visible at ~10 eV higher kinetic energy. The subsequent panels demonstrate application of Eq. (2.40) to the spectrum, thereby exposing the contribution from the different inelastic processes. The spectrum labelled "surface" represents the spectrum from which volume and surface excitations, as well as the ghost replicas have been eliminated. In the case of Si, the resulting spectrum shows weak loss peaks at the energy of the volume plasmons. These are attributed to the creation of intrinsic plasmons. The spectra of the bottom panel in (a) and (b) are shown in (c) together with a fit of the true spectra at the source to the theoretically expected line shapes.

In the case of insulators, where electrostatic screening is not very effective, electron–phonon coupling gives rise to a Gaussian broadening of the lineshape. In the sudden approximation, when the electronic transition proceeds sufficiently rapidly, the ground state zero point and thermal fluctuations are vertically projected onto the final (ionized) state in accordance with the Frank–Condon principle. The distribution of vibrational states follows a Poisson distribution which becomes Gaussian if the number of phonons is large enough [78].

In the case of metals or semiconductors with a sufficiently high density of states near the Fermi level, a possible response of the solid to the sudden appearance of the core–hole, in addition to the creation of intrinsic plasmons, is the creation of electron–hole pairs. The energy distribution of this process exhibits a $1/\omega$–singularity [79]. Taking into account lifetime broadening by convoluting the $1/\omega$ singularity with a Lorentzian, one obtains the Doniach–Sunjic lineshape for the source energy distribution $f_0(\omega)$ [80]:

$$f_0(\omega) \propto \frac{\cos[\frac{1}{2}\pi\alpha + (1-\alpha)\arctan(\omega/\gamma)]}{[\omega^2+\gamma^2]^{(1-\alpha)/2}} \tag{2.44}$$

Here α is the singularity index, that describes the asymmetry of the peak corresponding to the strength of the screening and γ is the lifetime broadening.

The solid curves in Fig. 2.12 represent the best fit to a Gaussian (the oxide data) accounting for the vibrational excitation of SiO_2 at room temperature and a Doniach-Sunjic lineshape (elemental data, see Equation (2.44)) accounting for the electron–hole pair creation in the course of the screening process. The singularity index α retrieved by the fitting procedure, provides valuable information on the screening of the core-hole [4]. Thus, elimination of multiple scattering can be used to expose the many body effects in photoemission spectra.

So far, the examples focussed on the aspect of the energy fluctuations during the transport using model calculations for the collision statistics (represented by the partial intensities). Although the examples indicate that the collision statistics are accurately understood (see e.g. Fig. 2.10), it seems worthwhile to present examples where the collision statistics have been experimentally investigated. Such a case is displayed in the Auger photoelectron coincidence spectra (APECS) shown Fig. 2.13. In APECS, pairs of Auger and photoelectrons emitted from the same atom are measured. Experimentally, this can be achieved in several different ways [81–83]. The partial intensities for the Auger-photoelectron pair are given by the product of the respective depth distribution functions $\phi_{n_A}(z)$ and $\phi_{n_X}(z)$, integrated over the depth:

$$C_{n_A,n_X} = \int_0^\infty \phi_{n_X}(z)\phi_{n_A}(z)dz \tag{2.45}$$

An example of the reduced APECS partial intensities $\gamma_{n_A,n_X} = C_{n_A,n_X}/C_{n_A=0,n_X=0}$ for Si2s photoelectrons and Si-LVV Augerelectrons excited by 300 eV photons is shown in Fig. 2.13b. The corresponding depth distribution functions are those shown in Fig. 2.7. The non-coincident (singles) Si2p partial intenities are shown for comparison as the dashed curve. The higher order Si2p partial intensities, measured in coincidence with the Auger peak (curve labelled $n_A = 0$) are significantly reduced compared to the singles intensities. The reason is that the Auger electrons (with $n_A = 0$) originate from shallow depths (see Fig. 2.7) and, since the photoelectrons are created at the same atom, they travel, on average, a shorter pathlength than the photoelectrons in the singles spectrum that originate from a much larger depth range. In other words, by measuring the photoelectrons in coincidence with Augerelectrons that have lost a certain fraction of their energy, one can select the depth range from which the detected coincident photoelectrons originate. Choosing $n_A = 1$

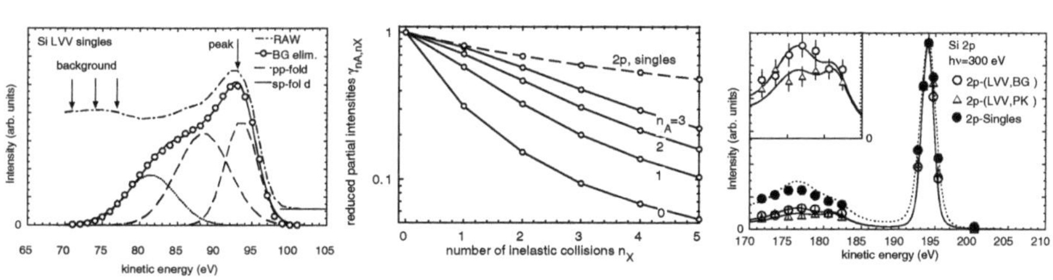

Fig. 2.13. (**a**) Experimental Si-LVV singles spectrum [51] (*dash-dotted line*); The same spectrum subjected to Eq. (2.40) is represented by *open circles.* (**b**) Reduced double differential partial intensities for bulk inelastic scattering $\gamma_{n_X,n_A} = C_{n_X,n_A}/C_{n_X=0,n_A=0}$ calculated from the curves in Fig. 2.7a and b using Eq. (2.45). The dashed curve represents the Si 2p singles partial intensities for bulk scattering. (**c**) Si 2p spectrum measured in coincidence with the background in the Si-LVV spectrum (*open circles*, see arrows marked "background" in (**a**), as well as with the peak of the Si-LVV Auger line (*open triangles*, arrow marked "peak" in (**a**). The *filled circles* are the corresponding singles intensities. The *solid* and *dotted lines* represent results of model calculations with the SESSA-Software (Simulation of Electron Spectra for Surface Analysis) [51] (see text). The data were normalized at the peak maximum. The inset shows an expanded view of the BG and PK spectra

consequently leads to an increase of the reduced partial intensities compared to the case $n_A = 0$, as shown by the corresponding curve in Figs. 2.13b and 2.13a shows the measured singles Si2p peak as chain dashed curve [84]. Elimination of the inelastic background due to surface and volume scattering using Eq. (2.40) results in the spectrum represented by the open circles. The solid curve is a fit of this spectrum to a linear combination of three Gaussians, indicated by the long-dashed and dotted curves, that compare well with the assignment of Pernaselci and Cini as the convolution of the p-p and s-p partial density of states of Si [85]. The filled circles (and dotted curve) in Fig. 2.13c represents the singles Si 2p spectrum, the triangles are the spectrum in coincidence with the peak in the Auger spectrum (see the arrow labelled "peak" in Fig. 2.13a). The bulk plasmon loss feature in this spectrum is significantly reduced compared to the plasmon in the singles spectrum. When the Si2p peak is measured in coincidence with the first loss in the SiLVV peak (see arrow labelled "background" in Fig. 2.13b), the plasmon loss in the Si2p peak (open circles in Fig. 2.13c) increases compared to the Si2p measured in coincidence with the Auger no–loss peak. These observations are in perfect agreement with the expected collision statistics shown in Fig. 2.13b.

The observed change of the plasmon intensity is thus a signature of the different depths sampled in the two coincidence spectra. An assessment shows that the electrons in the Si2p spectrum in coincidence with the Auger peak originate from an average depth of about 2.0 ± 2.1Å while the average emission depth of the spectrum measured in coincidence with the background amounts to about 4.7 ± 4.9Å. The singles spectrum, on the other hand, consists of electrons emitted from an average depth of 6.1 ± 6.5Å. Here the quoted uncertainties represent the fluctuations in the emission depth (see Fig. 2.7). Furthermore, these results provide direct experimental proof of the existence of extrinsic plasmons supporting the three step model for photoemission. However, the results also demonstrate the limited validity of the three step model in photoemission: It can be seen in Fig. 2.13c that the linewidth of the coincidence spectra is slightly smaller than for the singles spectrum. In early work on APECS, it was already anticipated that coincidence measurements would reduce the core-hole lifetime broadening and modify other contributions to line broadening. Indeed, line narrowing in photoemission spectra by APECS has been observed by several groups [86, 87] and is a clear indication for the one-step character of the Auger and photoelectron emission process [88].

The influence of surface excitations on the angular distribution of the zero order collision statistics is illustrated in Fig. 2.14, that shows the angular distribution of the Cu LMM and MVV Auger peaks emitted from a homogeneous Cu sample. Figure 2.14a shows the experimental configuration, employing a slanted sample holder, that is devised to separately study the excitation and emission part of the three step model in electron spectroscopy: When (I.) the azimuthal rotation axis points towards the electron gun, the emission angular distribution is obtained for a fixed angle of incidence by varying ϕ_s. On the other hand, when (II.) the azimuthal rotation axis is adjusted parallel to

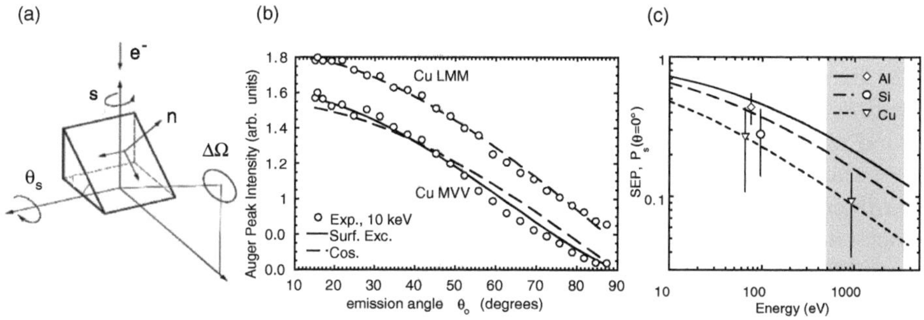

Fig. 2.14. (**a**) Schematic illustration of the experimental configuration to measure the emission angular distribution at fixed incidence. (**b**) Angular distribution of the Auger peak intensity for the Cu LMM (920 eV) and NVV (60 eV) transition for a primary energy of 3, 5 and 10 kV. *Data points*: experimental peak heights; *Solid lines*: fit of the data to Eq. (2.42); *dashed lines*: cosine distribution. (**c**) Survey of the results for the surface excitation parameter derived from Auger electron angular distributions for Al, Si and Cu (*data points*) and comparison with earlier results derived from REELS spectra [71] (*solid, dashed* and *dotted lines*). The *grey area* indicates the energy region where the earlier REELS measurements were performed

the analyzer axis, the emission angle is constant and varying ϕ_s, one obtains the incident angular distribution. In case (I.), all the processes playing a role in the excitation of the Auger electrons are the same for any value of ϕ_s. In consequence, any feature in the angular distribution is completely attributable to the escape process. Conversely, for case (II.), the angular distribution is entirely governed by the processes playing a role in the Auger electron excitation.

The emission angular distribution of the Auger peak intensities for a semi-infinite Cu sample are shown in Fig. 2.14b. The dashed line represents the cosine distribution. The distribution of the low energy Auger Cu MVV peak at ~60 eV is seen to deviate significantly from a cosine function. For the higher energy Auger transition a similar effect is seen, but it is much weaker. Similar results were obtained for Al and Si [89]. Since for a homogeneous sample the observed angular distribution cannot be explained by the effects of surface roughness or refraction by the inner potential [89], this is attributed to the influence of surface excitations. The solid line that represents a fit of these data to Eq. (2.42) is seen to describe this behaviour well. The resulting surface excitation parameter is shown in Fig. 2.14c as data points and compared with the empirical formula Eq. (2.42), based on measurements at higher kinetic energies (indicated by the grey area in Fig. 2.14c). This result shows that the range of validity of Eq. (2.42) extends down to rather low energies of ~50 eV.

2.3.3 Secondary Electron Emission

The examples in the previous two paragraphs demonstrate that electron reflection and emission phenomena at medium energies are generally understood quantitatively. Unfortunately, this is not the case for the emission of secondary electrons, i.e. emission of slow electrons induced by electron bombardment. Several factors complicate quantitative interpretation of secondary electron spectra. First of all, the definition of the term "secondary electron" is problematic in itself, since there is no experimental means of distinguishing a true secondary electron from a backscattered primary electron and model calculations indicate that the true secondary electron energy distribtution has a broad tail reaching up to energies just below the primary energy. In practice, however, only those electrons that are emitted from the sample with an energy of less than $\sim$50 eV are by convention designated as secondary electrons. Furthermore, experimental values for the secondary electron yield, i.e. the number of secondaries emitted per incident primary, exhibits significant scatter [90, 91]. Differences of up to 50% for any material and energy being the rule rather than the exception. Finally, the theory for the emission of secondary electrons is much more involved than for quasi–elastic electron reflection and emission discussed above.

Most theories for seondary electron emission are based on the three step model [92–97]. For the generation of secondaries one needs to consider the different mechanisms by which the primary electron can transfer energy needed to promote a secondary electron originally in the sea of loosely bound solid state electrons over the surface barrier. These mechanisms comprise single particle excitation by electron–electron scattering, ionization of core levels and decay of surface and bulk plasmons. For a detailed understanding of these excitation mechanisms, the band structure of the solid under consideration plays an important role. Furthermore, a primary electron can produce more than one secondary and these can in turn, provided they are sufficiently energetic, excite additional electrons of higher order in a cascade process.

As for the transport of the secondaries, the situation is more complicated than for quasi–elastic electron emission for several reasons. First of all, the interaction characteristics depend strongly on the actual energy of the electron. This not only holds for the total mean free paths (see Fig. 2.4), but also for the (normalized) differential mean free paths for elastic and inelastic scattering when the primary energy is smaller than 50 eV. As can be seen in Fig. 2.3a and b the shape of the differential mean free path for small energies losses is similar for primary energies above 200 eV, but for smaller energies it changes significantly, and for energies below 50 eV it changes dramatically. Note also that at such low energies, exchange and correlation effects need to be included in the calculation of the inelastic interaction characteristics, while for medium energies $\gtrsim$200 eV such effects are expected to be negligible.

For the elastic interaction the question of the potential to use in a partial wave analysis becomes more important at lower energies where differences between atomic potentials [12, 98, 99] and those applicable to the solid state, such as the muffin–tin potential [100] are larger than for medium energies. The main difference lies in the weakly bound electrons, corresponding classically to large impact parameters and quantum mechanically corresponding to high angular momentum. This main differences in the resulting elastic scattering cross sections concern small scattering angles. Therefore, the elastic mean free path in a sloid is significantly different from the gas phase (to which atomic potentials apply), while the transport mean free path, which emphasizes large scattering angles (see Eq. (2.2)), is practically unaffected by the scattering potential. Since the scattering parameter substantially exceeds unity in the relevant energy range (see Fig. 2.4), the transport in this energy regime is determined by momentum relaxation, allowing one to apply the so-called transport approximation [11, 38, 46], where the scattering cross section is replaced by the isotropic transport cross section and the transport mean free path replaces the elastic mean free path. In this case, the details of the interaction potential are of less importance. On the other hand, the assumption that appreciable deflections occur only during elastic collisions breaks down in the low energy range. Since the momentum relaxation is strong, this can be corrected for by replacing the usual (elastic) transport mean free path λ_{tr} by the total transport mean free path $\lambda_{tr,t}$:

$$\frac{1}{\lambda_{tr,t}} = \frac{1}{\lambda_{tr}} + \frac{1}{\lambda_{tr,i}}$$

where $\lambda_{tr,i}$ is the transport mean free path for inelastic scattering.

Summarizing the above statements on the electron transport in the low energy regime, it can be concluded that by virtue of the strong momentum relaxation, the secondary electron spectrum is a superposition of many different scattering orders for a complicated excitation function that is smeared out in the emitted energy distribution, making it very difficult to study the details of the excitation and inelastic scattering characteristics experimentally. This fact severely complicates a detailed comparison between theory and experiment.

A final essential aspect of secondary electron emission within the three step model is the escape over the surface barrier. The periodicity of the crystal potential is abruptly terminated at the surface giving rise to a surface barrier with a height of typically 10–20 eV, being of the order of, or larger than the energy of a secondary electron. The surface barrier is commonly taken to be the energetic distance from the bottom of the conduction band to the vacuum level, being equal to the electron affinity in the case of insulators, and being equal to the work function in the case of metals. During the passing of the surface barrier the energy component parallel to the surface is conserved, while the perpendicular component is reduced by the height of the barrier. Depending on the direction of surface crossing, the outgoing particle is either refracted at the surface or is reflected at the barrier. Thus, only a fraction of the internal angular distribution (within the so–called “escape–cone” de-

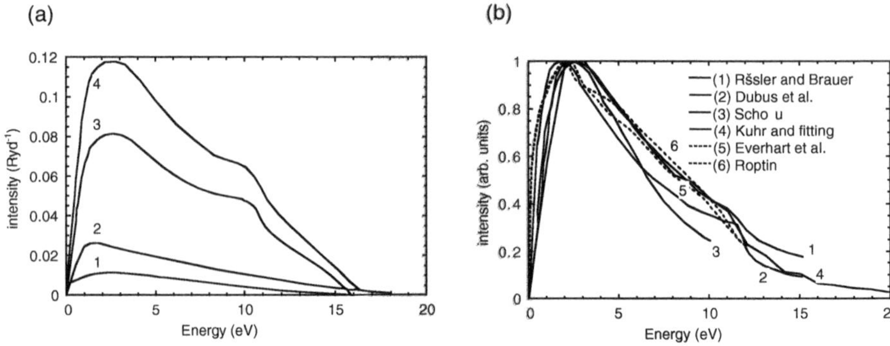

Fig. 2.15. Secondary electron spectrum of Al (in reciprocal Rydbergs) for a primary energy of 1 keV exhibiting contributions of different excitation mechanisms (after Ref. [101]). (1) excitation of core levels; (2) electron–electron scattering (single particle excitation); (3) decay of plasmons; The *curve* labelled (4) is the sum of these contributions. (**b**) comparison of different theoretical results for the shape of the secondary electron spectrum of Al, induced by 1 keV electrons [92–95] with experimental results [102, 103], normalized at the maximum of the peak

termined by the ratio of the energy and the barrier height) is transmitted through the barrier.

Experimental data on secondary electron emission scatter significantly from experiment to experiment (see [90, 91] for a review). One of the reasons for these disrepancies is that the surfaces for which values are quoted in the literature are not always measured under well-defined conditions. For example, a contaminated surface may have a significantly different barrier than its clean counterpart. This can lead to a dramatic change in the magntude of the escape cone and the probability for transmission over the barrier. Furthermore, the crystalline state of the surface influences the emitted spectrum not only because the surface barrier is different for different crystal orientations, but also since the states available for energy transfers from the primary electron to the electronic subsystem of the solid vary with orientation. Even on perfectly well defined surfaces, a measurement of the spectrum of low energy electrons is intrinsically difficult due to the presence of magnetic fields.

In Fig. 2.15a, the secondary electron energy spectrum for 2 keV incident electrons on an Al surface according to the detailed theory of Rösler and Brauer [101] is shown along with the contributions originating from the different excitation mechanisms: core level ionization (curve labelled 1), single particle excitations as a result of electron –electron scattering (curve labelled 2) and bulk plasmon decay (curve labelled 3). The curve labelled (4) is the sum of individual contributions. For Al, the main mechanism is the decay of plasmons that leads to the shoulder at about 10 eV, corresponding to the plasmon energy of Al of about 15 eV, reduced by the height of the surface barrier. Surface plasmon excitation is disregarded in this theory.

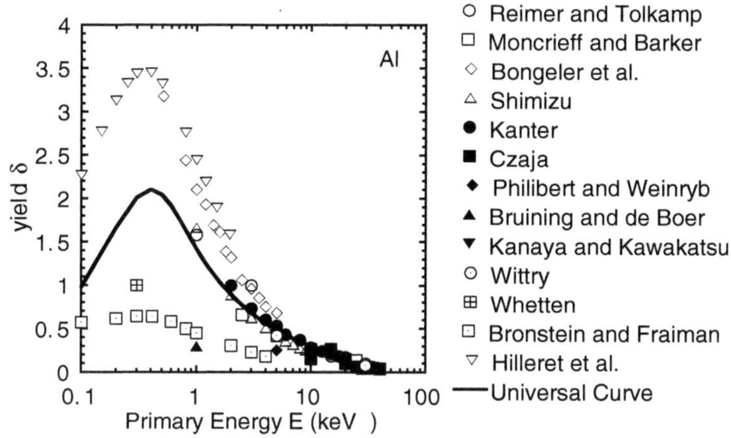

Fig. 2.16. Number of secondary electrons, i.e., the secondary electron yield δ, versus the primary energy for Al. The data points indicate experimental values taken from the sources indicated (Refs. [104–116]), the solid curve is the "universal" yield curve Eq. (2.46) using $E^m = 0.4$ eV and $\delta^m = 2.05$

A comparison of experimental results [102, 103] with different theoretical calculations [92–95] following the outline above is presented in Fig. 2.15b for the secondary electron spectrum of Al for a primary energy of 1 keV. These curves were normalized to unity at the peak position. Except for the semi-empirical approach of Schou, all considered theories predicts the appearance of a plasmon shoulder at around 10 eV, that is also seen in the experimental curves. The width and positions of the rising and falling edges of the spectrum are different for different theories. Note that the shoulder at ~5 eV, that approximately corresponds to the energy needed to excite a surface plasmon (reduced by the barrier height) is not reproduced by theory, although Kuhr and Fitting [95] explicitly consider decay of surface plasmons in their calculations.

The number of secondary electrons per incident primary, the secondary electron yield δ, is presented as a function of the energy in Fig. 2.16. The data points, that represent a collection of experimental data compiled in Ref. [90, 91], display substantial discrepancies, exceeding a factor of 2. The shapes of the measured yield curves is similar in most cases, however, and turns out to be very similar for any arbitrary material [91]. Several authors have noticed this similarity [91, 117–122] and developed empirical universal expressions for the secondary yield that gives a convenient description of the phenomenon of secondary electron emission. The form given by Lin and Joy [90] reads:

$$\delta(E) = 1.28\delta^m \left(\frac{E}{E_m}\right)^{-0.67} [1 - \exp\{-1.614 \left(\frac{E}{E_m}\right)^{-1.67})\}] \qquad (2.46)$$

Where the maximum yield δ_m and the position of the maximum E_m represent material dependent parameters. The merit of universal curves such as the one expressed in Eq. (2.46) lies in the fact that they can be used to characterize an experimental data set by a small number of parameters (2 parameters in the case of Eq. (2.46)). This greatly facilitates the finding and correcting of errors in individual data sets and allows one to identify trends in the dependence on target material etc. It is to hoped that in this way the discrepancy between theory and experiment can eventually be reduced to a tolerable level allowing to test the available theories and to deepen the understanding of the phenomenon.

2.4 Concluding Remarks

Phenomena involved in quasi–elastic electron emission and reflection have been reviewed and put into perspective with the emission of slow secondary electrons. While electron transport in the medium energy range turns out to be quantitatively understood, quantitative interpretation of secondary electron spectra is difficult since in such spectra many different processes are superimposed and the relevant source functions are obscured by intensive multiple scattering.

In the quasi–elastic case, reflection electron energy loss spectra can be used to study all relevant aspects of the signal electron transport: the angular distribution of the elastic peak can be used to study the shape of the elastic scattering cross section, while the energy loss portion of electron reflection spectra can be used to derive the dielectric function from which characteristics of the inelastic interaction of a particle with an arbitrary solid can be derived, not only for electrons but also for ions, photons and other particles. Indeed, such experiments have been used extensively in the past decades to unravel the different processes playing a role in electron emission in the medium energy range. Furthermore, it was shown that coincidence spectroscopy can be utilized to gain valuable information on the electron emission mechanism. In the case of the presented example this concerned the collision statistics in the emission of characteristic photoelectrons.

At low energies comparable to the energy of true secondaries, electron reflection measurements are very challenging experimentally. Furthermore, interpretation of reflection electron spectra is much more difficult at very low energies since the secondary electron spectrum and the loss features of the primaries overlap. In time resolved electron spectroscopy experiments, the resolution that can nowadays be achieved in practice is around several hundred picoseconds, which is clearly not good enough to completely resolve the secondary electron cascade, considering the fact that a 1 eV electron has a speed of about 6Å/fs. Nonetheless, coincidence experiments have a potential for the investigation of the phenomenon of secondary electron emission with a greater level of discrimination. For example, measuring thc secondary electron spectrum in coincidence with the loss features of a reflection spectrum should be feasible with state of the art experimental apparatus nowadays and could be used to expose –at least with a greater level of discrimination than with noncoincident measurements– the different excitation mechanism contributing to the peak of secondary electrons. Moreover, by selecting events in which more than one secondary electron is emitted as a result of a single impact of a projectile (be it an ion, an electron or an arbitrary other particle), valuable information on the collision cascade and the secondary electron cascade can be gained. Such experiments are presently being developed in the group of the author and it is hoped that in this way, with increasing time resolution of such experiments, essential aspects of the thermalization of slow electrons will become accesible to experiment in the near future.

Acknowledgments

Financial support of the present work by the Austrian Science Foundation FWF through Project No. P15938-N02 is gratefully acknowledged.

References

1. J. A. Venables, *Developments in Electron Microscopy and Analysis* (Academic Press, New York, 1995).
2. J. Goldstein, D. E. Newbury, P. Echlin, D. C. Joy, A. D. Romig, C. E. Lyman, C. Fiori, and E. Lifshin, *Scanning Electron Microscopy and X-ray Microanalysis* (Plenum, New York, London, 1992).
3. D. Briggs and J. Grant, eds., *Surface Analysis by Auger and X-Ray Photoelectron Spectroscopy* (IMPublications, Chichester, UK, 2003).
4. S. Hüfner, *Photoelectron spectroscopy* (Springer, Berlin, Heidelberg, New York, 1995).
5. H. Winter and F. Aumayr, in *Trapping Highly Charged Ions: Fundamentals and Applications*, edited by J. Gillaspy (NOVA Science Publisher Inc., New York, 1999).
6. H. S. W. Massey and E. Burhop, *Electronic and Ionic Impact Processes* (Clarendon Press, Oxford, 1952).
7. N. F. Mott and H. S. W. Massey, *The theory of atomic collisions* (Oxford University Press, Oxford, 1949).
8. A. Modinos, *Thermionic and Secondary Electron Spectroscopy* (Plenum, New York, 1984).
9. G. N. Fursey, *Field Emission in Vacuum Microelectronics, 1st Ed.* (Springer Verlag, Berlin–Heidelberg–New York, 2005).
10. W. Smekal, W. S. M. Werner, C. S. Fadley, and M. A. van Hove, J. Electron Spectrosc. Rel. Phen. **137**, 183 (2004).
11. W. S. M. Werner, Surf. Interface Anal. **31**, 141 (2001).
12. R. A. Bonham and T. G. Strand, J Chem Phys **39**, 2200 (1963).
13. A. Jabłonski and H. Ebel, Surf. Interface Anal. **11**, 627 (1988).

14. National Institute of Standards and Technology (NIST), NIST data bases SRDP 64, Gaithersburg MD (2003).
15. H. Raether, *Excitations of Plasmons and Interband Transitions by Electrons* (Springer, New York, 1980), vol 88 of Springer Tracts in Modern Physics.
16. P. Schattschneider, *Fundamentals of Inelastic Electron Scattering* (Springer, New York, Vienna, 1986).
17. R. F. Egerton, *Electron Energy Loss Spectroscopy in the Electron Microscope* (Plenum, New York and London, 1985).
18. L. D. Landau, E. M. Lifshitz, and L. P. Pitaevski, *Electrodynamics of Continuous Media* (Pergamom Press, Oxford, New York, 1984), 2nd edition, Translated by J. B. Sykes, J. S. Bell and M. J. Kearsley.
19. R. H. Ritchie, Phys. Rev. **106**, 874 (1957).
20. E. A. Stern and R. A. Ferrell, Phys. Rev. **120**, 130 (1960).
21. P. M. Echenique and J. B. Pendry, J. Phys. C: Solid State Phys **8**, 2936 (1975).
22. R. Nunez, P. M. Echenique, and R. H. Ritchie, J. Phys. C: Solid State Phys **13**, 4229 (1980).
23. F. Yubero and S. Tougaard, Phys. Rev. **B46**, 2486 (1992).
24. N. R. Arista, Phys. Rev. **A49**, 1885 (1994).
25. C. J. Tung, Y. F. Chen, C. M. Kwei, and T. L. Chou, Phys. Rev. **B49**, 16684 (1994).
26. C. Denton, J. L. Gervasoni, R. O. Barrachina, and N. R. Arista, Phys. Rev. **A57 , 4498** (1998).
27. M. Vicanek, Surf. Sci. **440**, 1 (1999).
28. K. L. Aminov and J. B. Pedersen, Phys. Rev. **B63**, 125412 (2001).
29. D. W. Lynch and W. R. Hunter, *Handbook of optical constants of solids*, Sect. 2.1.III., in [32] (1985).
30. K. Glantschnig and C. Ambrosch-Draxl, Phys. Rev. (2006).
31. C. Ambrosch-Draxl and J. O. Sofo, Comp. Phys. Comm. **17**, 1 (2006).
32. E. D. Palik, *Handbook of optical constants of solids* (Academic Press, New York, 1985).
33. E. D. Palik, *Handbook of optical constants of solids II* (Academic Press, New York, 1991).
34. B. L. Henke, E. M. Gullikson, and J. C. Davis, Atomic Data and Nuclear Data Tables **54**, 181 (1993).
35. W. S. M. Werner, Surf. Sci. **600**, L250 (2006a).
36. W. S. M. Werner and I. S. Tilinin, Surf. Sci. **268**, L319 (1992).
37. I. S. Tilinin and W. S. M. Werner, Phys. Rev. **B46**, 13739 (1992).
38. K. M. Case and P. F. Zweifel, *Linear Transport Theory* (Addison-Wesley, Reading, MA, 1967).
39. S. Goudsmit and J. L. Saunderson, Phys. Rev. **57**, 24 (1940).
40. L. D. Landau, J Phys (*Moscow*) **8**, 201 (1944).
41. W. S. M. Werner, Phys. Rev. **B71**, 115415 (2005a).
42. W. S. M. Werner and P. Schattschneider, J. Electron Spectrosc. Rel. Phen. **143**, 65 (2005).
43. W. S. M. Werner, Phys. Rev. **B55**, 14925 (1997).
44. A. Jabłonski, Surf. Interface Anal. **14**, 659 (1989).
45. S. Chandrasekhar, *Radiative Transfer* (Dover publications, New York, 1960).
46. V. V. Sobolev, *A Treatise on radiative transfer* (van Nostrand, Princeton, NJ, 1963).

47. I. S. Tilinin, Sov Phys JETP **55**, 751 (1982).
48. I. S. Tilinin, A. Jabłonski, and S. Tougaard, Phys. Rev. **B52**, 5935 (1995).
49. W. S. M. Werner, I. S. Tilinin, and M. Hayek, Phys. Rev. **B50**, 4819 (1994).
50. W. S. M. Werner, Surf. Interface Anal. **37**, 846 (2005b).
51. W. S. M. Werner, W. Smekal, and C. J. Powell, Simulation of Electron Spectra for Surface Analysis (SESSA), NIST data bases SRD–100, Gaithersburg MD (2005a).
52. W. S. M. Werner, H. Störi, and H. Winter, Surf. Sci. **518**, L569 (2002).
53. W. S. M. Werner, Surf. Interface Anal. **35**, 347 (2003a).
54. W. S. M. Werner, Surf. Sci. **526/3**, L159 (2003b).
55. W. S. M. Werner, Phys. Rev. **B74**, 075421 (2006b).
56. W. S. M. Werner, Surf. Sci. **588**, 26 (2005c).
57. W. S. M. Werner, C. Eisenmenger-Sittner, J. Zemek, and P. Jiricek, Phys. Rev. **B67**, 155412 (2003).
58. O. A. Baschenko, J. Electron Spectrosc. Rel. Phen. **57**, 297 (1991).
59. O. A. Baschenko and A. E. Nesmeev, J. Electron Spectrosc. Rel. Phen. **57**, 33 (1991).
60. P. Cumpson, J. Electron Spectrosc. Rel. Phen. **73**, 25 (1995).
61. S. Tougaard, Surf. Interface Anal. **8**, 257 (1986).
62. W. S. M. Werner, Surf. Interface Anal. **23**, 737 (1995a).
63. S. Tougaard and I. Chorkendorff, Phys. Rev. **B35**, 6570 (1987).
64. W. S. M. Werner, Surf. Sci., cond-mat/0611053 (2006c).
65. W. S. M. Werner, Appl. Phys. Lett. (2006d).
66. W. S. M. Werner, Phys. Rev. **B52**, 2964 (1995b).
67. R. Oswald, E. Kasper, and K. Gaukler, J. Electron Spectrosc. Rel. Phen. **61**, 251 (1993).
68. J. Zemek, P. Jiricek, W. S. M. Werner, B. Lesiak, and A. Jablonski, Surf. Interface Anal. **38**, 615 (2006).
69. W. F. Egelhoff, Phys. Rev. Lett. **71**, 2883 (1993).
70. W. S. M. Werner and M. Hayek, Surf. Interface Anal. **22**, 79 (1994).
71. W. S. M. Werner, W. Smekal, C. Tomastik, and H. Störi, Surf. Sci. **486**, L461 (2001a).
72. W. S. M. Werner, W. Smekal, H. Störi and C. Eisenmenger-Sittner, J. Vac. Sci. Technol. **A19**, 2388 (2001).
73. W. S. M. Werner, W. Smekal, T. Cabela, C. Eisenmenger-Sittner, H Störi, J. Electron Spectrosc. Rel. Phen. **114**, 363 (2001).
74. Y. F. Chen, Surf. Sci. **345**, 213 (1996).
75. S. Tanuma, C. J. Powell, and D. R. Penn, Surf. Interface Anal. **21**, 165 (1994).
76. C. J. Powell and A. Jabłonski, J Phys Chem Ref Data **28**, 19 (1999).
77. K. Goto, N. Sakakibara, Y. Takeichi, and Y. Sakai, Surf. Interface Anal. **22**, 75 (1994).
78. P. H. Citrin, G. K. Wertheim, and Y. Baer, Phys. Rev. **B16**, 4256 (1977).
79. P. W. Anderson, Phys. Rev. Lett. **18**, 1049 (1967).
80. S. Doniach and M. Sunjic, J Phys C **3**, 285 (1970).
81. H. W. Haak, G. A. Sawatzky, and T. D. Thomas, Phys. Rev. Lett. **41**, 1825 (1978).
82. S. M. Thurgate, Surf. Interface Anal. **20**, 627 (1993).
83. E. Jensen, R. A. Bartynski, S. L. Hulbert, and E. D. Johnson, Rev. Sci. Instrum. **63**, 3013 (1992).

84. W. S. M. Werner, W. Smekal, H. Störi, H. Winter, G. Stefani, A. Ruocco, F. Offi, R. Gotter, A. Morgante, and F. Tomasini, Phys. Rev. Lett. **94**, 038302 (2005b).
85. A. Pernaselci and M. Cini, J. Electron Spectrosc. Rel. Phen. **82**, 79 (1996).
86. E. Jensen, R. A. Bartynski, S. L. Hulbert, E. D. Johnson, and R. Garrett, Phys. Rev. Lett. **62**, 71 (1989).
87. S. M. Thurgate and Z. T. Jang, Surf. Sci. **466**, L807 (2000).
88. M. Ohno, J. Electron Spectrosc. Rel. Phen. **124**, 53 (2002).
89. W. S. M. Werner, H. Trattnik, J. Brenner, and H. Störi, Surf. Sci. **495**, 107 (2001b).
90. D. C. Joy, Scanning **17**, 270 (1995).
91. D. C. Joy, http://pciserver.bio.utk.edu/metrology/download/E-solid/ database.doc (2003).
92. M. Rösler and W. Brauer, in *Particle Induced Electron Emission I*, edited by G. Höhler (Springer, Heidelberg, 1992), vol. 122, pp. 1–65.
93. A. Dubus, J. Devooght, and J. C. Dehaes, Phys. Rev. **B36**, 5110 (1987).
94. J. Schou, Phys. Rev. **B22**, 2141 (1980).
95. J. C. Kuhr and H. J. Fitting, J. Electron Spectrosc. Rel. Phen. **105**, 257 (1999).
96. J. P. Ganachaud and M. Cailler, Surf. Sci. **83**, 498 (1979a).
97. J. P. Ganachaud and M. Cailler, Surf. Sci. **83**, 519 (1979b).
98. A. C. Yates, Comp Phys Comm **2**, 175 (1971).
99. A. Jabłonski, F. Salvat, and C. J. Powell, J. Phys. Chem. Ref. Data **33**, 409 (2004).
100. J. B. Pendry, *Low Energy Electron Diffraction* (Academic Press, London, New York, 1974).
101. M. Rösler and W. Brauer, Phys Stat Sol **148**, 213 (1988).
102. D. Roptin, Ph.D. thesis, University of Nantes, Nantes, France (1975).
103. T. E. Everhart, N. Saeki, R. Shimizu, and T. Koshikawa, Journal of Applied Physics **47**, 2941 (1976).
104. L. Reimer and C. Tolkamp, Scanning **3**, 35 (1980).
105. D. A. Moncrieff and P. R. Barker, Scanning **1**, 195 (1976).
106. R. Bongeler, U. Golla, M. Kussens, L. Reimer, B. Schendler, R. Senkel, and M. Spranck, Scanning **15**, 1 (1993).
107. R. Shimizu, J. Appl. Phys **45**, 2107 (1974).
108. M. Kanter, Phys. Rev. **121**, 1677 (1961).
109. W. Czaja, Journal of Applied Physics **37**, 4236 (1966).
110. J. Philibert and E. Weinryb, C. R. Acad. Sci. **256**, 4535 (1964).
111. H. Bruining and J. M. de Boer, Physica **V**, 17 (1938).
112. K. Kanaya and M. Kawakatsu, Journal of Applied Physics **D5**, 1727 (1972).
113. D. B. Wittry, Proc. 4th Conf. on X-ray Optics and Microanalysis, ed. R. Castaing, (Hermann:Paris) p. 168 (1966).
114. N. R. Whetten, *Methods in Experimental Physics IV* (Academic Press, New York, 1962).
115. I. M. Bronstein and B. S. Fraiman, *Vtorichnaya Elektronnaya Emissiya* (Nauka, 1969), (in Russian).
116. N. Hilleret, J. Bojko, O. Grobner, B. Henrist, C. Scheuerlein, and M. Taborelli, Proc. 7th European Particle Accelerator Conference, Vienna. p. 127 (2000).
117. V. V. Makarov and N. N. Petrov, Sov. Phys. Sol. State **23**, 1028 (1981).
118. E. M. Baroody, Phys. Rev. **78**, 780 (1950).

119. J. L. H. Jonker, Philips Research Repts. **7**, 1 (1952).
120. A. J. Dekker, Solid State Physics **6**, 251 (1958).
121. G. F. Dionne, Journal of Applied Physics **46**, 3347 (1975).
122. H. Seiler, Journal of Applied Physics **54**, R1 (1983).

3

Potential Electron Emission from Metal and Insulator Surfaces

Friedrich Aumayr and Hannspeter Winter

3.1 Potential Electron Emission – Basic Mechanisms

Impact of slow ions (impact velocity < 1 a.u. $= 25$ keV/amu) on solid surfaces is of genuine interest in plasma- and surface physics, and related applications. Nature and intensity of the resulting inelastic processes depend both on the kinetic and the potential (= internal) ion energy carried toward the surface. For most practical applications the kinetic projectile energy is of foremost relevance as, e.g., in ion-induced kinetic electron emission (KE [1–4]), ion-surface scattering and kinetic sputtering [5, 6]. However, ion-induced processes can also depend on the internal (potential) energy of the projectile, especially if the latter exceeds the kinetic projectile energy, resulting in additional electron emission or sputtering (potential electron emission PE [4, 7–10], potential sputtering [11–13]. Considerable potential energy will be stored in a multiply charged ion (MCI) Z^{q+} during its production where q electrons are removed from an originally neutral atom. The same potential energy will become again available if the MCI encounters a solid surface.

PE due to impact of slow singly, doubly and multiply charged ions on atomically clean metal surfaces has been thoroughly studied by Hagstrum [7, 8, 14, 15] who measured the yields and energy distributions of ion-induced slow electrons. From these studies he concluded that PE arises from relatively fast electronic transitions (rates $\geq 10^{14}$ s^{-1}) from the surface into empty projectile states, which require no minimum impact velocity and already start before the ion has entered the surface selvedge. PE yields increase strongly with the projectile's potential energy, i.e. their charge state. At higher impact velocity also kinetic electron emission (KE [1–4]) will produce slow electrons (see also Sect. 3.2 in the contribution of C. Lemell and J. Burgdörfer and the contribution by Helmut Winter et al. to this book), the fraction of which cannot simply be distinguished from the one due to PE. Based on theoretical studies on "radiationless" electronic transitions between a metal surface and a slow ion or excited atom by Massey [16–20], Hagstrum identified four types of one- and two-electron transitions (a–d) as being relevant for PE (see Fig. 3.1).

F. Aumayr and H. Winter: *Potential Electron Emission from Metal and Insulator Surfaces*,
STMP **225**, 79–112 (2007)
DOI 10.1007/3-540-70789-1_3

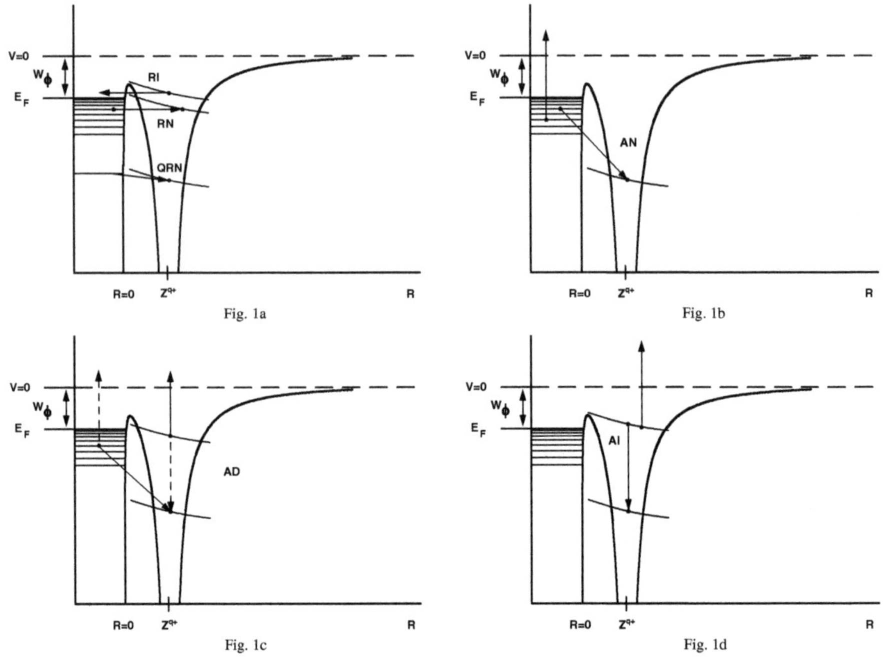

Fig. 3.1 a,b. Electron energy level diagrams showing resonant electronic transitions (**a**) and the Auger neutralisation process (**b**) for ions in front of a metal surface (R: particle-surface distance). The shaded region indicates the occupied part of the conduction band (W_ϕ: work function, E_F: Fermi energy) **c, d:** Electron energy diagrams showing Auger de-excitation (**c**) and autoionization processes (**d**) for projectiles in front of a metal surface. With decreasing distance to the surface the levels are shifted due to image charge effects

(a) *Resonant neutralisation* (RN, Fig. 3.1a) transfers an electron from the surface into unoccupied states of the approaching ion which overlap filled surface valence band states. RN itself does not give rise to electron emission but acts as a precursor for subsequent electron emitting transitions (see below). Arifov et al. [21] noted that for MCI impact a sequence of RN processes can take place, which generates a short-lived multiply-excited particle, later termed "hollow atom" [22].

(b) *Resonant ionization* (RI, Fig. 3.1a) is inverse to RN and transfers an electron from the projectile into an empty surface state with binding energy less than the surface work function W_ϕ.

(c) *Auger neutralisation* (AN, Fig. 3.1b; sometimes named *Auger capture*) can cause electron ejection from the surface valence band if the available potential energy exceeds twice the surface work function W_ϕ. One surface electron is captured by the ion and another one ejected with a kinetic energy $E_e \leq W_i' - 2W_\phi$.

The respective electron energy distribution reflects a self-convolution of the surface-electronic-density-of-states (S–DOS).

(d) *Auger de-excitation* (AD, Fig. 3.1c) of the projectile takes place if the latter, after an RN- or AN transition, carries an excitation energy still larger than W_ϕ. The excited projectile electron interacts with a surface electron such that the latter becomes ejected and the former demoted, or another surface electron is captured into the projectile and the initially excited electron ejected. In contrast to AN, energy distributions of electrons from AD are directly reflecting the S–DOS.

Hagstrum [7, 8, 15] regarded electronic transitions (a–d) in an adiabatic model (no coupling between electronic and nuclear motion) in order to derive the total slow electron yield. The respective transition rates increase exponentially with decreasing ion-surface distance according to the overlap between the S-DOS and the projectile-based electronic wave functions. Consequently, these transitions start most probably from the Fermi edge of the S-DOS. By assuming these transition probabilities as independent on the ion impact velocity, neutralisation of a singly charged ion is found to take place most probably at distances of a few Angstroms, whereas neutralisation of a MCI, depending on its charge state q, can already start at a comparably much larger distance (cf. below).

By generalizing this scheme, Arifov et al. [21] proposed that a MCI can (resonantly) capture a number of electrons from the surface within a rather short time, which leads to

(e) *Autoionisation* (AI, Fig. 3.1d) of the thereby produced multiply-excited particle (see Fig. 3.1d). AI is the intra-projectile AD of transiently formed doubly or multiply excited particles, where one or more electrons are ejected into vacuum, with other electron(s) in the projectile becoming demoted to lower states. Projectile AI was first found as Auger de-excitation of doubly-excited projectiles following the impact of He^{2+} or metastable He^{+} [23]. The energy distributions of slow electrons resulting from AI are therefore not directly reflecting the target S-DOS.

(f) *Quasi-resonant neutralisation* (QRN, Fig. 3.1a) is a near-resonant electron transition between target- and projectile core states which can only take place in close collisions where sufficient overlap of inner electronic orbitals will be achieved. Such QRN is of interest in the later stage of the interaction of MCI with a surface and within the bulk.

(g) *Radiative de-excitation* (RD) of excited projectile states after RN or AN of singly charged ions is much less probable than Auger de-excitation, since the respective transition rates will be about six orders of magnitude smaller than for Auger transitions. However, since radiative transition rates increase with about the fourth power of the projectile core charge [24], whereas Auger transition rates are not strongly affected by electron-core interaction, the latest steps of deexcitation of MCI projectiles which involve recombination of inner-shell vacancies may also give rise to soft X-ray emission (see below).

During the last two decades novel powerful MCI sources and easily accessible ultrahigh vacuum equipment have remarkably improved the conditions for studying such MCI-surface interaction processes also for higher primary ion charge states. For example, impact of a slow Th^{80+} ion on a clean gold surface can produce the tremendous amount of more than 300 electrons [25, 26] (cf. Sect. 3.1). Slow electron emission has been investigated in conjunction with other experimental signatures as the outgoing trajectories, final charge states and kinetic energy loss of MCI scattered on flat single-crystalline target surfaces, and with emission of projectile-characteristic fast Auger electrons and soft X-rays (for a general review see Arnau et al. [9]). The thereby collected expertise has formed our present understanding of MCI – surface interaction, which is characterized by the sketch shown in Fig. 3.2.

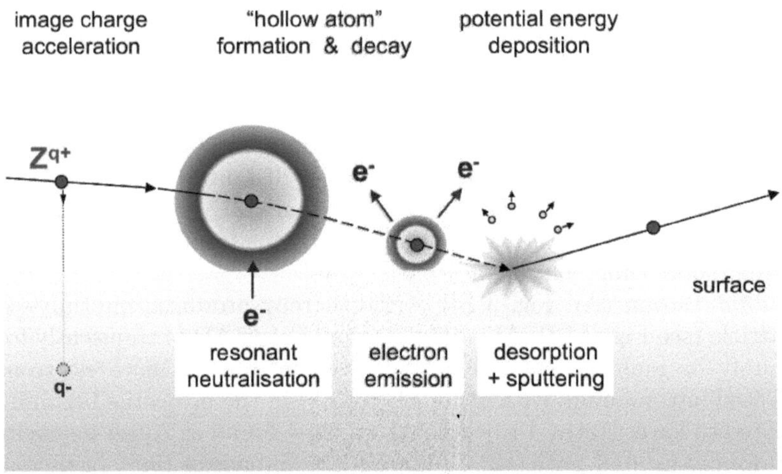

Fig. 3.2. Hollow atom formation and -decay during slow MCI impact on a metal surface [27]

The neutralisation of the MCI starts with the formation of a transient multiple-excited projectile which carries empty inner shells and has thus been called "Hollow Atom" (HA). This name was introduced by Briand et al. [22] in order to explain the appearance of projectile-characteristic soft X-ray emission from impact of fast MCI on gas-covered metals. We now know that such X-rays arise primarily in the late stage of HA decay in the target bulk, whereas slow electrons can already be emitted before the HA-turned MCI has touched the surface [9]. Upon its impact the MCI reclaims its missing electrons from the surface in order to become neutral, which leads to the emission of many slow electrons via AI and gives rise to strong electronic excitation at the surface where the potential energy is deposited during a short time (typically less than hundred femtoseconds) within a small area (typically one nanometer squared). The above surface stages can be well described by the so-called "classical over-the-barrier model" [28] (for details see the contribution to this book by C. Lemell and J. Burgdörfer, Sect. 3.4). As soon as the HA approaches the surface more closely, it will become screened by the metal electron gas which further accelerates its deexcitation. Eventually, inside the solid the so far remaining inner shell vacancies of strongly screened HA will be filled, which causes emission of projectile-characteristic fast Auger electrons and/or soft X-rays (see above), depending on the respective fluorescence yield. The just mentioned projectile recombination- and relaxation processes cannot be precisely separated from each other, since fast Auger electron emission may already occur before the close surface contact, and slow electron emission also after penetration of the surface. Anyhow, the slow electrons provide information on the situation of HA above and at the target surface, and fast Auger electron- and/or soft X-ray spectra on the HA development mainly below the surface [9].

For ions with a kinetic energy well above the respective KE threshold [1], the total electron yield results from both PE and KE, with the relative importance of both processes being difficult to assess. A detailed description of KE with its underlying mechanisms can be found in the contributions of J. Burgdörfer and C. Lemell as well as the contribution by H. Winter to this book. In our review KE will only be treated as far as it is a non-negligible competitor for PE and has to be separated from the latter. A clear-cut separation of PE and KE, however, is only possible under certain well-defined circumstances:

(a) for projectile ions with kinetic energies well below the KE threshold (exclusive PE)
(b) for slow HCI in very high charge states where the potential energy greatly exceeds the ions kinetic energy (dominant PE)
(c) for projectiles of any velocity in their neutral ground state (exclusive KE).

In Sects. 3.2 and 4.2 we demonstrate that the coincident measurement of electron emission and projectile energy loss can help to separate PE and KE also in situations where neither contribution is negligible.

Sometimes kinetic and potential effects are almost indistinguishable. E.g., recently a further kind of electron emission process has been observed for singly charged ion impact on MgO. It is interpreted either (3.1) as creation of a hole in the valence band by resonant electron capture, followed by an Auger transition (i.e. a PE effect) [29], or (2) as the production of a surface exciton via electron promotion in projectile collisions with O^- target-ions (a kinetic process) followed by exciton autoionization (an Auger-type effect) [30] (see also the contribution of P. Zeijlmans van Emmichoven to this book).

Impact of fast heavy particles on metal surfaces can give rise to plasmon excitation [31]. It has been found that such plasmons can also be excited by slow ions [32, 33] via their potential energy ("potential excitation of plasmons"). Clear signature for excitation of such plasmons is their subsequent one-electron decay which manifests itself as a characteristic feature in the electron energy distribution [31–34]. Slow-ion induced plasmons are thus PE-related phenomena (see our Sect. 3.3 and the much more detailed treatment by R. Baragiola and C. Monreal in this book).

3.2 Experimental Methods

In this chapter we will shortly outline the experimental techniques, which are nowadays available to study ion-induced electron emission. We describe methods for the production of MCI, the preparation of surfaces, the detection, number counting and energy-analysis of ion-induced slow electrons, and present some more sophisticated coincidence techniques, in particular to separate PE from KE in grazing incidence experiments.

3.2.1 Production of Slow Multiply Charged Ion Beams

The production of ions, their selection with respect to charge, mass and momentum and their transport to the experimental area of interest are well-established experimental techniques. Of special interest in the present context are multicharged ion (MCI) sources which nowadays can deliver beams of Z^{q+} ions (atomic number Z, charge state q) for virtually any chemical species up to the fully stripped ("naked") ions ($q = Z$). In the following we describe some MCI sources, which naturally can also deliver slow singly charged ions.

ECRIS – Electron Cyclotron Resonance Ion Source

An ECRIS [35–37] consists of the main components as shown in Fig. 3.3a. A discharge chamber filled with the working gas (pressure typically $\leq 10^{-5}$ mbar; ion production from non-volatile substances see below) is immersed in a "min-B" magnetic field geometry (i.e., the magnetic field strength increases from the plasma center outward) providing the necessary ion confinement. Such magnetic field configurations are generated by solenoids (which for larger set-ups may be superconducting) and multipole magnetic fields produced by permanent magnets (Sm-Co, Fe-Nd-B). Microwave radiation with a frequency up to 28 GHz is fed into the discharge chamber to produce a plasma wherein the electrons are heated by electron cyclotron resonance (ECR) when drifting through the so-called ECR zone (cf. Fig. 2.1), where the electron cyclotron frequency $\nu_{ec} = e \cdot B/2\pi.m_e$ matches the applied microwave frequency (e and m_e are electron charge and -mass, respectively, and B is the local magnetic field). The electrons can acquire hundreds of keV and thus ionize the magnetically confined ions which stay rather cold. This permits efficient step-by-step electron impact ionization up to high q values and extraction of high-quality (low-emittance) MCI beams. The energy spread of extracted ions is typically between 5 and 10 eV times charge state q.

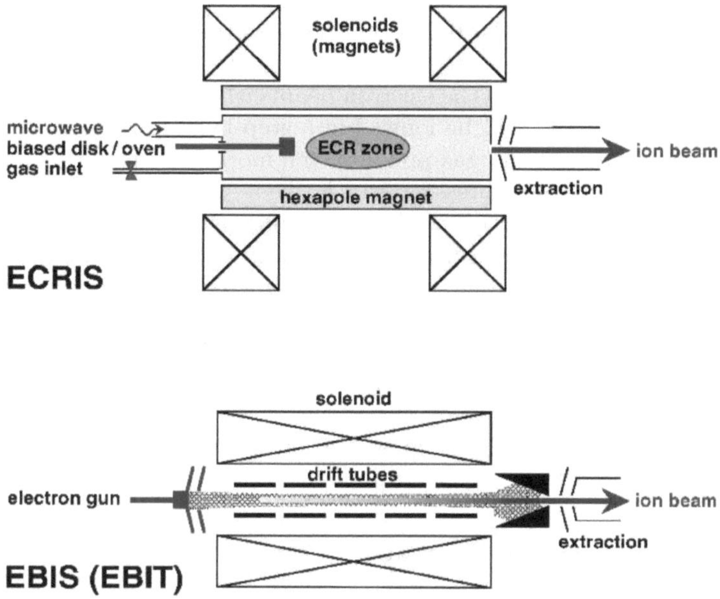

Fig. 3.3. (**a**) Main components of an ECRIS for MCI production (cf. text) (**b**) Main components of an EBIS/EBIT for MCI production (cf. text)

Production of ions from solid compounds can be achieved via sputtering or vaporisation from suitable electrodes brought near to the ECR zone, or by evaporisation from a crucible inside the discharge chamber. Stable operation of the ECRIS plasma requires additional electrons which can be delivered from a hot filament or by ion-induced electron emission from oxide-coated inner walls of the discharge vessel, or from probes biased negative with respect to the ECR plasma. Since its invention in the late sixties ECRIS have undergone rapid development. Of special interest for the experimenter is the actual trend to rather compact all-permanent magnet ECRIS which require astonishingly low microwave power.

EBIS – Electron Beam Ion Source

EBIS utilizes a high-density electron beam for both ion confinement and step-by-step ionization [38]. An intense electron beam with typical length of about one meter, a diameter of about 0.1 mm and a current density of up to several thousands A/cm^2 inside a series of drift tubes which can be connected to different potentials is focused by a strong magnetic field of typically one Tesla produced by a long (usually superconducting) solenoid (cf. Fig. 3.3b). The electron beam, by virtue of its strong negative space charge, provides strong confinement for positive ions as long as appropriate potential barriers are applied on both ends. The ion confinement can last from a few milliseconds up to many seconds until the full space-charge compensation of the electron beam is reached.

The ions are brought into the electron beam either by gas injection or from external ion sources and will be ionised in a step-by-step manner. The lower the surrounding background gas pressure, the more efficiently this ionization proceeds. The background pressure can be kept well below 10^{-10} mbar by cryogenic pumping on the inner walls of the superconducting solenoid. At any time during the ion confinement only a narrow group of MCI charge states is present. In the short-pulsed mode, ions are extracted by rapidly lowering one of the axial potential barriers (extraction times typically 5–50 μs). However, the EBIS may also be operated c.w. in the so-called "leaky mode" which delivers less highly charged ions than in the (short) pulsed mode. EBIS are capable of producing up to fully stripped Xe^{54+} ions (about 10^4 per pulse) or fully stripped Ar^{18+} (about 10^8 per second; [39]). Typically, the emittance is $\leq 10\ \pi$.mm.mrad and thus smaller than for ECRIS.

EBIT – Electron Beam Ion Trap

EBIS and EBIT have as common working principle the step-by-step ionization of ions which are trapped in the space charge of a dense electron beam. The EBIT concept has been developed by Marrs et al. [40, 41]). It involves a much shorter electron beam (only a few cm) than the EBIS, which is easier to realise and probably more stable. Originally, the EBIT was devised for studying soft X-ray radiation from trapped MCI subjected to electron impact excitation and -ionization for up to H-like U^{91+} [42]. However, the MCI can also be extracted along the electron beam direction [43], in which case similar ion charge spectra and MCI yields are obtained as with EBIS. For production of MCI from non-gaseous compounds, singly charged ion species can be injected in a similar way as for the EBIS, e.g., from a metal vapour vacuum arc source ("MEVVA"; [44]). EBIT are relatively small devices [45], comparable in size, technology and maintenance costs to medium-performance electron microscopes. Recently, a very compact EBIT has been developed with the magnetic field configuration for electron beam confinement produced from permanent magnets [46], which therefore needs no cryogenic components.

3.2.2 Preparation of Target Surfaces

A basic requirement for reproducible experimental results to be compared meaningfully to theory are well defined target surfaces. Here we distinguish polycrystalline and single crystalline surfaces. In case of polycrystalline materials the sample cleanliness is important but structural aspects cannot be neglected. Especially for low ion energies possible preferential orientation of the grains of a polycrystal may cause channeling leading to structural effects in particle penetration, scattering and sputtering. For more advanced electron emission- and especially ion scattering experiments single crystal surfaces are necessary. In general, the target preparation follows conventional recipes [47–49]. *Metal single crystals* are polished and oriented before being transferred into the UHV chamber. Target preparation consists of sputtering and annealing cycles. The target cleanliness is usually checked by Auger-electron-spectroscopy (AES) or ion scattering spectrometry (ISS) [47, 50].

Annealing temperatures for single-crystalline surfaces can be kept below 0.5 T_m where T_m is the bulk melting temperature. Higher temperatures may cause severe segregation of impurities. In cases of obnoxious impurities suitable gas-surface reactions, e.g. with oxygen to remove carbon or with hydrogen to remove oxygen, may be necessary. The structural quality of the surface is controlled in many cases by low electron energy diffraction (LEED) or by surface channeling effects [47, 50]. For low index directions of a single crystal surface characteristically shaped spatial ion distributions are observed [51, 52], whereas for high index or "random" directions the angular peak width of reflected ions is a good indicator for the "flatness" of a surface [10, 49].

For *semiconductor* surfaces the sputtering and annealing procedure is less effective or simply destructive. Due to the covalent bonding, annealing of a sputtered semiconductor surface is poor. Semiconductor surfaces are prepared by following chemical recipes in case of Si and a heating process in vacuum [48]. In some cases, i.e. III–V semiconductors, cleaning in vacuum is possible. In case of oxides the extreme sensitivity to ion bombardment causes problems. Beside structural damage by preferential sputtering, i.e. the depletion of oxygen in the near surface region, defect creation causes severe changes of surface properties. Thin oxide layers can be grown *in situ* on the proper metal substrate [53, 54].

For *insulator* surfaces a severe difficulty in ion beam experiments is caused by the generation of electrical charges at the surface. Both the impact of positively charged ions and the consequent electron emission contribute to a positively charged surface layer. This will not only change the impact energy and beam geometry, but also the energy distribution of emitted charged particles. Since their energies are usually very low, even a charge-up by less than one Volt can severely influence measured total yields of charged secondary particles. Charge building-up at the surface can be overcome in several ways.

- Flooding the target with charge carriers of appropriate polarity.
- Deposition of insulating target materials as thin (μm) or ultra thin films (nm) on metal substrates, to reduce the electrical resistance of the surface layer.
- Heating of a sample up to temperatures where it becomes a good ionic conductor (e.g. for alkali halide targets).

3.2.3 Techniques for Studying Ion–Induced Electron Emission

Total Yield and Number Statistics of Slow Electrons

The total electron yield γ (mean number of electrons emitted per single projectile impact) can be determined from the fluxes of the incoming projectiles I_p and the emitted electrons $I_e = \gamma \cdot I_p/q$ (so-called current measurement [1]). With charged projectiles this can be accomplished in a straightforward manner by measuring target currents with and without permitting the electrons to leave the target, which is done by appropriate target biasing. Precautions have to be taken against possible disturbances from charged particle reflection, secondary ion emission and, especially, spurious electron production due to impact of reflected or scattered projectiles or electrons, all such effects possibly causing additional electron emission from the target region [1]. For current measurements in general ion currents of at least nA are necessary.

A superior technique for determination of total electron yields γ involves the electron emission statistics (ES), i.e. the probabilities W_n for ejection of $1, 2, \ldots, n$ electrons per incident projectile, from which the total yield is obtained as the first moment of the W_n [55–59].

Figure 3.4 shows an appropriate set-up where the incoming ions can be accelerated or decelerated by a four-cylinder lens assembly to any desired nominal impact energy, before hitting a target surface under normal incidence. In some experiments the ion impact energy was only limited by the projectile ion image charge interaction with the surface [25]. Practically all electrons ejected from the target with energies smaller than 60 eV into the full 2π solid angle are deflected by a highly transparent (96%) conical electrode and then, after extraction from the target region, accelerated and focused onto a surface barrier detector connected to $U_e \geq +20$ kV with respect to the target. The resulting ejected electron trajectories have been indicated in Fig. 3.4. The probability W_0 that no electron is emitted cannot be determined directly, but may practically be neglected for yields $\gamma \geq 3$. This ES technique requires ion fluxes at the target surface of less than 10^4 projectile/s only and is therefore ideally suited for the comparably weak MCI beams from EBIS and EBIT (see above). Additionally, charging-up of insulator surfaces under HCI bombardment will be avoided [60, 61]. Furthermore, since MCI-induced potential electron emission depends strongly on the MCI charge state, ES measurements can be utilized to distinguish between different MCI species with equal or nearly equal charge – to – mass ratios present in “mixed” ion beams [62].

Apart from giving access to precise total electron yields, the emission number statistics is also of genuine interest by itself since it can be analyzed with respect to the total number of electrons involved in particular emission processes and the mean “single electron emission probability”: Only a fraction of the electrons excited in an ion-induced event can actually escape into vacuum [5, 63, 64].

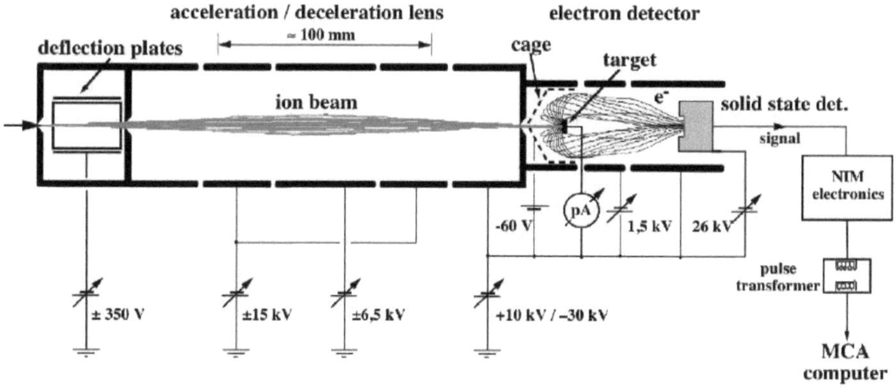

Fig. 3.4. Experimental setup for measuring the electron emission statistics [59]

Ejected Electron Energy Distribution

For electron energy analysis various types of spectrometers have been applied. Electron spectrometers consist essentially of electrodes producing a well defined electric field. The notation of such spectrometers refers primarily to the shape of these electrodes (parallel plate-, cylindrical mirror-, cylindrical- or spherical spectrometer). In Fig. 3.5 we show a setup with a single parallel plate electron spectrometer [34].

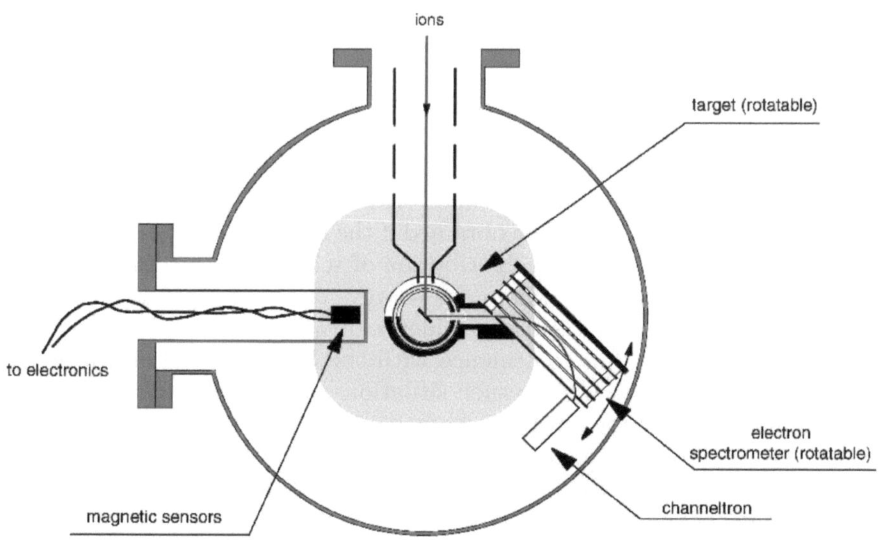

Fig. 3.5. Experimental setup for measuring electron energy distributions [34]

Electrons from the target surface enter the electric field through an entrance slit and, after appropriate deflection, leave the field by passing the exit slit. An important quantity of the spectrometer is the solid angle defined as the product of the polar and azimuthal acceptance angles. Another important property is its (relative) energy resolution $\Delta E/E$ defined as the accepted energy interval ΔE divided by the electron energy E. The quality of a spectrometer is also determined by its focusing power since such electron spectrometers can be constructed to focus in different orders. Various aspects have to be considered if a choice is to be made between different types of spectrometers. Favourable features of the spectrometers are a high resolution, large transmission and simplicity of design. High resolution is required when individual Auger lines are to be measured. Natural line width and separation of adjacent lines are of the order of 1 eV. Accordingly, resolutions $\Delta E/E$ of 10^{-4} to 10^{-3} are required using, e.g., 1000 eV electrons. Such a resolution may in principle be realised by reduction of the spectrometer slits or by deceleration of the electrons. The latter method takes advantage of the fact that $\Delta E/E$ is usually constant so that a reduction of E will also decrease ΔE. The deceleration method is advantageous, since the loss in the spectrometer efficiency is relatively small. Furthermore, the deceleration method allows for varying ΔE during the measurements. For high resolution measurements one has to shield against the earth magnetic field. A reduction factor of about 100 can be achieved by mumetal shielding inside the scattering chamber. Spurious electric fields must also be avoided which are generally produced by electrons collected at insulating surfaces.

Coincidence Techniques

The relative importance of ion induced PE and KE from solid surfaces is not easy to determine. Measurements performed under grazing angles of incidence are of particular interest here, since then the projectiles interaction with the surface proceeds along a well-defined trajectory (surface channeling [10]). More detailed information can be obtained if the electron emission is observed in coincidence with the angular distribution of scattered projectiles [66–68].

Figure 3.6 shows a setup where the energy- and angular distributions of projectiles impinging under grazing incidence on a flat monocrystalline target surface can be observed in coincidence with the ES of ejected electrons (for further details, see Sect. 4.2). In such situations neutral scattered projectiles are often much more abundant than charged ones. Kinetic energy distributions of both neutral and charged particles can be determined by means of time-of-flight (TOF) techniques which, however, requires a well-defined time structure (short pulsing) of the projectile beam.

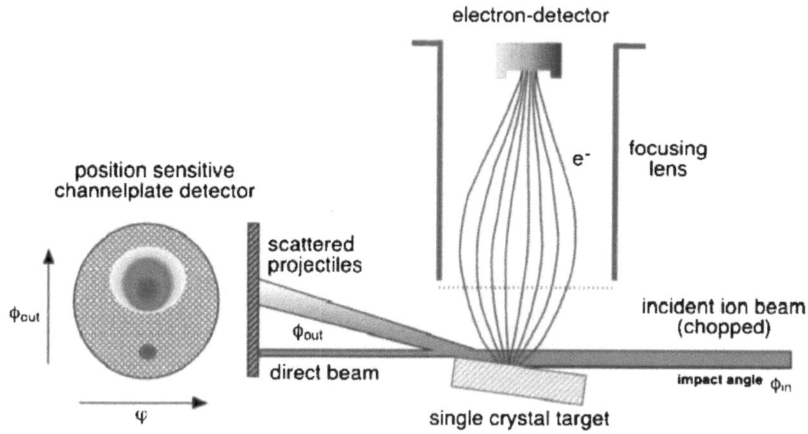

Fig. 3.6. Experimental setup (schematic) for measuring ES of ejected electrons from grazing incidence of slow MCI on a flat target surface in coincidence with scattered projectiles [68]

3.3 Potential Electron Emission from Metal Surfaces

3.3.1 MCI Impact Under Normal Incidence on Metal Surfaces

Although potential electron emission is already possible for singly charged ions or even excited atoms [7, 8, 23, 69, 70], a substantial number of PE electrons can only be emitted during the interaction of MCI with surfaces. From early yield measurements [21] a linear dependence of the total electron yield γ on the projectiles potential energy (i.e. total MCI recombination energy) has been assumed in accordance with multi-step RN/AI relaxation cascades in front of the surface. This linear relationship between the total electron yield and the MCI potential energy breaks down for higher MCI charge states [71–75], but no saturation of the total electron yield as a function of charge state could be observed for a metal surface even at very high MCI charge states [25, 26], Fig. 3.7). Total electron yields as derived from ES measurements for impact of slow MCI up to q = 80 are shown in Fig. 3.7.

If we regard a larger impact energy range, such data exhibits the following general behaviour. Starting at the lowest possible impact velocity caused by mirror-charge attraction of the projectile [26], the electron yield drops continuously towards a minimum and then rises again due to the onset of KE (cf. Fig. 3.8). For a clean gold surface this KE starts at a threshold impact velocity of about 2.10^5 m/s [76, 77] (see also Sect. 3.2 in the contribution of C. Lemell and J. Burgdörfer to this book). In the exclusive PE regime (i.e. below the onset of KE), according to [25, 26, 57, 58, 78] γ can be approximated by the empirical relation

$$\gamma_{PE}(v) \approx \frac{c}{\sqrt{v}} + \gamma_\infty \tag{3.1}$$

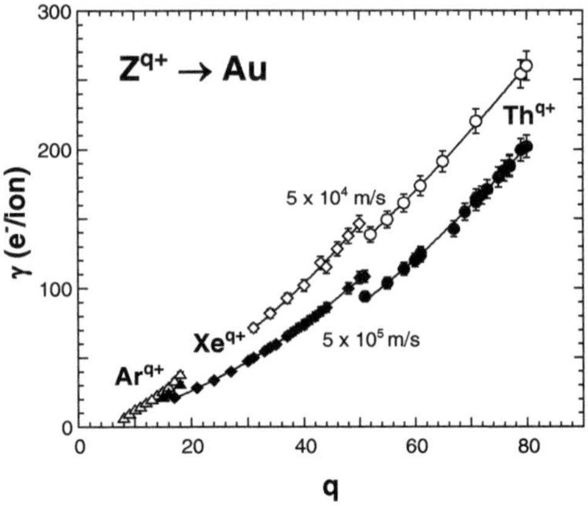

Fig. 3.7. Charge-state dependent total electron yields for impact of Ar^{q+}, Xe^{q+}, Th^{q+} on clean polycrystalline Au measured at two impact velocities [26]

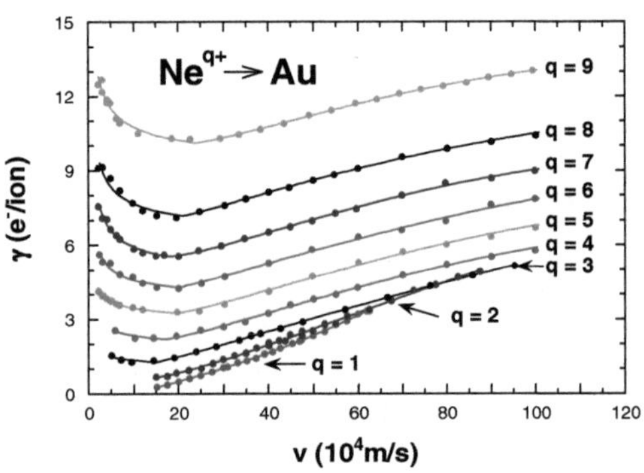

Fig. 3.8. Total electron yields γ vs. ion velocity v measured for impact of singly and multiply charged Ne ions on atomically clean polycrystalline gold [76]. Observed electron yields result jointly from KE (which from the threshold around $2 \cdot 10^5$ m/s monotonically increases with impact velocity) and PE (which increases with ion charge state)

c and γ_∞ are constants depending on the collision system under consideration, and the impact velocity v has to be taken as the effective projectile velocity, i.e. including the appropriate image charge acceleration.

In order to extract information on the above-surface neutralisation phase, measured total yields have been compared with COB model calculations [57, 78]. Increase of the electron yield towards low collision velocity (cf. Fig. 3.8) is described by the first term in Eq. (3.1), which is attributed mainly to AI cascades which deliver the more electrons the more time is left in front of the surface. The velocity-independent part of the PE yield as denoted by the constant γ_∞ in Eq. (3.1) stands for very fast electron emission in the surface selvedge due to screening. Here we can no more distinguish between conduction band electrons and electrons bound to the projectile; in other words, the projectile core has become "dressed" by conduction band electrons [79–81]. Further insight into the origin of the slow electron emission is provided if we compare measured ES with model probability distributions [82]. We assume that these electrons are emitted independently of each other with an about constant chance p. Then the probability for emission of n electrons follows from the binomial distribution for p and an electron ensemble of size N as its second parameter. By best-fitting binomial distributions with different values p and N to a measured ES we can evaluate both parameters. ES for slow MCI impact on metal- and insulator surfaces follow closely binomial distributions [59] which indicates that then the electrons have mainly been ejected above and at the surface. On the other hand, electron emission mainly from below the surface due to impact of faster MCI which also induce KE results in ES which clearly deviate from a binomial distribution and fit much better to Poisson- [59] or Polya distributions [64, 82].

3.3.2 MCI Impact Under Grazing Incidence on Metal Surfaces

Coincidence measurements beween ES and scattered projectiles have been performed for grazing impact of slow MCI on clean monocrystalline Au(111) with an experimental setup as sketched in Fig. 3.6 [67, 68]. Figure 3.9 shows an intensity distribution of scattered projectiles as recorded with a position sensitive detector for 0.45 keV/amu Ar^{8+} ions directed at an angle of incidence of 5° onto a Au(111) surface [66]. The peaked feature on the top right hand side represents a small fraction of the primary ion beam that has passed above the target (see Fig. 3.6), whereas the broad peak results from scattered projectiles. Specularly reflected projectiles contribute to the central peak, while scattering from surface imperfections (e.g. steps) is responsible for the tail of the scattering distribution. On the bottom of Fig. 3.9 we show the mean number of electrons emitted in coincidence with different parts of the angular distribution shown on top. Apparently, less electron emission results from specularly reflected projectiles than for scattering under larger angles.

In Fig. 3.10 we show various ES from these measurements. In the upper panel there is a "non-coincident ES" resulting from all impinging projectiles without selection. The "coincident ES" is obtained in coincidence with projectiles from the complete scattering distribution shown on top of Fig. 3.9. From the difference we see that a considerable fraction of projectiles is not specularly reflected, which apparently produces a higher electron yield. In the middle of Fig. 3.10 we show (1) an ES coincident with truly specularly reflected projectiles (i.e. for the central peak on top of Fig. 3.9), and (2) another ES measured coincidently with projectiles scattered out of the specular direction (i.e. in the tail on top of Fig. 3.9). ES (2) clearly corresponds with a higher electron yield than ES (1) [66]. Finally, on the bottom of Fig. 3.10 an ES is shown for Ar^{8+} impact under normal incidence on polycrystalline Au with an impact energy which was equal to the kinetic energy component

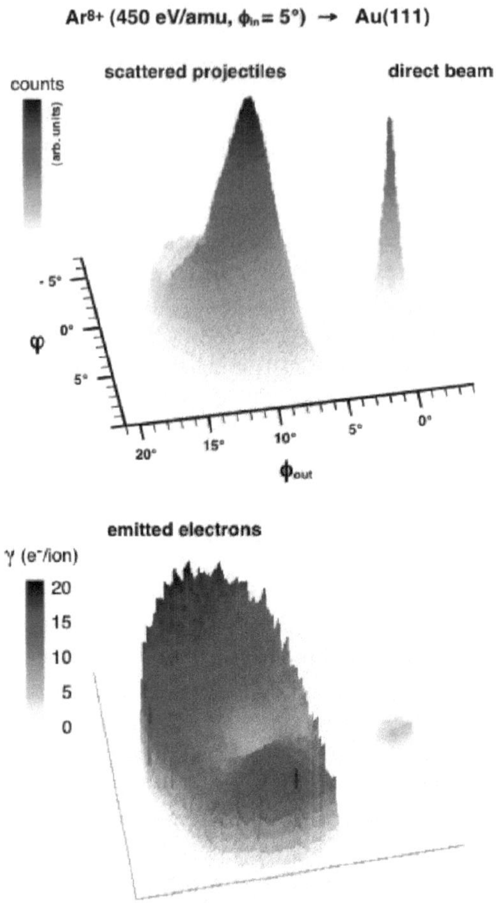

Fig. 3.9. *Top*: Intensity distribution of scattered projectiles as recorded on a position sensitive detector for the case of 0.45 keV/amu Ar^{8+} ions impinging under a grazing angle $\phi_{\text{in}} = 5°$ onto a Au(111) surface [66]. *Bottom*: Mean number of emitted electrons measured in coincidence with projectiles scattered into different exit angles (positions correspond to top of this figure)

normal to the surface in the case of grazing incidence [58]. Since 100 eV Ar projectiles ($v = 2.2 \times 10^4$ m/s) can hardly produce any KE, the ES on the bottom of Fig. 3.10 results exclusively from PE by neutralization of Ar^{8+} ions, which release a total potential energy of about 600 eV upon surface impact.

The similarity of this ES and the one labeled (1) in the middle of Fig. 3.10 shows that in the case of metal surfaces specularly scattered projectiles provide approximately the same PE yield as the much slower ones impinging under a much larger angle. ES (1) can therefore be related to PE by projectiles which did not approach the top-most surface layer any closer than about 1 a.u. and,

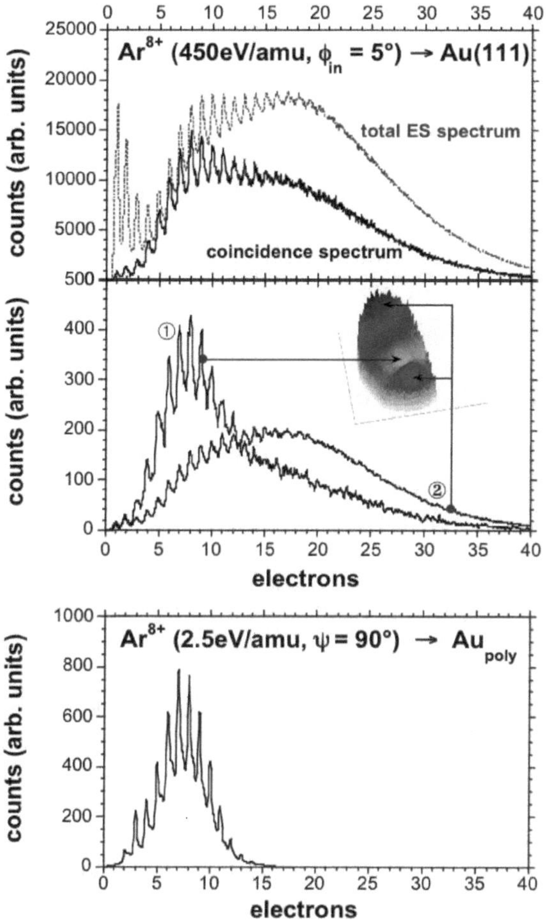

Fig. 3.10. *Top*: ES for 0.45 keV/amu Ar^{8+} ions impinging under a grazing angle $\phi_{in} = 5°$ onto a Au(111) surface, measured non-coincident and in coincidence with all scattered projectiles [66], respectively. *Middle*: ES for 0.45 keV/amu Ar^{8+} ions impinging under a grazing angle $\phi_{in} = 5°$ onto a Au(111) surface, measured in coincidence with the central part (1) and with the wings (2) of projectile scattering distribution shown on top of Fig. 3.9, respectively. *Bottom*: ES for 2.5 eV/amu normal incidence of Ar^{8+}on polycrystalline Au [58]

as to another conclusion from this observation, the above-surface PE part from hollow-atom relaxation above a metal surface is shown to depend on the perpendicular impact velocity component only.

Two notes need to be added here. First, for grazing scattering of MCI on metal surfaces, the exit angle of specularly reflected projectiles is increased via image charge attraction on the incoming trajectory [25, 83–85]. For example,

Ar^{8+} ions will gain near a Au surface (work function $W_\phi = 5.1$ eV) about 30 eV (see also Sect. 3.4 in the contribution of C. Lemell and J. Burgdörfer to this book) which is a non-negligible fraction of the initial perpendicular energy of 100 eV for the above discussed case. Actually, for very high ion charges q this image charge acceleration will make the grazing incidence regime difficult to access.

Secondly, sufficiently fast grazing incident projectiles can produce KE by elastic collisions with quasi-free electrons above the surface (see contribution by H. Winter to this book). However, Ar projectiles with a velocity of 0.1 a.u. produce a KE yield of less than 2% which is negligible for the present considerations.

3.3.3 Potential Excitation of Plasmons

As discussed in more detail in the contribution of R.A. Baragiola and C. Monreal to this book, electron spectra for impact of slow singly [32, 86] and multiply charged ions [86] on quasi-free electron-metals may contain contributions from one-electron decay of plasmons. The respective signature appears at electron energies $E_p - W_\phi$ ($E_p \ldots$ target bulk- or -surface plasmon energy). Socalled "potential excitation of plasmons" ("PEP") arises if the ion potential energy exceeds $E_p + W_\phi$, but plasmons can also "indirectly" be excited by sufficiently fast electrons from KE.

A special situation was encountered for H^+ impact on Al(111) where a conspicuous structure in the respective electron spectra has originally been related to plasmon decay [87], but is probably caused by the diffraction of multiply-scattered KE electrons [34].

According to theoretical predictions [88] for impact of slow ions ($v < 1$ au) on clean metal surfaces an important de-excitation channel for the projectile potential energy should be due to plasmon excitation (i.e. collective oscillations of the quasi-free electron gas in the solid). Electron spectra for impact of singly [32, 86] and multiply charged ions [86] on quasi-free electron metal surfaces show fairly weak peaks at electron energies $E_p - W_\phi$ ($E_p \ldots$ bulk- or surface plasmon energy). Consequently, for clean Al ($W_\phi \approx 4.3$ eV) the one-electron decay of bulk plasmons ($E_p \approx 15.3$ eV) should leave an electron peak at 11 eV and for surface plasmons at 6.6 eV. Bulk plasmon excitation in aluminum due to the kinetic projectile energy can only proceed beyond ca. 40 keV/amu [32, 86, 89]. Therefore, slower ions can only excite plasmons due to their potential energy, in competition with Auger electron emission. Such "potential excitation of plasmons – PEP" requires an ion potential energy exceeding the sum of $E_p + W_\phi$.

However, "indirect" excitation of plasmons can take place by sufficiently fast electrons (≥ 35 eV) which are produced from KE in the same collision process [90]. Guided by recent studies on PEP [32, 86], we have measured electron spectra for impact of $\leq$10 keV H^+ H_2^+, He^+, Ne^{q+} and Ar^{q+} ($q = 1, 2$)

on atomically clean poly- and monocrystalline Al. In these electron spectra the following features have been observed (see Fig. 3.11) [91]:

(a) At ca. 63 eV the Al-LMM Auger electron peak from sputtered neutral Al atoms, and minor peaks on its left side from Auger electron emission from sputtered Al^+ ions.
(b) Doppler-shifting autoionisation lines from doubly excited projectiles (He^+ and Ne^{q+}).
(c) A broad peak at about 11 eV which is related to bulk-plasmon decay, and in some cases also a considerably weaker surface plasmon peak at 6.6 eV.

Figure 3.12 shows electron spectra for impact of 5 keV H^+ (impact angle $\psi = 5°$ with respect to the surface) on poly- and monocrystalline Al.

For polycrystalline Al only a weak (indirectly excited [90]) bulk plasmon peak (c) can be recognized, whereas for the monocrystalline Al(111) surface a considerably more prominent peak is seen which however moves if the electron emission angle α is changed with respect to the surface, instead of staying at 11 eV as expected for bulk-plasmon decay (see above). Such "moving peaks" have already been observed in [87] and were there ascribed to plasmon decay. In contrast to this interpretation, we explain this feature by the diffraction of slow electrons from KE which have undergone multiple scattering near the surface of the monocrystalline target [34]. Variation of total electron yields when changing the target crystal orientation vs. the incident ion beam can be ascribed to changing conditions for projectile channeling (see, e.g., [92]). However, in the present measurements the ion incidence angle ψ as kept fixed and the angle of electron emission α varied. Similar diffraction features as for 5 keV H^+ have been found for 10 keV H_2^+ impact on Al(111) [34], but not for any other projectile ions. This can probably be understood from the fact that projectiles heavier than protons cause considerably stronger sputtering and thus more surface roughening than impact of H^+ or H_2^+.

Another remarkable result concerns the relation between the importance of potential excitation of bulk plasmons and the projectile potential energy. Comparing relative plasmon-excitation probabilities for different primary ions suggests an excitation process which is quasi-resonant with respect to the available potential energy [34]. PEP can of course not proceed at too low potential energy (e.g., for H^+ or Ar^+, where the observed plasmons are excited by fast electrons from KE). PEP is apparently most probable for Ne^+ where the ion neutralisation energy practically matches the potential energy which is required for bulk plasmon excitation (see above), and it becomes clearly less probable for higher potential energy (He^+, Ar^{2+}). In particular, the significance of bulk plasmon peaks in the electron spectra for Ne^+ and Ne^{2+} is not much different for poly- and monocrystalline Al surfaces, which can be explained in the following way [34]. The potential energy arising from the first neutralization step $Ne^{2+} \Rightarrow Ne^{+*}$ is too low for excitation of bulk plasmons, and the subsequent de-excitation of the intermediate singly charged excited ion $Ne^{+*} \Rightarrow Ne^+$ provides an already "too large" potential energy. There-

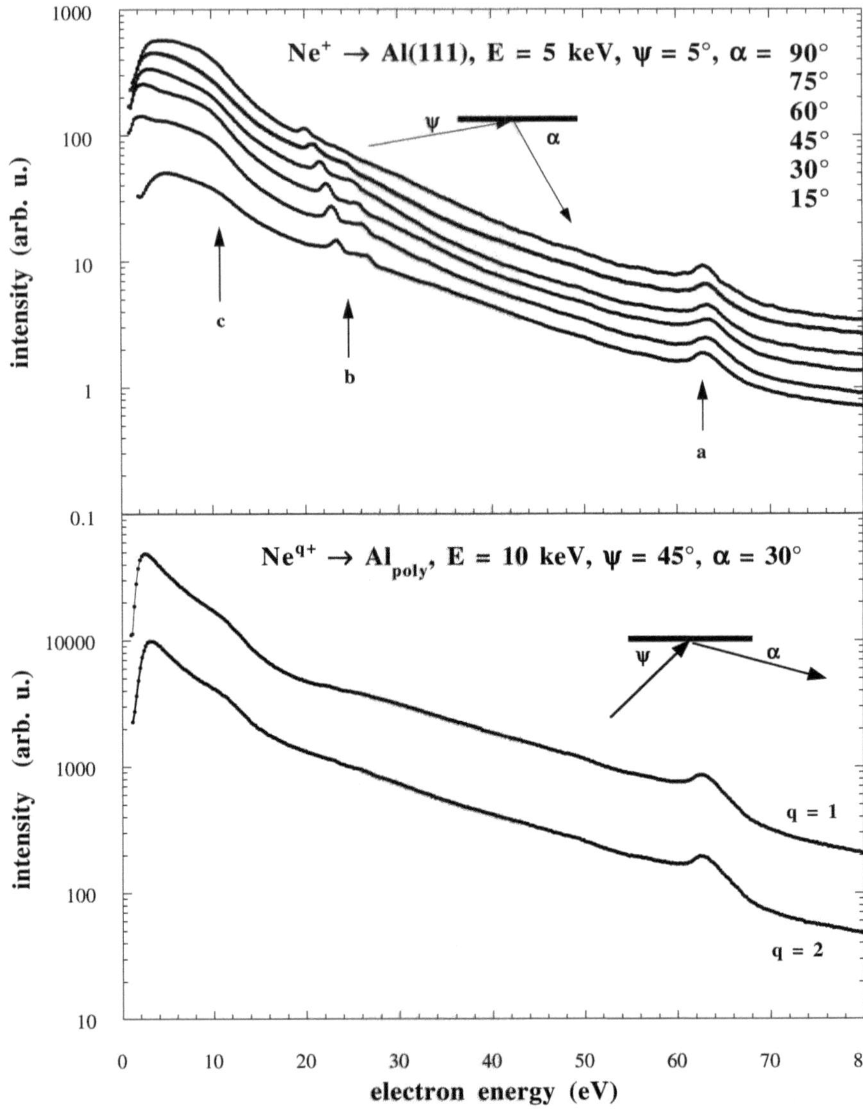

Fig. 3.11. (**a**) Electron spectra (in arbitrary units) for impact of 5 keV Ne^+ on clean Al(111) for 5° ion incidence- and various electron emission angles. The different spectra have been set-off with respect to each other for better comparability. (**b**) Electron spectra (in arbitrary units) for impact of 10 keV Ne^+ and Ne^{2+} on clean polycrystalline Al for 45° ion incidence- and 30° electron emission angles. The two spectra have been set-off with respect to each other for better comparability (from [91])

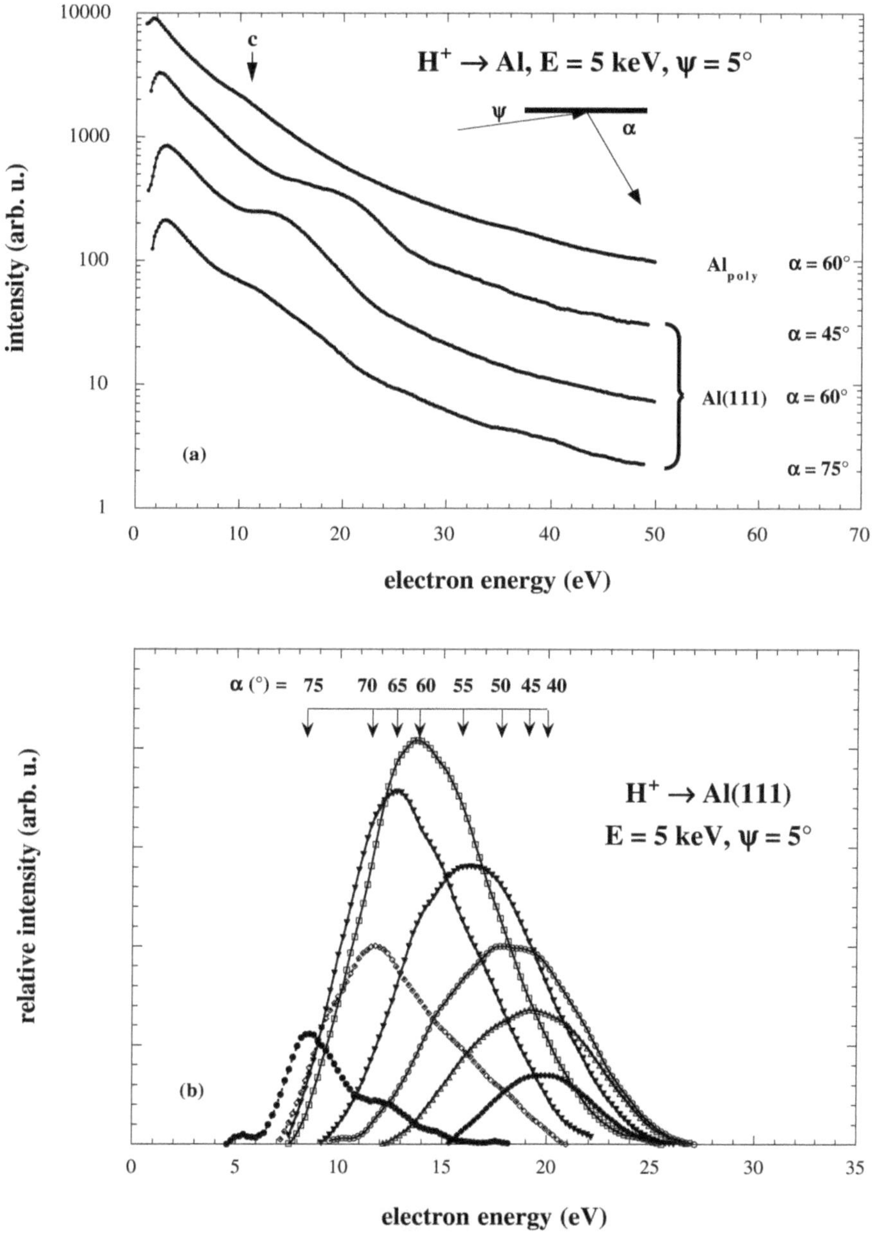

Fig. 3.12. (*top*) Comparison of electron spectra (in arbitrary units) for impact of 5 keV H^+ on poly- and monocrystalline aluminum. Spectra have been set-off with respect to each other for better comparison. (*bottom*) Relative size and electron-energy position observed for different electron emission angles α for the "moving peaks" in electron spectra induced by impact of 5 keV H^+ on Al(111) (from ref. [34])

fore, for impact of Ne^{2+} the probability for bulk plasmon excitation will be dominated by the potential energy of the intermediate Ne^+ ground state ion, and it should therefore not be significantly more important than for impact of Ne^+ (see Fig. 3.11).

3.4 Potential Electron Emission from Insulator Surfaces

3.4.1 MCI Impact Under Normal Incidence on Insulator Surfaces

Investigations of electron emission from insulator surfaces are more difficult to perform than for metal surfaces (c.f. Sect. 2.2). Reliable experimental evidence on MCI-induced PE from insulator surfaces is therefore scarce and also mainly limited to low charge state ions ($q \leq 10$) [60, 61, 93]. Considerable differences can be expected to the case of PE from metal surfaces (see outlook).

In the following we report on measurements of total electron yields for impact of N^{q+} ($q = 1, 5, 6$) and Ar^{q+} ($q = 1, 3, 6, 9$) ions on clean, polycrystalline lithium-fluoride LiF [60, 61]. The results differ considerably from the ones measured for metal surfaces (Fig. 3.13). Because of the larger binding energy for LiF valence band electrons, the neutralization sequence starts closer to the surface than for metal targets and less time is available for above-surface AI. For projectiles without inner shell vacancies (e.g. N^{5+}), a considerably smaller electron yield is therefore found than for a Au surface, whereas for N^{6+} (one K-shell vacancy) the yields for both target species become about the same (Fig. 3.13). This has been attributed to the comparably more efficient secondary electron emission in LiF which is induced by the fast

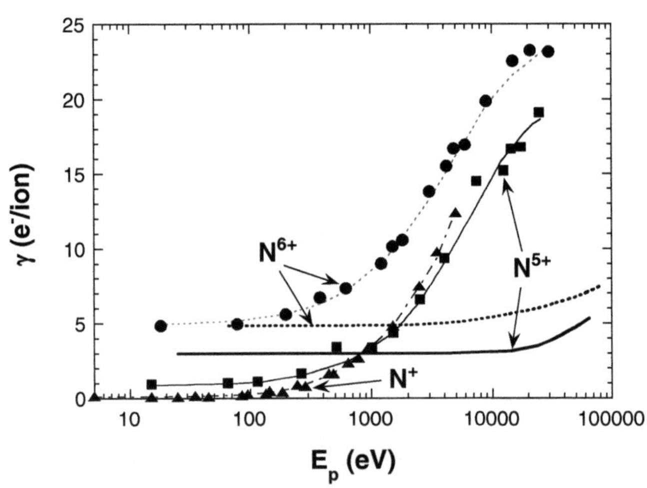

Fig. 3.13. Total slow electron yields γ for impact of N^{5+} and N^{6+} on clean polycrystalline Au (*lines only*) and LiF (*lines through data points*) vs. impact energy [60]

($E_e \approx 350$ eV) KLL-electrons from decay of N^{6+} projectile K-shell vacancies, and to a considerably larger inelastic mean free path of these secondary electrons (see contribution by *W*. Werner to this book). In other words, for impact of N^{6+} on LiF, in comparison with an Au surface the less efficient "above-surface" electron production is compensated by a more efficient below-surface secondary electron emission [60].

However, for LiF the PE contribution to the total electron yield is only dominant over the respective KE at considerably low ion impact energy (Figs. 3.13 and 3.14), due to the low KE threshold energy (c.f. the contribution on KE by Helmut Winter et al. in this book). In contrast to metal surfaces, KE from LiF starts at much smaller impact energy (lower KE threshold than for Au), increases faster with projectile velocity and shows an "inverse" dependence on incident charge state q of the Ar^{q+} projectile ions (less emission for higher q) [61, 94]. A better separation of PE from KE therefore requires slow MCI projectiles in higher charge states (work in progress; see outlook)

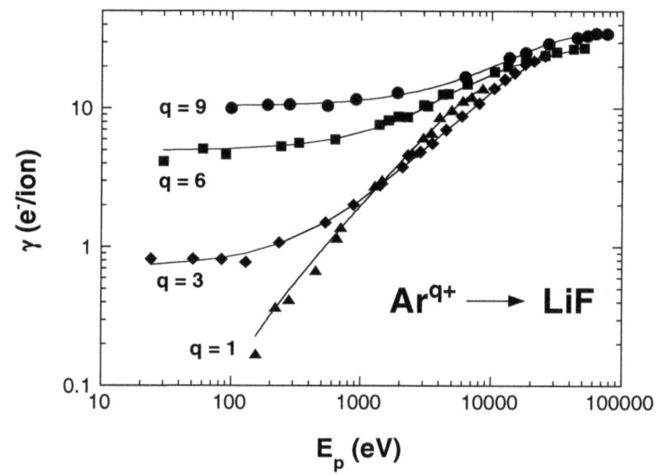

Fig. 3.14. Total slow electron yields γ for impact of Ar^{q+} ($q = 1, 3, 6, 9$) on LiF vs. impact energy [61]

3.4.2 MCI Impact Under Grazing Incidence on Insulator Surfaces

A first attempt to separate PE and KE in coincidence measurements of ES and angular distributions of scattered projectiles as described in [66] was unsuccessful due to the stong KE contribution present [95]. In order to separate KE and PE contributions, we attempted to use the close relationship between KE and the inelastic energy loss of scattered projectiles. A time-of-flight (TOF) unit added to the setup (Fig. 3.6) eventually allowed us to perform

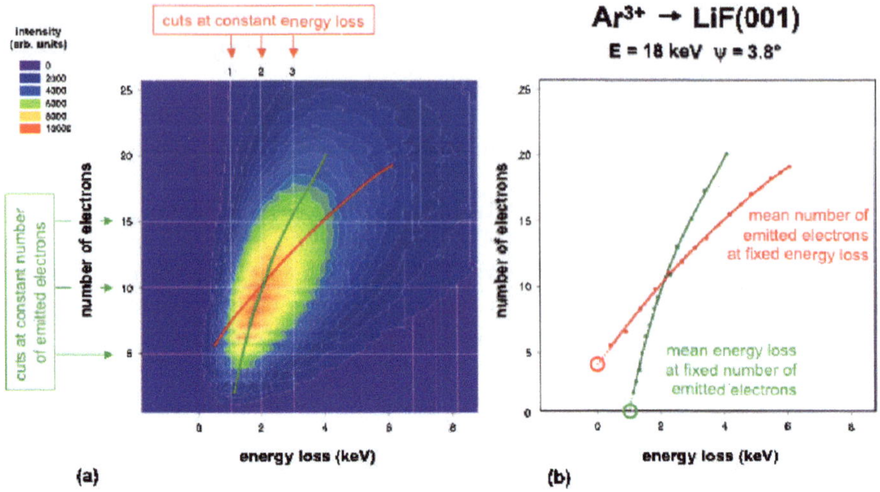

Fig. 3.15. (**a**) Coincidence spectra of number of emitted electrons vs. projectile energy loss for 18 keV Ar^{3+} impact on a LiF(001) surface (angle of incidence 3.8°) [68]. (**b**) Cuts through these coincidence spectra at constant energy losses provide related mean numbers of emitted electrons, and cuts for given numbers of emitted electrons show related mean energy losses. The two curves have been extrapolated to zero energy loss and zero number of emitted electrons (*circles*), respectively (for further details cf. text)

ES measurements in coincidence with projectile energy loss [68] under grazing incidence conditions.

Figure 3.15 shows as a general trend for those studies a direct correlation of the mean number of emitted electrons with projectile energy loss. Extrapolation of the resulting curve to the hypothetical case of projectiles with no energy loss at all (not directly observable in our experiment) leads to an electron emission yield which cannot be caused by kinetic energy loss of the projectile [68]. Since such electrons are not emitted at the expense of the projectile's kinetic energy, they can only result from deposition of projectile potential energy E_{pot}, i.e. they correspond to the "pure" potential electron emission yield $\gamma_{PE}(\Delta E \to 0)$.

Plotting such extrapolated $\gamma_{PE}(\Delta E \to 0)$ values for different Ar^{q+} projectiles as a function of the related potential energy does support this interpretation. In Fig. 3.16 we find a linear relationship between the "pure" PE yield and the potential energy brought towards the surface by the different MCI, with no dependence on the kinetic energy which had been varied between 18 and 54 keV [68]. Most notably, the data points are close to the limit of potential energy conservation (solid line in Fig. 3.16).

Auger processes leading to PE (see Chap. 1) require a potential energy of at least twice the minimum electronic binding energy W_ϕ at the surface (corresponding to the work function for metal targets). The maximum possible

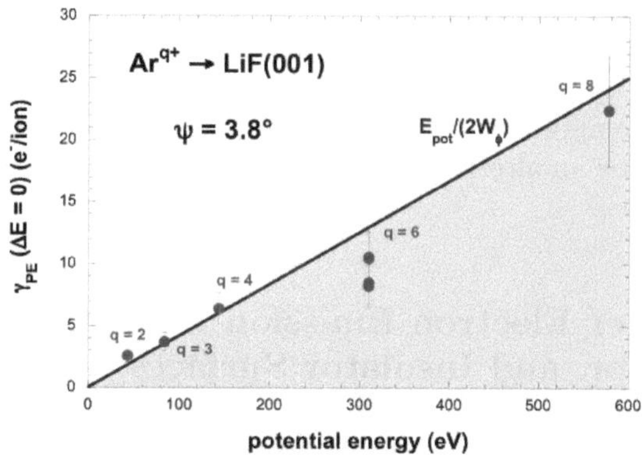

Fig. 3.16. "Pure" PE yields vs. MCI potential energy (data points) for grazing impact of Ar^{q+} (angle of incidence 3.8°) compared with theoretical prediction (*solid line*, cf. text) [68]

number of electrons emitted via PE is therefore given by $n_{\text{max}} = E_{\text{pot}}/2W_{\phi}$. This maximum possible number of PE electrons is indeed obtained from our experimental data, taking into account a binding energy of about 12 eV [96] for the highest occupied states in the $F^{-}(2p)$ valance band of LiF (solid line in Fig. 3.16). Furthermore, we have assumed 100% probability for electron escape from the surface which seems reasonable considering the large band gap of LiF(001) which extends to vacuum energies and prevents accommodation of the slow electrons inside the solid.

This remarkable finding suggests that up to the highest ion charge state/potential energy which we have applied (so far Ar^{8+}, $E_{\text{pot}} \approx 580$ eV), the electronic properties of the alkali halide (limited hole mobility, possible reduction of the electron capture rate due to hole formation and eventually necessary capture of more tightly bound electrons, see, e.g., [97]) impose no limitations on its ability to provide sufficient electrons for complete neutralization and de-excitation within the (limited) surface interaction time. We note that these PE yields for grazing Ar^{q+} impact are by more than a factor of 2 larger than PE yields by Vana et al. [61] for normal incident Ar^{q+} on LiF (c.f. Fig. 3.14). It is important to stress that in grazing collisions the projectiles interact with many different (F^{-}) sites over a rather large lateral extension. In contrast to that Wirtz et al. [97] have shown that for normal impact on a LiF surface the rates for capture from neighboring sites are considerably (typically one order of magnitude) smaller than for capture from the Fluorine atom closest to the projectile impact site. Further neutralization from the same site is also less probable because it requires capture of a more tightly bound electron. Our findings are also consistent with observations on the image charge acceleration of multiply charged ions in grazing scattering on a LiF surface [98]. There the interaction energies gained by the projectiles also point to a complete projectile neutralization along the grazing scattering flight path.

Since surface-channeled projectiles interact with the surface along well defined and calculable trajectories [10], the here presented technique in principle also allows to investigate PE yields as a function of the closest distance of projectile approach towards the surface. For not too high ion charge states this could be an alternate way to determine distance dependent Auger rates [99–101].

3.5 Fast Auger Electron Emission for Metal, Semiconductor, and Insulator Surfaces

Recombination of inner shell vacancies accompanied by emission of fast Auger electrons (or soft X-rays) takes place mainly in the latest stage of the HA relaxation cascade. High energy Auger electron spectra show profound structures which, by means of Hartree-Fock atomic structure calculations, can serve for identifying the HA configuration at the moment of the particular Auger decay [102–104].

Whether this emission occurs still above or already below the surface has been a controversial issue in the field [9, 105–107].

In Fig. 3.17 we compare high resolution KLL Auger spectra obtained for N^{6+} ion impact on different target surfaces [108]. While results for semiconducting p-doped Si(100) are very similar to the ones for Al(110), a pronounced difference can be seen for the insulating LiF(100) surface; the conspicuous absence of a peak on the low energy side of the KLL spectrum, which for Si and Al arises from slow L-shell filling via Auger cascades, has been explained by Limburg et al. [108] by the much smaller electron mobility in LiF. Because of the large LiF band gap L-shell filling can probably only start after the projectile has entered the close collision regime at and below the surface.

Furthermore, blocking of resonance ionisation (RI) due to the existence of the LiF band gap has been made responsible for delayed lower-shell population in HA which are in front of the surface. Khemliche et al. [109] have compared results for O^{7+} induced KLL Auger electron emission for a clean LiF surface and a Au target surface covered by up to a single monolayer of LiF. The authors argued that it is not the large band gap which only in the case of bulk LiF could restrain RI back into the target, but rather the high LiF work function that hampers neutralization of MCI in front of the surface.

As mentioned before, recombination of inner-shell vacancies can also proceed by characteristic X-ray emission. From highly resolved X-ray spectra produced in HCI-surface collisions, the number of spectator electrons residing in higher n-shells at the moment of the radiative inner shell transition and the time scale for such inner-shell filling in an HA can be estimated. Such measurements give direct evidence for the transient formation of HA [22]. However, as for the corresponding Auger electron spectroscopy, the question to what extent such X-ray emission can occur still "above" or only "below" the surface is not yet settled [111]. In any case, X-ray spectra reported by Briand

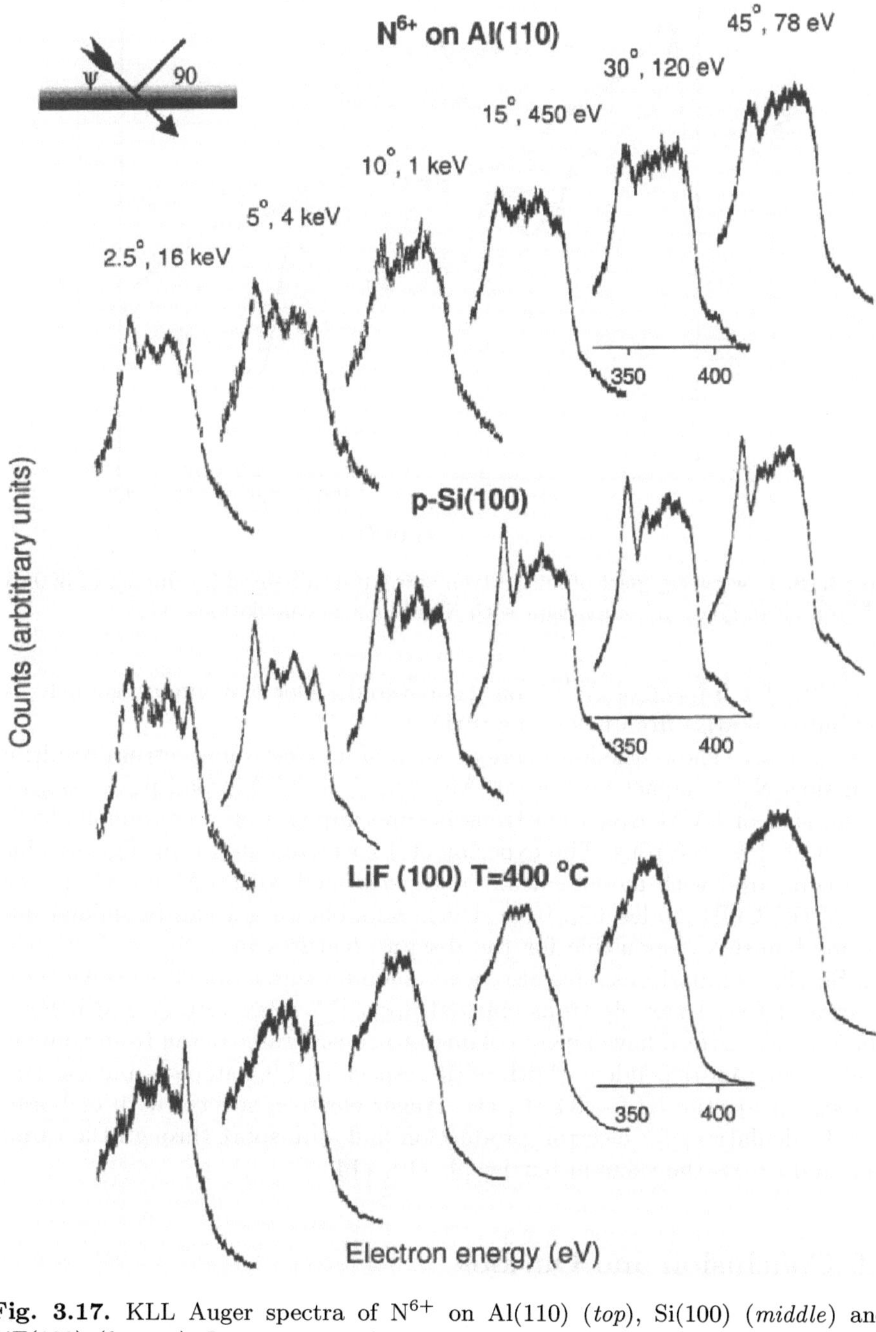

Fig. 3.17. KLL Auger spectra of N^{6+} on Al(110) (*top*), Si(100) (*middle*) and LiF(100) (*bottom*). Ion energy and incidence angle have been varied in order to keep constant the ion velocity component normal to the surface [110]

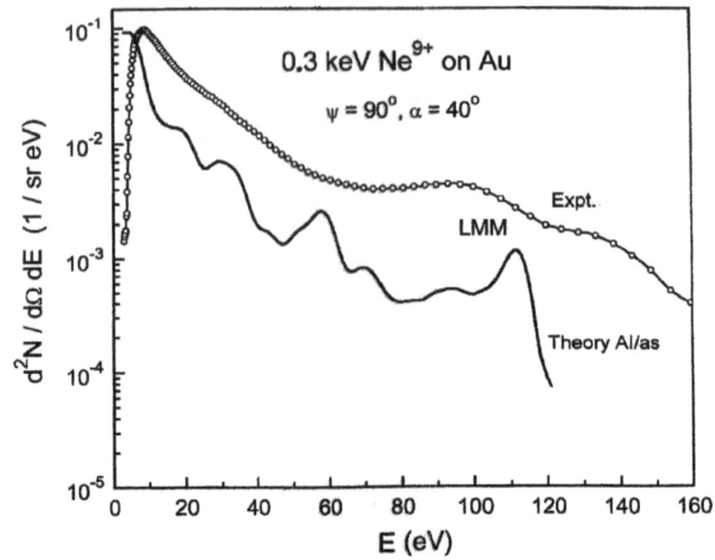

Fig. 3.18. Low energy part of an electron spectrum measured for impact of 300 eV Ne^{9+} on clean gold, in comparison with Monte Carlo calculations [112]

et al. [22] for impact of Ar^{17+} on gas-covered silver can unambiguously be attributed to arise from below the surface.

Figure 3.18 shows the low energy region of an electron spectrum resulting from slow Ne^{9+} impact on a clean Au surface [112]. A broad peak assigned to emission of LMM Auger electrons is superimposed on a continuous background of slow electrons. The experimental spectrum shown in Fig. 3.18 has been compared with model calculations performed with a Monte Carlo version of the COB model [65], from which valuable insight can be gained into the mechanisms responsible for the discrete features in such electron spectra. Similar comparisons have also been made for impact of N^{6+} on Au [113]. Spectra of fast Auger electrons emitted upon HA relaxation at and below a metal target surface have been explained to considerable detail from comparisons of angular-dependent (both with respect to MCI impact and electron take-off) projectile KLL-, KLM-, etc. Auger electron spectra with elaborate model calculations for electron production and -transport through the target bulk and across the vacuum barrier [9, 112, 114].

3.6 Conclusion and Outlook

Although studied since the early fifties of the past century, electron emission due to the potential energy of ions is still an active field of research. For slow singly and multiply charged ions impinging on a metal surface experimental data are abundant and the basic processes are meanwhile well understood. For the case of MCI-induced PE from insulator surfaces the experimental evidence is more scarce and limited to low charge state ions. There are a number of reasons why our present knowledge on PE by ion – metal interaction cannot be extrapolated to collisions with insulating surfaces.

(i) The dielectric response of insulating surfaces (image charge acceleration, hole mobility, etc.) is quite different from that of metallic targets, resulting in new phenomena observable in experiment and because of their complexity posing a considerable challenge to theory.
(ii) Electron transfer from an insulator surface to a MCI can lead to local charging-up of the surface. The image charge acceleration usually observed for metal surfaces could thus be altered or even be overcompensated, leading to a repulsion of the projectile by the positive hole charges on the surface ("trampoline effect" [97, 115]). If such a trampoline effect exists, a pure electronic interaction of the projectile ion with the target without any significant momentum transfer to individual target nuclei would be possible. This offers important implications for using slow highly charged ions as a gentle tool for nanostructuring of surfaces [9, 13, 116, 117].
(iii) The wide electronic band gap will affect and alter electronic transfer, emission and excitation processes as compared to conducting surfaces.

By relating the projectile energy loss to kinetic electron emission we have started to determine contributions from potential electron emission even in the presence of a considerable KE yield. Our results suggest a practically complete use of the available potential energy for electron emission during grazing scattering, in sharp contrast to findings for the normal incidence case. Considerably smaller PE yields for perpendicular MCI impact on LiF suggest further search for the so-called trampoline effect.

In order to understand the difference between interaction of MCI with insulator and metal surfaces, we presently study PE due to impact of slow MCI in very high charge states ($30 < q < 70$) on insulating surfaces. These experiments are currently in progress using MCI beams from the Heidelberg EBIT (MPI for Nuclear Physics Heidelberg, Germany). For normal ion impact the effects described above might lead to a leveling-off of γ above a certain projectile charge state. The point of saturation will most likely critically depend on the details of the hole dynamics and therefore give important information on the latter. By systematic variation of the ion impact angle we hope to clarify the different mechanisms of hollow atom formation near and potential electron emission from insulating target surfaces.

Acknowledgments

This work has been supported by Austrian Science Foundation FWF, Austrian Academy of Sciences ÖAW (Project KKKÖ 1-2005) and was carried out within Association EURATOM-ÖAW. Partial support by the European Project RII3 026015 is also acknowledged.

Abbreviations

AD	Auger de-excitation
AES	Auger electron spectroscopy
AI	auto-ionization
AL	Auger loss to (empty states of) the conduction band
amu	atomic mass unit
AN	Auger neutralization
a.u.	atomic unit
arb.	units arbitrary units
COB	classical over-the-barrier (model)
EBIS	electron beam ion source
EBIT	electron beam ion trap
ECR	electron cyclotron resonance
ECRIS	electron cyclotron resonance ion source
ES	electron emission statistics
FWHM	full width at half maximum
HA	hollow atom
HCI	highly charged ion
KE	kinetic electron emission
MCI	multi-charged ion
PE	potential electron emission
PO	peeling - off
QRN	qusi-resonant neutralization
RD	radiative de-excitation
RI	resonant ionization
RN	resonant neutralization
S-DOS	surface density-of-states
SIMS	secondary ion mass spectrometry
SS	screening shift (of energy levels)
UHV	ultra high vacuum

References

1. D. Hasselkamp, in Particle Induced Electron Emission II, edited by G. Höhler (Springer, Heidelberg, 1992), Vol. **123**, p. 1.
2. J. Schou, Scanning Microsc. **2**, 607 (1988).
3. M. Rösler and W. Brauer, in Particle Induced Electron Emission I, edited by G. Höhler (Springer, Berlin, 1991), Vol. 122.
4. R. Baragiola, in Chap. IV in Low energy Ion-Surface Interactions, edited by J. W. Rabalais (Wiley, 1993).

5. Fundamental Processes in Sputtering of Atoms and Molecules (SPUT 92), edited by P. Sigmund (Mat. Fys. Medd., Copenhagen, 1993).
6. H. Gnaser, Low-Energy Ion Irradiation of Solid Surfaces (Springer Berlin, 1999).
7. H. D. Hagstrum, Phys. Rev. **96**, 325 (1954).
8. H. D. Hagstrum, Phys. Rev. **96**, 336 (1954).
9. A. Arnau, et al., Surf. Sci. Reports **27**, 113 (1997).
10. H. Winter, Physics Reports **367**, 387 (2002).
11. T. Neidhart, F. Pichler, F. Aumayr, HP. Winter, M. Schmid, and P. Varga, Phys. Rev. Lett. **74**, 5280 (1995).
12. M. Sporn, G. Libiseller, T. Neidhart, M. Schmid, F. Aumayr, HP. Winter, P. Varga, M. Grether, and N. Stolterfoht, Phys. Rev. Lett. **79**, 945 (1997).
13. F. Aumayr and HP. Winter, Phil. Trans. Roy. Soc. (London) **362**, 77 (2004).
14. H. D. Hagstrum, Phys. Rev. **91**, 543 (1953).
15. H. D. Hagstrum, Phys. Rev. **104**, 672 (1956).
16. H. S. W. Massey, Proc. Cambridge Phil. Soc. **26**, 386 (1930).
17. H. S. W. Massey, Proc. Cambridge Phil. Soc. **27**, 469 (1931).
18. A. Cobas and W. E. Lamb, Phys. Rev. **65**, 327 (1944).
19. M. L. E. Oliphant and P. B. Moon, Proc. Roy. Soc. (London) A **127**, 388 (1930).
20. S. S. Shekter, J. Exp. Theoret. Phys. (USSR) **7**, 750 (1937).
21. U. A. Arifov, E. S. Mukhamadiev, E. S. Parilis, and A. S. Pasyuk, Sov. Phys. Tech. Phys. **18**, 240 (1973).
22. J. P. Briand, L. de Billy, P. Charles, S. Essabaa, P. Briand, R. Geller, J. P. Desclaux, S. Bliman, and C. Ristori, Phys. Rev. Lett. **65**, 159 (1990).
23. H. D. Hagstrum and G. E. Becker, Phys. Rev. B **8**, 107 (1973).
24. H. A. Bethe and E. E. Salpeter, Quantum Mechanics of One- and Two Electron Systems (Academic Press, New York, 1957).
25. F. Aumayr, H. Kurz, D. Schneider, M. A. Briere, J. W. McDonald, C. E. Cunningham, and HP. Winter, Phys. Rev. Lett. **71**, 1943 (1993).
26. H. Kurz, F. Aumayr, D. Schneider, M. A. Briere, J. W. McDonald, and HP. Winter, Phys. Rev. A **49**, 4693 (1994).
27. HP. Winter and F. Aumayr, Euro. Phys. News **33**, 215 (2002).
28. J. Burgdörfer, P. Lerner, and F. W. Meyer, Phys. Rev. A **44**, 5674 (1991).
29. Y. T. Matulevich and P. A. Zeijlmans van Emmichoven, Phys. Rev. B **69**, 245414 (2004).
30. P. Riccardi, M. Ishimoto, P. Barone, and R. A. Baragiola, Surf. Sci. **571**, L305 (2004).
31. H. Raether, Surface Plasmons, Berlin, 1988).
32. R. A. Baragiola and C. A. Dukes, Phys. Rev. Lett. **76**, 2547 (1996).
33. N. Stolterfoht, D. Niemann, V. Hofmann, M. Rösler, and R. A. Baragiola, Phys. Rev. Lett. **80**, 3328 (1998).
34. H. Eder, F. Aumayr, P. Berlinger, H. Störi, and HP. Winter, Surf. Sci. **472**, 195 (2001).
35. R. Geller, B. Jacquot, and M. Pontonnier, Rev. Sci. Instrum. **56**, 1505 (1985).
36. B. Wolf, Handbook of Ion Sources (CRC Press, Boca Raton, New York, London, Tokyo, 1995).
37. P. Sortais, Nucl. Instrum. Methods B **98**, 508 (1995).
38. E. D. Donets, in The Physics and Technology of Ion Sources, edited by I. G. Brown (John Wiley, New York, 1989), p. Chap. 12.

39. E. D. Donets, in HCI-92, edited by P. Richard, M. Stöckli, C. L. Cocke and C. D. Lin (AIP Conf. Proc., Manhattan, Kansas, 1992), Vol. **274**, p. 663.
40. R. E. Marrs, M. A. Levine, K. D. A, and J. R. Henderson, Phys. Rev. Lett. **60**, 1715 (1988).
41. M. A. Levine, R. E. Marrs, J. R. Henderson, D. A. Knapp, and M. B. Schneider, Phys. Scr. T**22**, 157 (1988).
42. R. E. Marrs, P. Beiersdorfer, and D. Schneider, Physics Today **10**, 27 (1994).
43. D. Schneider, et al., Phys. Rev. A **42**, 3889 (1990).
44. I. G. Brown, The Physics and Technology of Ion Sources (John Wiley, New York, 1989).
45. J. R. Crespo López-Urrutia, B. Bapat, B. Feuerstein, D. Fischer, H. Lörch, R. Moshammer, and J. Ullrich, Hyperfine Interactions 146/147, 109 (2003).
46. V. P. Ovsyannikov and G. Zschornack, Rev. Sci. Instrum. **70**, 2646 (1999).
47. S. Speller, W. Heiland, and M. Schleberger, Exper. Meth. in Phys. Sci. **38**, 1 (2001).
48. E. Taglauer, Appl. Phys. A **51**, 238 (1990).
49. H. Winter, Progr. Surf. Sci. **63**, 177 (2000).
50. H. Niehus, W. Heiland, and E. Taglauer, Surf. Sci. Reports **17**, 213 (1993).
51. L. Folkerts, S. Schippers, D. M. Zehner, and F. W. Meyer, Phys. Rev. Lett. **74**, 2204 (1995).
52. A. Niehof and W. Heiland, Nucl. Instrum. Meth. Phys. Res. B **48**, 306 (1990).
53. U. Diebold, J. M. Pan, and T. E. Madey, Surf. Sci. **331–333**, 845 (1995).
54. R. M. Jaeger, H. Kuhlenbeck, H. J. Freund, M. Wuttig, W. Hoffmann, R. Franchy, and H. Ibach, Surf. Sci. **259**, 235 (1991).
55. G. Lakits, F. Aumayr, and HP. Winter, Rev. Sci. Instrum. **60**, 3151 (1989).
56. F. Aumayr, G. Lakits, and HP. Winter, Appl. Surf. Sci. **47**, 139 (1991).
57. H. Kurz, F. Aumayr, C. Lemell, K. Töglhofer, and HP. Winter, Phys. Rev. A **48**, 2182 (1993).
58. H. Kurz, K. Töglhofer, HP. Winter, F. Aumayr, and R. Mann, Phys. Rev. Lett. **69**, 1140 (1992).
59. H. Eder, M. Vana, F. Aumayr, and HP. Winter, Rev. Sci. Instrum. **68**, 165 (1997).
60. M. Vana, F. Aumayr, P. Varga, and HP. Winter, Europhys. Lett. **29**, 55 (1995).
61. M. Vana, F. Aumayr, P. Varga, and HP. Winter, Nucl. Instrum. Meth. Phys. Res. B **100**, 284 (1995).
62. F. Aumayr, H. Kurz, HP. Winter, D. Schneider, M. A. Briere, J. W. McDonald, and C. E. Cunningham, Rev. Sci. Instrum. **64**, 3499 (1993).
63. C. Lemell, HP. Winter, F. Aumayr, J. Burgdörfer, and C. Reinhold, Nucl. Instrum. Meth. Phys. Res. B **102**, 33 (1995).
64. M. Vana, F. Aumayr, C. Lemell, and HP. Winter, Intern. J. Mass Spectr. Ion Proc. 149/150, 45 (1995).
65. C. Lemell, HP. Winter, F. Aumayr, J. Burgdörfer, and F. W. Meyer, Phys. Rev. A **53**, 880 (1996).
66. C. Lemell, J. Stöckl, J. Burgdörfer, G. Betz, HP. Winter, and F. Aumayr, Phys. Rev. Lett. **81**, 1965 (1998).
67. C. Lemell, J. Stöckl, HP. Winter, and F. Aumayr, Rev. Sci. Instrum. **70**, 1653 (1999).
68. J. Stöckl, T. Suta, F. Ditroi, HP. Winter, and F. Aumayr, Phys. Rev. Lett. **93**, 263201 (2004).

69. H. D. Hagstrum, Phys. Rev. **119**, 940 (1960).
70. H. D. Hagstrum and G. E. Becker, Phys. Rev. Lett. **26**, 1104 (1971).
71. S. T. de Zwart, PhD thesis, University of Groningen, (1987).
72. S. T. de Zwart, A. G. Drentje, A. L. Boers, and R. Morgenstern, Surf. Sci. **217**, 298 (1989).
73. M. Delaunay, M. Fehringer, R. Geller, D. Hitz, P. Varga, and HP. Winter, in XIV ICPEAC Conference, edited by M. J. Coggiola et al., Palo Alto, CA, 1985), p. 477.
74. M. Delaunay, M. Fehringer, R. Geller, D. Hitz, P. Varga, and HP. Winter, Phys. Rev. B **35**, 4232 (1987).
75. M. Delaunay, M. Fehringer, R. Geller, P. Varga, and HP. Winter, Europhys. Lett. **4**, 377 (1987).
76. H. Eder, F. Aumayr, and HP. Winter, Nucl. Instrum. Meth. B **154**, 185 (1999).
77. G. Lakits, F. Aumayr, M. Heim, and HP. Winter, Phys. Rev. A **42**, 5780 (1990).
78. H. Kurz, F. Aumayr, C. Lemell, K. Töglhofer, and HP. Winter, Phys. Rev. A **48**, 2192 (1993).
79. R. Diez Muino, A. Arnau, and P. Echenique, Nucl. Instrum. Meth. Phys. Res. B **98**, 420 (1995).
80. R. Diez Muino, A. Salin, N. Stolterfoht, A. Arnau, and P. Echenique, Phys. Rev. A **57**, 1126 (1998).
81. R. Diez Muino, N. Stolterfoht, A. Arnau, A. Salin, and P. Echenique, Phys. Rev. Lett. **76**, 4636 (1996).
82. HP. Winter, M. Vana, C. Lemell, and F. Aumayr, Nucl. Instrum. Meth. Phys. Res. B **115**, 224 (1996).
83. C. Lemell, HP. Winter, F. Aumayr, J. Burgdörfer, and F. W. Meyer, Phys. Rev. A **53**, 880 (1996).
84. F. W. Meyer, L. Folkerts, H. O. Folkerts, and S. Schippers, Nucl. Instrum. Meth. Phys. Res. B **98**, 441 (1995).
85. H. Winter, Europhys. Lett. **18**, 207 (1992).
86. D. Niemann, M. Grether, M. Rösler, and N. Stolterfoht, Phys. Rev. Lett. **80**, 3328 (1998).
87. B. van Someren, P. A. Zeijlmans van Emmichoven, I. F. Urazgil'din, and A. Niehaus, Phys. Rev. A **61**, 032902 (2000).
88. P. Apell, J. Phys. B: At. Mol. Opt. Phys. **21**, 2665 (1988).
89. H. Raether, Surface Plasmons,Springer Tracts in Modern Physics **111**, Springer, Berlin (Springer, Berlin, 1988).
90. E. A. Sanchez, J. E. Gayone, M. L. Martiarena, O. Grizzi, and R. A. Baragiola, Phys. Rev. B **61**, 14209 (2000).
91. HP. Winter, H. Eder, F. Aumayr, J. Lörincik, and Z. Sroubek, Nucl. Instrum. Meth. Phys. Res. B **182**, 15 (2001).
92. N. Benazeth, Nucl. Instrum. Meth. Phys. Res. B **194**, 405 (1982).
93. S. N. Morozov, D. D. Gurich, and T. U. Arifov, Izv. Akad. Nauk SSSR Ser. Fiz. **43**, 137 (1979).
94. M. Vana, H. Kurz, HP. Winter, and F. Aumayr, Nucl. Instrum. Meth. Phys. Res. B **100**, 402 (1995).
95. J. Stöckl, C. Lemell, HP. Winter, and F. Aumayr, Phys. Scr. T92, 135 (2001).
96. D. Ochs, M. Brause, P. Stracke, S. Krischok, F. Wiegershaus, W. Maus-Friedrichs, V. Kempter, V. E. Puchin, and A. L. Shluger, Surf. Sci. **383**, 162 (1997).

97. L. Wirtz, C. O. Reinhold, C. Lemell, and J. Burgdörfer, Phys. Rev. A **67**, 12903 (2003).
98. C. Auth, T. Hecht, T. Igel, and H. Winter, Phys. Rev. Lett. **74**, 5244 (1995).
99. T. Hecht, H. Winter, and R. W. McCullough, Rev. Sci. Instrum. **68**, 2693 (1997).
100. R. C. Monreal, L. Guillemot, and V. A. Esaulov, J. Phys.: Condens. Matter **15**, 1165 (2003).
101. Y. Bandurin, V. A. Esaulov, L. Guillemot, and R. C. Monreal, Phys. Rev. Lett. **92**, 017601 (2004).
102. J. Limburg, J. Das, S. Schippers, R. Hoekstra, and R. Morgenstern, Surf. Sci. **313**, 355 (1994).
103. J. Limburg, J. Das, S. Schippers, R. Hoekstra, and R. Morgenstern, Phys. Rev. Lett. **73**, 786 (1994).
104. S. Schippers, J. Limburg, J. Das, R. Hoekstra, and R. Morgenstern, Phys. Rev. A **50**, 540 (1994).
105. J. Das, L. Folkerts, S. Bergsma, and R. Morgenstern, in 6th HCI Conference, edited by P. Richard, M. Stöckli, C. L. Cocke and C. D. Lin (AIP, Manhattan, KS, 1993), Vol. **274**, p. 563.
106. R. Köhrbrück, M. Grether, A. Spieler, N. Stolterfoht, R. Page, A. Saal, and J. Bleck-Neuhaus, Phys. Rev. A **49**, R1529 (1994).
107. F. W. Meyer, S. H. Overbury, C. C. Havener, P. A. Zeijlmans van Emmichhoven, and D. M. Zehner, Phys. Rev. Lett. **67**, 723 (1991).
108. J. Limburg, S. Schippers, R. Hoekstra, R. Morgenstern, H. Kurz, F. Aumayr, and HP. Winter, Phys. Rev. Lett. **75**, 217 (1995).
109. H. Khemliche, J. Villette, P. Roncin, and M. Barat, Nucl. Instrum. Meth. Phys. Res. B **164**, 608 (2000).
110. J. Limburg, S. Schippers, R. Hoekstra, R. Morgenstern, H. Kurz, M. Vana, F. Aumayr, and HP. Winter, Nucl. Instrum. Meth. Phys. Res. B **115**, 237 (1996).
111. F. Aumayr, HP. Winter, J. Limburg, R. Hoekstra, and R. Morgenstern, Phys. Rev. Lett. **79**, 2590 (1997).
112. N. Stolterfoht, D. Niemann, M. Grether, A. Spieler, A. Arnau, C. Lemell, F. Aumayr, and HP. Winter, Nucl. Instrum. Meth. Phys. Res. B **124**, 303 (1997).
113. D. Niemann, M. Grether, A. Spieler, N. Stolterfoht, C. Lemell, F. Aumayr, and H. Winter, Phys. Rev. A **56**, 4774 (1997).
114. J. Thomaschewski, J. Bleck-Neuhaus, M. Grether, A. Spieler, and N. Stolterfoht, Phys. Rev. A **57**, 3665 (1998).
115. J. P. Briand, S. Thuriez, G. Giardino, G. Borsoni, M. Froment, M. Eddrief, and C. Sebenne, Phys. Rev. Lett. **77**, 1452 (1996).
116. F. Aumayr and HP. Winter, e-J. Surf. Sci. Nanotech. **1**, 171 (2003).
117. N. Nakamura, M. Terada, Y. Nakai, Y. Kanai, S. Ohtani, K. Komaki, and Y. Yamazaki, Nucl. Instrum. Meth. Phys. Res. B **232**, 261 (2005).

4

Kinetic Electron Emission for Grazing Scattering of Atoms and Ions from Surfaces

Helmut Winter

4.1 Introduction

Electron emission induced by impact of atomic particles on solid surfaces is of substantial interest in fundamental research and technological applications. As examples we mention particle detectors, surface analytical tools, display technology, or plasma wall interactions. As a consequence a substantial body of literature has been devoted to this field. In general, two different regimes for the emission processes are of particular relevance: (1) kinetic emission (KE) where kinetic energy of impinging projectiles is transferred to electrons bound in a solid, and (2) potential emission (PE) where internal excitation energies of atomic projectiles is converted into electronic excitations of the target. The contributions of both mechanisms to the emission of electrons from a solid target depend on specific conditions of the collision, in particular, the kinetic energy of the projectiles and their internal electronic excitation energies. In a simple picture, the ratio of kinetic energy and internal potential energy provides an estimate on possible contributions of the two different processes for the ejection of electrons. In this chapter we will concentrate our discussion on recent progress in studies on kinetic emission (KE) phenomena. Potential emission (PE) of electrons induced by multiply and highly charged ions will be outlined in the Chap. 3 by F. Aumayr and HP. Winter.

The importance of electron emission induced by particle impact for a variety of different fields and applications has led over the last century to a large amount of experimental and theoretical studies. For detailed information we refer to some articles which review the developments in this field [1–5]. In general, the emission of electrons is an intricate problem, since different mechanisms contribute to the final ejection of electrons from the solid into vacuum. The emission process starts with the primary excitation of electrons within a solid target, is followed by electron transport to the surface region (accompanied by excitations of secondary electrons), and eventually the crossing of the solid-vacuum boundary. All these mechanisms have been investigated in detail and modeled in the framework of theories taking into account the

H. Winter: *Kinetic Electron Emission for Grazing Scattering of Atoms and Ions from Surfaces*, STMP **225**, 113–151 (2007)
DOI 10.1007/3-540-70789-1_4

electronic excitation and transport phenomena [6, 7]. However, despite substantial efforts in exploring the basic interaction mechanism, important basic features concerning the interactions mechanisms are only vaguely understood. As an example, we mention here the emission of electrons at low kinetic energies, where the small projectile energy leads to a substantial decrease for the yields of emitted electrons. Experimental difficulties made studies on the resulting threshold behaviour of electron emission quite challenging [8, 9] so that the knowledge on this important issue of the emission processes is fairly incomplete.

A further example is the fairly different electron emission yields observed for impact of atoms and ions on insulator and metal surfaces made use of in particle detection. It is known for a long time that electron yields for impact of atomic particles on insulating materials can be substantially higher than for impact on metal targets [10, 11]. In order to illustrate this, at first glance surprising feature, we have sketched in Fig. 4.1 an energy scheme for a metal (left panel) and an insulator target (right panel). It is observed that tightly bound valence electrons from insulators (binding energy typically 10 eV) can be more efficiently extracted from the solid than conduction electrons from metals with a work function of typically 4 to 5 eV. Arguments for this finding were based on the different electron transport affected by the band gap of the insulator, however, details on the interaction sequence, in particular the primary excitation process was not clear until recently.

In this chapter we will discuss recent progress in studies on KE for impact of atomic particles on metal and insulator surfaces. We will focus on specific aspects related to the impact of projectiles on the target surface under a grazing angle of incidence, i.e. fast scattering of atoms or ions with the

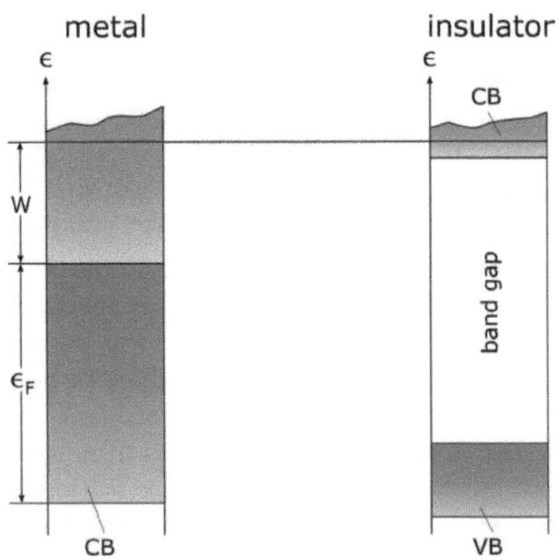

Fig. 4.1. Sketch of electronic energy diagram for metal (*left panel*) and insulator (*right panel*)

solid proceeds in the regime of surface channeling [12–17]. Under those conditions, scattering of atomic projectiles from the surface is characterized by a sequence of events with small angular deflections resulting in well defined trajectories for ensembles of projectiles. For sufficiently small grazing angles of incidence, projectiles will not penetrate into the bulk and interact with the selvedge of the surface only. Since impact parameters with individual surface atoms amount to typically some a.u. (atomic unit of length 1 a.u. = 0.529 nm, a.u. are used throughout this paper unless otherwise stated) electron excitation and emission phenomena during the interaction of atomic projectiles with an electron gas can be investigated in some detail. It is this specific feature which makes grazing ion/atom surface scattering from solid surfaces particularly attractive for studies on the elementary mechanisms for the emission of electrons induced by atomic particles.

In a general scenario of particle induced electron emission, projectiles penetrate the bulk of the solid, may induce there a variety of excitation processes, and are eventually absorbed in deeper layers of the target. For grazing scattering most projectiles are scatterly specularly from the topmost layer of the surface and can analyzed then with respect to their energy. As a consequence it is possible to apply concepts of translation energy spectroscopy – well established in collisions with atoms in the gas phase [18] – to atom-surface scattering. Coincident detection of electrons with the energy loss of the specific particle which has excited the electrons allows one to map the interaction scenario in very detail. We will outline in the next section basic issues of the experimental techniques and discuss the new physical insights into the electronic excitation and emission phemomena obtained by this method. We will show that with this experimental technique absolute total electron emission yields can be reliably measured as low as about 10^{-5}. This property is the basis of detailed studies on the threshold behaviour as outlined by examples for different sorts of projectile-target combinations.

4.2 Concepts and Experimental Techniques

A basic mechanism for the transfer of kinetic energy from a fast atomic projectile to electrons is the elastic collision between these two partners of very different masses. The energy transfer can simply be estimated from conservation of energy and momentum. In head-on binary collisions, electrons are scattered from the projectiles as being reflected from a moving wall so that initial electron velocity (momentum) vectors $\mathbf{v}_{\text{in}} = (v_x\ ,\ v_z)$ are inverted with respect to the direction of the projectile velocity vector $\mathbf{v}_p$ and enlarged according to a velocity (momentum) transfer 2 $\mathbf{v}_p$. In the xz-frame with the x axis along $\mathbf{v}_p = (v_p, 0)$ we have for the final electron velocity or momentum

$$\mathbf{v}_e = (-v_x, v_z) + \mathbf{v}_p \tag{4.1}$$

Making use of eq. (4.1), the outcome of such collisions on the initial momentum distribution of electrons can be visualized in momentum space as illustrated in Sect. 3.2.1. For an estimate on the electron emission process from a solid target we consider the maximum energy transfer to an electron of kinetic energy E_e and mass m_e ($m_e = 1$ a.u.) in a collision with an atomic projectile of mass M. For an electronic binding energy of electrons with respect to vacuum E_b Baragiola et al. [2, 8] derived a threshold for electron emission at a projectile energy

$$E_{\mathrm{th}} = \frac{M}{2m_e}\left(E_e - \sqrt{E_e\left(E_e + E_b\right)} + \frac{E_b}{2}\right) \tag{4.2}$$

This amounts in collisions of, e.g., H atoms with an Al surface ($E_e =$ 10.6 eV = Fermi energy, $E_b = W = 4.3$ eV = work function) to $E_{\mathrm{th}} = 168$ eV ($v_{\mathrm{th}} = 0.082$ a.u.), whereas for LiF ($E_e \approx 4$ eV = width of F2p valence band, $E_b = 12$ eV) we find $E_{\mathrm{th}} = 1836$ eV ($v_{\mathrm{th}} = 0.271$ a.u.). Note that the resulting E_{th} for the insulator is about one order of magnitude higher than for the metal surface which is in contrast with the experimental findings (see, e.g., Fig. 4.6).

Investigations on the threshold behaviour of KE have been performed until recently under large angle impact of ions on metal surfaces [8, 9]. Since reliable measurements of very small total electron yields are an important prerequisite for studies in the threshold regime, those studies were affected in this respect by the limited performance of established experimental methods as current measurements or detection of electron number distributions. Furthermore, other contributions as potential emission (PE, see contribution of Aumayr and Winter in Chap. 3) for incident ions or promotion processes related to close encounters of projectiles with atoms of the crystal lattice could not be excluded. As a consequence, important aspects of the threshold behaviour was not clear and the role of subthreshold mechanisms remained uncertain.

In recent years, substantial progress in studies on KE and, in particular, its threshold behaviour can be stated. These developments are closely related to the experiments performed for impact under a grazing angle of incidence where projectiles are reflected specularly in front of the surface [13, 19–21]. The new type of experiments makes use of this property and provides information on the final energies of projectiles after the collision with the target. Additional detection of emitted electrons allows one furthermore to relate the emission of electrons to excitations of the target during the scattering event, in particular, for the coincident recording of the two channels [22–25].

A typical setup used for such studies is sketched in Fig. 4.2 [26]. In order to reduce contributions of PE to a negligible level, neutral projectiles have to be used. Then the energy loss of projectiles scattered from the target surface (in sketch: LiF(001) mono-crystalline surface) can only be obtained using time-of-flight (TOF) techniques. Chopped beams of neutral atoms are produced via field plates biased with voltage pulses of ns rise time and subsequent near resonant charge transfer in a gas cell. TOF studies of the chopped incident beams provide a direct control on possible occupations of metastable levels for

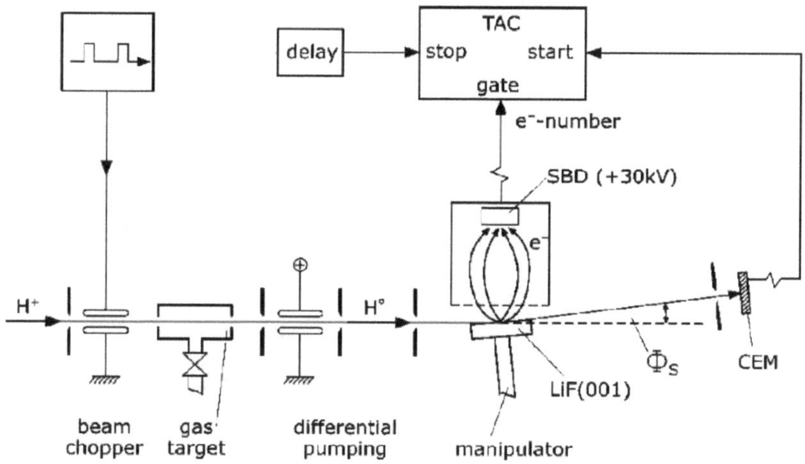

Fig. 4.2. Sketch of experimental setup for coincident TOF-electron number studies [26]

the neutralized atoms which might contribute via PE to the overall emission of electrons.

The TOF part of the setup is of standard design where the detector (channelplate electron multiplier, CEM in sketch) provides the "start" signal. The "stop" signal is obtained from the delayed output signal of the chopper generator. In the studies, a stable overall energy resolution of a few eV for 1 keV pojectiles has to be achieved by electronics comprising a digital delay with ns stability and resolution [26]. Furthermore an ion source with a low energy spread as well as a good emittance is needed [27, 28].

The detection of electrons can be performed in different ways. Electrons emitted from the surface during ion/atom impact might be recorded by means of a channelplate detector [29] or an array of channelplates [23, 30]. A detector pulse may serve to monitor the presence of electrons and used as gating signal for the TOF branch. The pulse heights of the CEM can be used to estimate the electron number distribution. Roncin and coworkers [23] made use of a CEM array in order to derive information on the spatial distribution of emitted electrons. A TOF analysis for the electron channel allowed those authors to measure the electron energy. As an example for results obtained with this setup we display in Fig. 4.3 TOF spectra for 600 eV H° ions scattered from a LiF(001) surface under a grazing angle of incidence $\Phi_{\text{in}} = 1.8°$. The data were recorded in LIST mode and analyzed in terms of events for a specific number of emitted electrons (here: no electron, one electron, two electrons). The spectra reveal peaked structures with defined energy shifts for different number of emitted electrons. We will analyze and discuss these spectra in the next section.

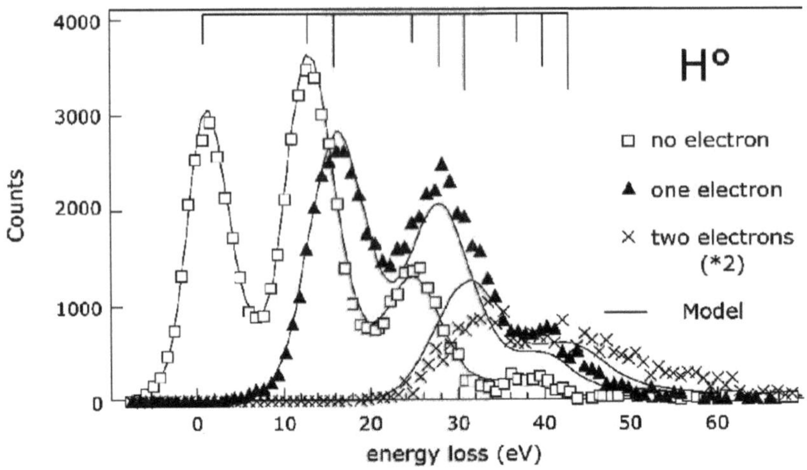

Fig. 4.3. Energy loss spectra for impact of 600 eV H^+ ions on LiF(001) under $\Phi_{in} = 1.8°$. Open symbols denote data coincident with no electrons [23]

The setup for the coincident recording of TOF spectra and number of emitted electrons as displayed in the sketch of Fig. 4.2 is operating in the laboratory of the author and was developed in collaboration with groups from the TU Vienna (F. Aumayr and HP. Winter) and FHI Berlin (J. Viefhaus and U. Becker) [26]. Electrons emitted from the target are collected by a small attractive electric field of about 100 V/cm and recorded by a surface barrier detector (SBD) at a potential of about 25 kV where the pulse heights are proportional to the number of electrons emitted during single collision events [31–33]. Measurements are triggered by a TOF event where the two pulse heights of a time-to-amplitude converter (TAC) and of the SBD are stored in a two dimensional array. In Fig. 4.4 we show as an example a spectrum obtained for scattering of 1 keV H° atoms from LiF(001) under $\Phi_{in} = 1.8°$. This spectrum is obtained for similar conditions as for the data shown in Fig. 4.3 and reveals also defined peaks. Note that the ordinate is converted from flight times to energy loss and the abscissa is the SDB pulse height proportional to the number of emitted electrons. The noise signal of the SBD represents events without emission of electrons.

Since in our method SBD pulse heights are recorded only when an event is registered by the channelplate detector, one can also obtain accurate information on events related to the emission of no electron (noise level of SBD). This is the basis for precise measurements of low total electron yields γ from measured probabilities W_n for a specific number n of emitted electrons [9] via

$$\gamma = \sum_0^\infty nW_n \Big/ \sum_0^\infty W_n \tag{4.3}$$

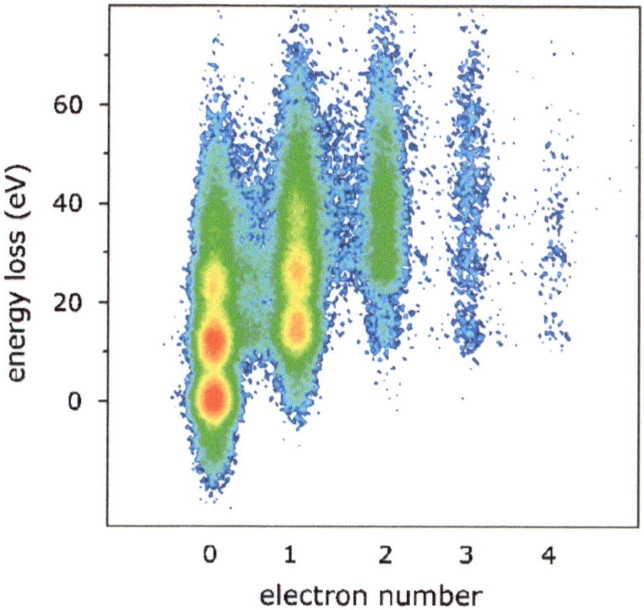

Fig. 4.4. 2D plot of coincident TOF (vertical axis, "energy loss") and SBD (horizontal axis, "electron number") spectra for scattering of 1 KeV hydrogen atoms from LiF(001) under $\Phi_{\text{in}} = 1.8°$

For low γ W_o will dominate the electron number spectrum and can reliably be derived from a coincident electron number spectrum as inferred from the spectrum in Fig. 4.5 for the scattering of 2 keV H° atoms from an Al(111) surface. The non-coincident spectrum of a free running SBD is dominated by the noise of the detector, and information on contributions of W_o can be obtained only via statistical arguments from W_n for $n \neq 0$.

For precise measurements of the electron number distributions (END) and of total electron yields one has to take into account the collection efficiency of the setup for electrons (including transmission of a highly transparent grid at the entrance of the electron detector which shields high electric fields from the target region) and the reflection of fractions of electrons (typically 10 %) from the surface of the SBD. The latter effect leads to additional intensity shifted from main peaks in the spectra; correction of this effect is straightforward [34]. The sensitivity of the method to record extremely low total electron yields is limited by the properties of the chopped projectile beam. In order to keep the data acquisition time in reasonable bounds, one has to enhance the primary beam intensity. Then, however, the probability for finding more than one projectile in a bunch increases and might affect the measurements, since the second projectile also participates to the overall electron signal. This additional projectile needs not be detected by the CEM, could undergo large

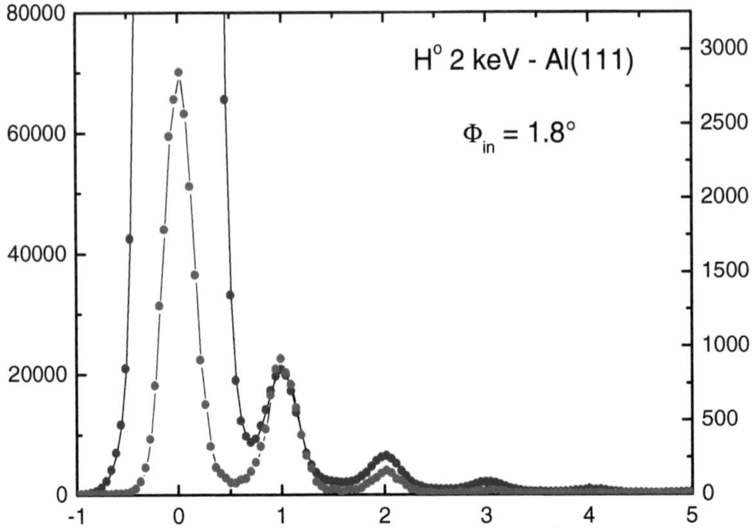

Fig. 4.5. Coincident (*full red/grey circles*) and non-coincident (*full blue/black circles*) electron number spectra for scattering of 2 keV H° from Al(111) under $\Phi_{\mathrm{in}} = 1.8$ deg

angle collisions with defects at the target surface, and results in enhanced probabilities for emission of electrons. It turns out that this feature limits the detection of reliable total electron yields to about 10^{-5}. The sensitivity to quantitative electron yields is not achieved by other experimental techniques and makes this scheme of detection very attractive for studies of threshold phenomena.

4.3 Studies on Electron Emission Phenomena at Insulator and Metal Surfaces

In the following we discuss studies on KE for grazing scattering of fast atomic projectiles from metal and insulator surfaces. Since the lengths of trajectories for particle impact scale as $1/\sin\Phi_{\mathrm{in}}$, experiments performed under glancing angles of incidence are particularly sensitive to the defect structure of the target surface. Therefore the elaborate preparation of atomically clean and, in particular, flat surfaces in an UHV chamber at a base pressure of some 10^{-11} mbar is substantial for such studies [13].

With the experimental procedures outlined in the preceding section total electron yields were measured for grazing scattering of fast H° atoms from insulator and metal surfaces. From the energy scheme sketched in Fig. 4.1 and the discussion above, one would expect a more efficient emission of elec-

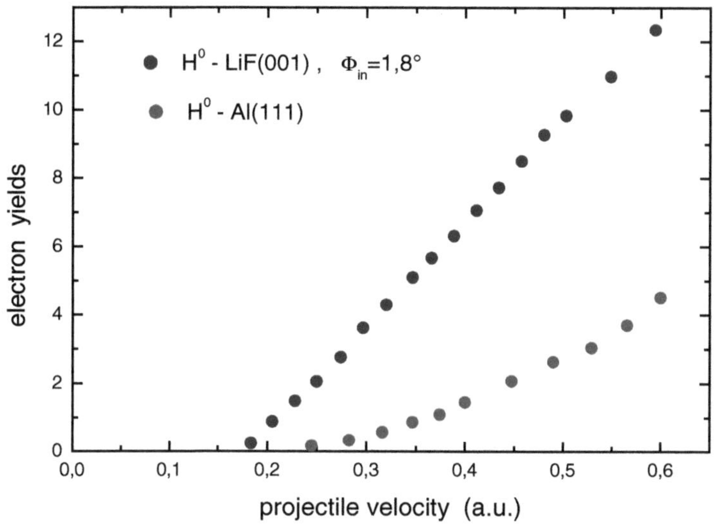

Fig. 4.6. Total electron emission yields as function of projectile velocity for scattering of hydrogen atoms from LiF(001) (*blue/black full circles*) and Al(111) (*red/grey full circles*) under a grazing angle of incidence $\Phi_{\text{in}} = 1.8°$

trons from the metal. However, the experimental data in Fig. 4.6 shows the opposite behaviour and implies that different mechanisms are responsible for the ejection of electrons for the two different classes of materials. The yields observed for scattering from LiF(001) (blue/black dots) are clearly larger than for impact on Al(111) (red/grey dots) and show a smaller velocity for the onset of electron emission. In this respect we mention that enhanced total yields are well established for emission of electrons during large angle impact of particles [10].

4.3.1 Insulator Surfaces

For a detailed investigation of the electron emission process we make use of coincident translation energy spectra as shown in Figs. 4.3 and 4.4. Important information on specific excitation mechanisms during particle impact follows directly from the 2D-representation of the TOF and electron number spectra in Fig. 4.4. This spectrum is dominated by events which do not result in the emission of an electron. The discrete peak structures present in the spectra reflect the broad band gap of ionic crystals (about 14 eV for LiF). The intense peak in the lower left corner with an energy loss of less than 1 eV stems from projectiles which are elastically scattered under channeling conditions in front of the topmost layer of surface atoms with a negligible transfer of energy to the crystal lattice [35]. The second peak in the left column (0 electrons) shows an additional energy loss of 12 eV; the resulting excitation process proceeds without emission of an electron and was first ascribed by Roncin et al. [23]

to the production of a surface exciton, a local excitation of an electron-hole pair (see below). Also the production of a second exciton can be identified in the spectrum. In the second column (1 electron) we find events combined with the emission of one electron, emission of one electron and production of one and more excitons. The mean energy loss for emission of an electron is slightly larger than for the production of an exciton and amounts to 14 eV. In the third column (2 electrons) we find events related to the emission of two electrons, emission of two electrons plus production of one exciton, etc.

From the data in Fig. 4.4 we derive relative intensities and probabilities for the emission of 0, 1, and 2 electrons as well as the production of a specific number of excitons. The bars in Fig. 4.7 represent the experimental data (full bars) and the result of a statistical analysis (open bars) in terms of an interaction model sketched in Fig. 4.8.

Key feature of this model is the formation of a negative ion in a local capture event of an electron from a halide site (here: F^- ions) which is part of the "flat" band of partly localized valence electrons. Since this "active site" is embedded in the lattice of an ionic crystal an additional binding by surrounding positive charges ("Madelung potential") results in a substantial increase of binding energies of negative ions and a reduction in the energy defect in collisions of atoms with anions. As a consequence, one finds an enhanced probability P_{bin} for electron capture. In the further sequence of the collision, this level crosses the exciton level with transition probability P_{LZ}.

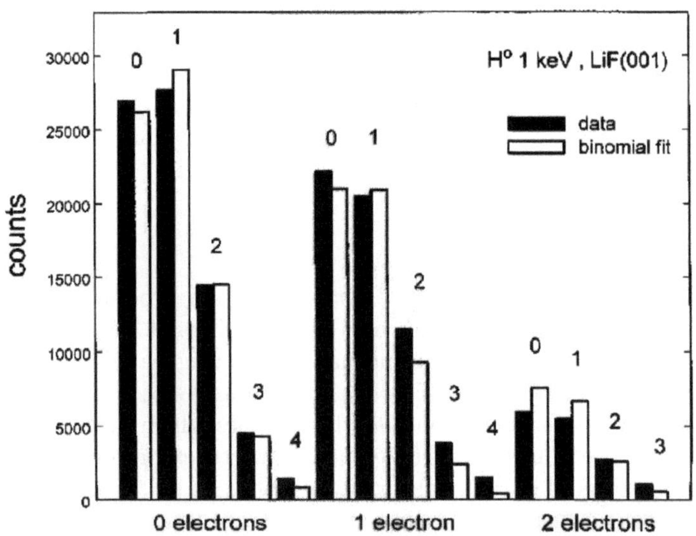

Fig. 4.7. Bar graph of intensities for spectra shown in Fig. 4.4 for emission of specific number of electrons and production of excitons (number at bar). Full bars: experiment, open bars: analysis of data in terms of binomial statistics with parameters derived from interaction model [36]

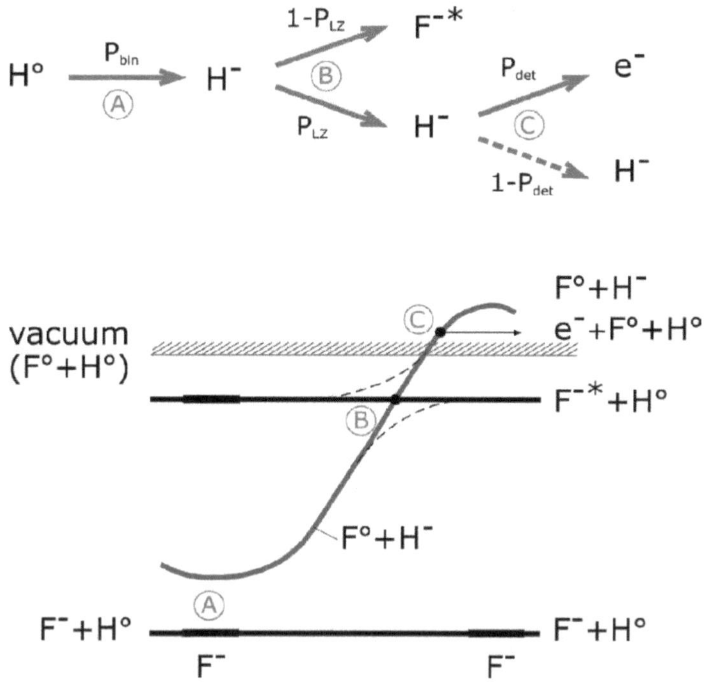

Fig. 4.8. Sketch of interaction path (*upper part*) and interaction model (*lower part*) comprising potential energy curves for scattering of hydrogen atoms from LiF(001) under grazing angle of incidence

Electrons emitted into vacuum result from the detachment of the (transient) negative ion with probability P_{det}. The experimental distribution of electrons (n_e) and excitons (n_{ex}) shown in Fig. 4.7 is well reproduced for a number of collisions n_{coll} by the binomial distribution

$$P^{(n_{\mathrm{coll}})}_{n_{\mathrm{ex}}\,n_e} = \binom{n_{\mathrm{coll}}}{n_{\mathrm{ex}}+n_e}(1-P_{\mathrm{bin}})^{n_{\mathrm{coll}}-(n_{\mathrm{ex}}+n_e)} P_{\mathrm{bin}}^{n_{\mathrm{ex}}+n_e}\binom{n_{\mathrm{ex}}+n_e}{n_e}(1-P_{\mathrm{LZ}})^{n_{\mathrm{ex}}}P_{\mathrm{LZ}}^{n_e} \tag{4.4}$$

with parameters $n_{\mathrm{coll}} = 12$, $P_{\mathrm{bin}} = 0.138$, and $P_{\mathrm{LZ}} = 0.418$ [36].

Studies on Threshold Behaviour for H Atoms Scattered from a LiF(001) Surface

The coincident detection of TOF and electron number spectra provides direct information on the probability W_o for the emission of no electron. This makes this technique a powerful tool for studies of very low γ as present near the threshold for KE.

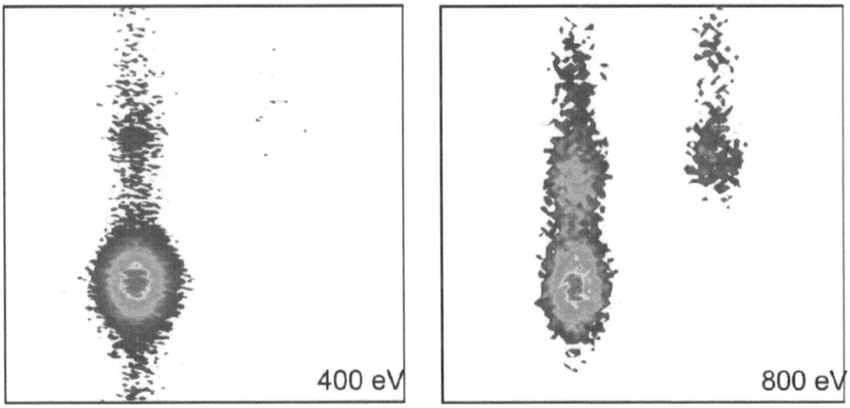

Fig. 4.9. 2D plot of coincident TOF and SBD spectra for scattering of 400 eV (*left*) and 800 eV (*right*) hydrogen atoms from LiF(001) under $\Phi_{\text{in}} = 1.8$ deg

In Fig. 4.9 we show 2D spectra for H atoms of energy 400 eV (left panel) and 800 eV (right panel) scattered from a LiF(001) surface. Comparison with the spectrum displayed in Fig. 4.4 indicates that probabilities for electronic excitations are strongly reduced at lower projectile energies. At 400 eV the spectrum is completely dominated by the peak from elastically scattered projectiles and only a small fraction of excitons (small peak above prominent peak) is produced. The signal from emission of electrons is extremely weak and shows only a few events.

The resulting probabilities for exciton production (full circles) and electron emission (full triangles) are displayed in Fig. 4.10 as function of projectile velocity for constant normal energy $E_z = 0.6$ eV. In addition, fractions of negative H^- ions in the scattered beam (full squares) were measured. The probabilities are generally small (< 0.1) and grow with increasing projectile energy (velocity). The solid curves represent results from an analysis of data in terms of the model for the electronic interaction mechanisms described above.

In more detail, the formation of a negative ion takes place at "active sites" mediated by the confluence of levels for the two collision partners (labeled "A" in the sketch of potential energy curves in Fig. 4.8). The probability for a transition in this binary collision can be estimated from Demkov theory [37]

$$P_{\text{bin}} = \frac{1}{2} \sec h^2 \left(\frac{\pi \alpha \Delta E}{2v} \right) \tag{4.5}$$

with $1/\alpha = (E_1^{1/2} + E_2^{1/2})/2^{1/2}$, E_1, E_2 being electron binding energies of the collision partners and ΔE the effective energy difference between potential curves (energy defect). In passing we note that this approach was first used in the field of atom surface interactions for the description of negative ion formation during grazing scattering from halide surfaces [38, 39]. After formation of the negative ion the diabatic potential curve for $(H^- + F^\circ)$ crosses

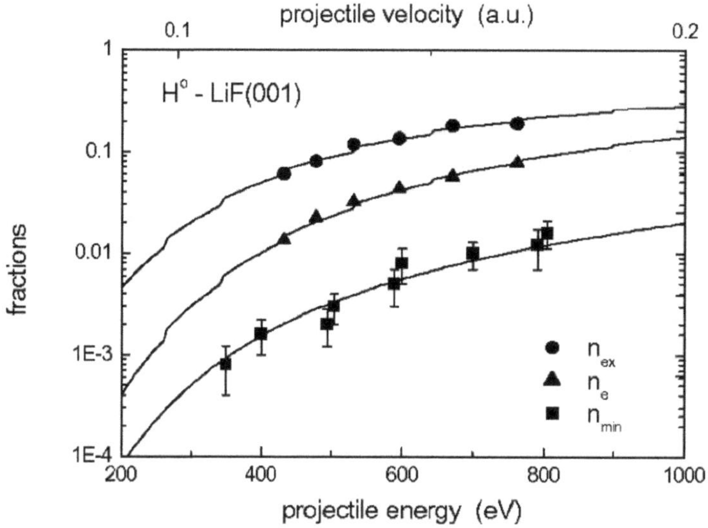

Fig. 4.10. Exciton (*full circles*), electron (*full triangles*), and H^- fractions (*full squares*) as function of projectile energy for scattering of hydrogen atoms from LiF(001) with constant normal energy $E_z = 0.6$ eV

the exciton level ($H^\circ + F^{-*}$) with a transition probability approximated by Landau-Zener theory [40]

$$P_{LZ} = \exp\left(\frac{\pi \Delta\varepsilon_x^2}{2\,(dV/dR)_x\, v}\right) \tag{4.6}$$

with $\Delta\varepsilon_x$ being the energy gap of the adiabatic potential curves, $(dV/dR)_x$ the difference of slopes of the diabatic potential energy curves at the diabatic crossing point. Negative ions surviving the crossing with probability P_{LZ} may detach with probability P_{det} before the next active site is reached. The complete reaction scheme for scattering from LiF is presented in the upper panel of Fig. 4.8.

Based on this reaction scheme, we consider only the occupation of a neutral atom n_o and negative ion n_{min} (normalization $n_o + n_{min} = 1$) which is modified (n'_o and n'_{min}) after one interaction cycle according to

$$n'_o = n_o\,(1 - P_{bin}) + n_o P_{bsin}\,[(1 - P_{LZ}) + P_{LZ} P_{det}] + n_{min}\, P_{det} \tag{4.7}$$

$$n'_{min} = n_o P_{bin}\, P_{LZ}\,(1 - P_{det}) + n_{min}\,(1 - P_{det}) \tag{4.8}$$

The probability for emission of an electron n_e and production of a surface exciton n_{ex} is given by

$$n_e = n_o P_{bin} P_{LZ} P_{det} + n_{min}\, P_{det} \tag{4.9}$$

$$n_{ex} = n_o P_{bin}\,(1 - P_{LZ}) \tag{4.10}$$

Total electron yields and production of excitons are obtained from the iteration over n_{coll} effective collisions (number of active sites being passed by projectiles). Since potential energy curves for the present case and details on the detachment mechanisms are presently not available, relevant quantities enter as free parameters, i.e. $\alpha\ \Delta E$, $\beta = \pi \Delta\varepsilon_x^2 / 2(dV/dR)_x$, and P_{det} .

The solid curves in Fig. 4.10 represent best (correlated) fits to the data using eqs. (4.5–4.8). Variation of n_{coll} yields lowest χ^2 for about 10 collisions, and with $\alpha\ \Delta E = 0.22$ a.u., $\beta = 0.21$ a.u., and $P_{\text{det}} = 0.50$ the data in Fig. 4.10 are fairly well reproduced. This holds also for the statistical analysis of data shown in Fig. 4.7. From the fractions of negative H^- ions in the scattered beam one can derive P_{det} as function of projectile velocity and finds from the analysis for low velocities about 0.5 with a pronounced increase at higher velocities. Calculations by Borisov and Gauyacq [41] using a wave-packet-propagation method show similar values and trends for electron detachment of H^- ions in front of a LiF surface.

Studies on Electron Emission for Scattering of H Atoms from a KI(001) Surface

For scattering from a KI(001) surface a similar interaction scenario takes place. However, the lower binding energy of valence band electrons in KI (8eV) compared to LiF (12 eV) and the smaller band gap (KI: 6 eV, LiF: 14 eV) lead to substantially modified electronic interactions. In Fig. 4.11 we show a 2D-spectrum for the energy loss coincident with the number of emitted electrons after scattering of 400 eV H° atoms from KI(001) under $\Phi_{\text{in}} = 1.8°$. Comparison with the spectrum for scattering from LiF(001) (cf. Fig. 4.9) reveals a pronounced increase in excitation probabilities, manifested in much higher probabilities for production of surface excitons and emission of one electron.

The origin for this observation is based on the clearly smaller binding energy for I^- ions forming the valence band of KI. Then the energy defect ΔE in the collision is reduced which leads according to eq. (4.5) to higher capture probabilities P_{bin}. The energy loss spectrum from Fig. 4.11 coincident with the emission of no electron is displayed in Fig. 4.12. The solid curve represents a best fit to the data assuming discrete energy losses owing to the production of surface excitons and the emission of electrons to vacuum with small kinetic energies. The dashed curves denote additional contributions necessary to reproduce the spectrum and which are ascribed to loss of electrons to the conduction band of KI. Since, different from LiF, the bottom of the valence band starts about 2 eV below the energy for vacuum, this channel for electron loss is expected to play a role here. In a detailed analysis [42] it turns out that electron loss to the conduction band has an increasing contribution for larger projectile energies and explains that less electronic excitation energy results in the emission of electrons than for, e.g. LiF. With this property one can also understand, why the yields of H^- ions are not higher for scattering

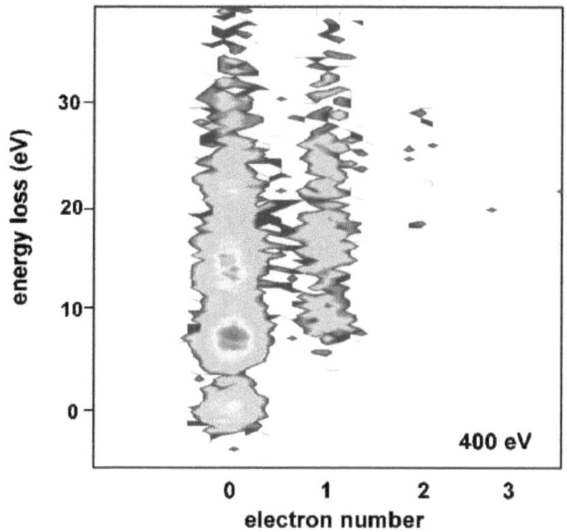

Fig. 4.11. 2D plot of coincident TOF and SBD spectra for scattering of 400 eV hydrogen atoms from KI(001) under $\Phi_{in} = 1.8$ deg

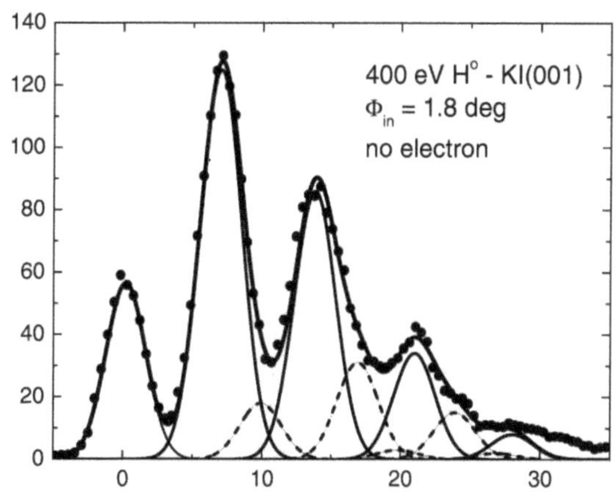

Fig. 4.12. Energy loss spectra for scattering of 400 H° atoms from KI(001) under $\Phi_{in} = 1.8°$. *Full curves*: contributions from elastic scattering and production of surface excitons; dahed curves: contributions from excitation to the conduction band of KI

from KI than from LiF. From the significantly smaller energy defect ΔE for KI and the resulting larger P_{bin} , one would expect larger negative ion fraction here than for LiF. A discussion on electron phenomena from surfaces of oxides (thin films) is given by Zeijlmans van Emmichoven and Matulevich in Chap. 7 of this book.

4.3.2 Metal Surfaces

In Fig. 4.13 we show a 2D-spectrum for scattering of 1 keV hydrogen atoms from an Al(111) surface under a grazing angle of incidence $\Phi_{in} = 1.8°$. The electronic structure of the aluminum target is fairly well described by the approximation of a free-electron gas. The scattering conditions are the same as for the data shown in Fig. 4.4 for scattering from LiF(001). The spectrum for the metal reveals substantial differences compared to the insulator: (1) instead of discrete peaks only a broad peak structure with substantially larger energy loss is found for the metal; (2) the signal owing to emission of electrons is clearly reduced, only a weak peak related to the emission of one electron can be detected here. Comparison of spectra in Figs. 4.2 and 4.13 implies that the interaction mechanisms for electronic excitation and emission from insulator and metal target have to be substantially different.

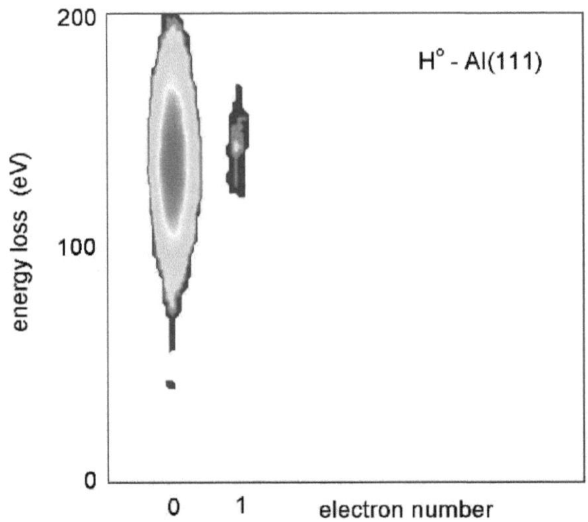

Fig. 4.13. 2D plot of coincident TOF (energy loss) and SBD spectra for scattering of 1 keV hydrogen atoms from Al(111) under $\Phi_{in} = 1.8°$

Model for Energy Transfer in Collisions of Atoms with Conduction Electrons of Metal Targets

For grazing collisions of atomic projectiles with surfaces of metal targets elastic encounter with (free) electrons is the dominant mechanism for electronic excitations. The momentum and energy transfer in such collisions can be deduced from eq. (4.1). A straightforward visualization of the momentum transfer and the relevant occupied and unoccupied phase space for conduction electrons can be achieved by the scheme of a shifted Fermi sphere in momentum space [43].

As follows from eq. (4.1), the component of momentum for the (light) electron parallel to the direction of the incident (heavy) atom is inverted and enhanced by $q = 2m\ v_p$. Since the distribution of occupied states for a free electron metal is represented by the Fermi sphere in momentum space, the final momentum distribution can be visualized by a Fermi sphere shifted by momentum q (see Fig. 4.14). The effect of the collision on a Fermi momentum vector $\mathbf{k}_F$ is sketch for a final electron momentum (red/grey arrow in figure) obtained by inversion with respect to the direction of the incident projectile and adding of $\mathbf{q}$.

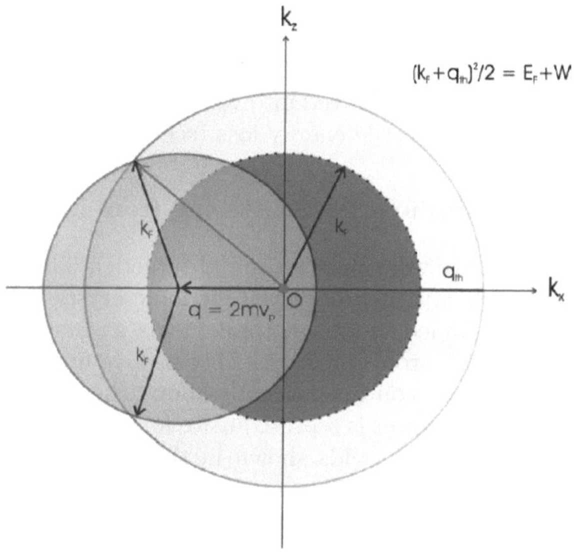

Fig. 4.14. Visualization of momentum transfer in collision of atoms with conduction electrons by concept of shifted Fermi sphere in momentum space

Excitations of the electron gas can take place only, if the final electron momentum is outside the initial Fermi sphere (section in $k_x k_z$-plane is drawn in dark blue). Emission of electrons to vacuum is possible, if the final kinetic energy is larger than the surface potential, i.e. $(k_F + q)^2/2 > E_F + W$ with E_F being the Fermi energy (10.6 eV for Al) and W the work function (4.3 eV for Al(111)). From this condition follows for the emission of electrons a threshold of minimum energy transfer $q_{\mathrm{th}} = 2mv_{\mathrm{th}}$ with the projectile threshold velocity (see also eq. (4.2))

$$v_{\mathrm{th}} = \frac{v_F}{2}\left[\sqrt{1+W/E_F} - 1\right] \tag{4.9}$$

For v_F and E_F for the bulk of Al one derives $v_{\mathrm{th}} = 0.088$ a.u.

The phase space for excited electrons with sufficient energy (momentum) to overcome the surface barrier is represented by the cap resulting from the section of the shifted Fermi sphere with the sphere of radius $k_F + q_{\text{th}} = (2(W + E_F))^{1/2}$. From geometrical arguments follows in lowest order for the volume of the sphere cap a proportionality to $(v - v_{\text{th}})^2$, i.e. a quadratic threshold law. In neglecting transport and passage of the solid-vacuum boundary one expects in close vicinity of v_{th} a quadratic threshold behaviour of KE. In passing we note that for low q (low projectile velocities) the volume of a cap with respect to the static Fermi sphere increases linearly with q or v_p. This is the established v_p-proportionality for electronic stopping via weak excitations of Fermi electrons which are, however, not sufficient to excite electrons to vacuum energies.

In the framework of this simple model, it is also possible to calculate from the momenta of initial and final electronic states excitation energies and the energy transfer to electrons. This is of particular interest for those electrons which are eventually emitted into vacuum and which can be detected in coincidence with the projectile energy loss (see below).

KE Studies Near Threshold for Scattering from Metal Surfaces

From the discussion above we expect for metal targets a different dependence of KE on projectile velocity than observed for ionic crystals. In Fig. 4.15 we show results from a detailed study on total yields as function of projectile velocity for impact of He° atoms on an Al(111) surface under a grazing angle of incidence of $\Phi_{\text{in}} = 1.9°$. As outlined already above, the reliable measurement of total yields of 10^{-4} and lower is a prerequisite for performing those studies. We stress that the maximum yields shown in the plot amount to 4 percent only. Near threshold the total yields can be fitted fairly well by a quadratic dependence on velocity. From the fit a threshold velocity $v_{\text{th}} = 0.112$ a.u. is obtained which is clearly larger than the value estimated in the previous section for bulk properties of Al ($v_{\text{th}} = 0.088$ a.u.). The value of v_{th} derived from the experiment corresponds to collisions with electrons of velocity of about 0.65 v_F. This corresponds to an effective electron density of 25 percent of the bulk value as present at a distance of about 3 a.u. in front of the topmost layer of surface atoms. Therefore the enhanced classical threshold is ascribed to grazing surface collisions, where projectiles are specularly reflected with a distance of closest approach in front of the surface plane. This feature of the collision geometry can be made use of in detailed studies on the distribution of Fermi momenta in the selvedge of metal surfaces.

In the inset of Fig. 4.15 the yields near the classical threshold are plotted on an enlarged scale (maximum $2 \cdot 10^{-3}$). We find small fractions of emitted electrons above the fit curve. These yields are attributed to collisions of projectiles with imperfections of the target surface (primarily steps terminating terraces formed by surface atoms). Then projectiles perform close encounter collisions with the target and probe regions of higher electron densities, i.e. close to bulk values. The data shown at the enlarged scale is indeed consistent with v_{th} calculated for bulk parameters of the electron gas.

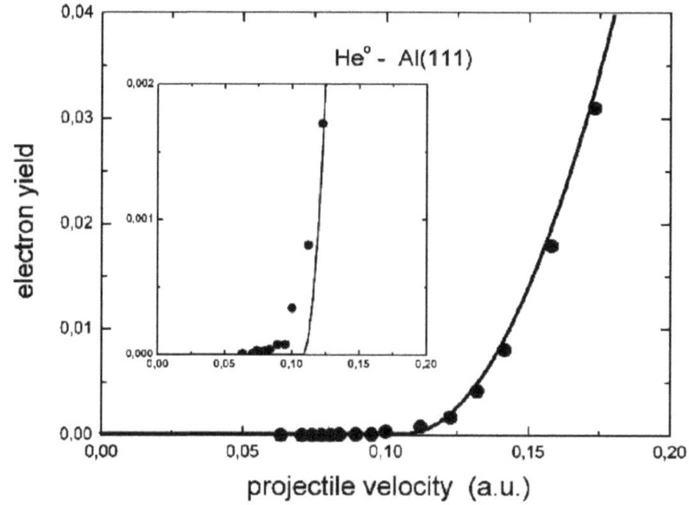

Fig. 4.15. Total electron yields as function of projectile velocity for He° atoms (*full circles*) scattered from Al(111) under 1.9°. *Solid curve*: fit to quadratic threshold law. Inset: vertical scale enlarged by factor of 20

In a more detailed study on the threshold behaviour, the variation of the projectile velocity for different angles of incidence leads to trajectories which probe the largest momenta of conduction electrons at different distances from the surface. One expects to find smaller Fermi momenta for increasing distances from the surface, resulting according to eq. (4.9) in enhanced v_{th}. This feature is observed in experiments performed for different Φ_{in}.

In Fig. 4.16 we show a study on total electron yields for He° atoms scattered from Al(111) under $\Phi_{\mathrm{in}} = 2.2°$ and $\Phi_{\mathrm{in}} = 5.5°$ [44]. Note that for a quadratic threshold law a plot of the square root of γ vs. projectile velocity gives a linear dependence which can easily be analyzed concerning the kinetic threshold (see also "Fowler-plot" for analysis of work function from photo emission near threshold [45]). The plots in Fig. 4.16 are consistent with a linear dependence and reveal for the two angles very different velocities for the onset of electron emission.

Under the assumption that the dispersion for large electron momenta can be approximated by $E_{\max} = k_{\max}^2/2$ (in bulk: $E_{\max} = E_F$) one derives from eq. (4.9)

$$k_{\max} = \left(W - 2v_{\mathrm{th}}^2\right) / 2v_{\mathrm{th}} \tag{4.10}$$

From the work function W of the target and the measured threshold velocity, the maximum electronic momenta are obtained as plotted in Fig. 4.17 as function of distance from the topmost layer of surface atoms. The distance of closest approach is deduced from experiments on rainbow scattering under axial surface channeling conditions [46]. The curves in the figure represent simple estimates on the maximum momentum of electrons parallel to the surface

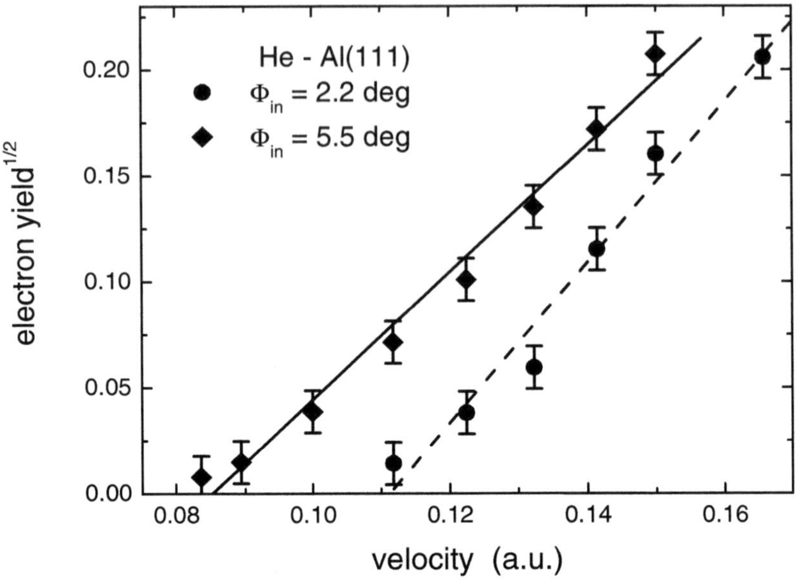

Fig. 4.16. (Electron yields)$^{1/2}$ as function of velocity for He° atoms scattered from Al(111) under $\Phi_{\text{in}} = 2.2°$ (*full circles*) and $\Phi_{\text{in}} = 5.5°$ (*diamonds*) [44]

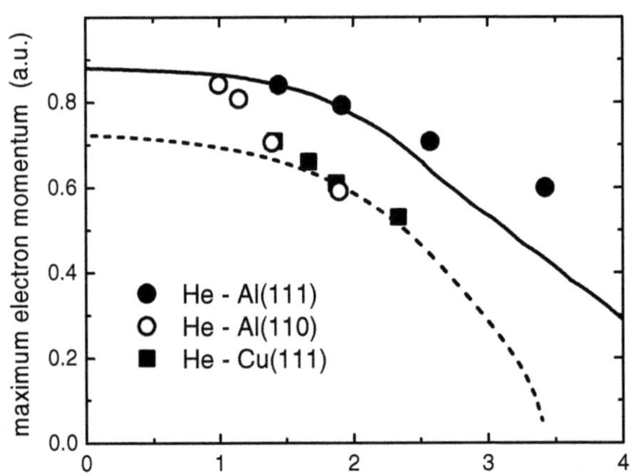

Fig. 4.17. Maximum of electron momentum as function of distance from surface for scattering of He° atoms from Al(111) (*full circles*), Al(110) (*open circles*), Cu(111) (*full squares*) [44]

in the effective surface potential according to $k_{\text{max}}^2/2 = k_F^2/2 - k_z^2/2$ with k_z being the component of momentum normal to the surface plane.

This simple estimate is fairly consistent with the experimental data and in accord with features expected from the electronic structure of the different target surfaces. For example, the momenta for Cu(111) (full squares) compared to Al(111) (full circles) are systematically smaller owing to the smaller Fermi energy/momentum in copper. The clear-cut shift of the momenta for Al(111) and Al(110) is consistent with the shift of the reference plane ("jellium edge") between the two different faces of the crystal. The effective electronic surface potential $V_{\text{surf}}(z)$ derived from this work is in fair agreement with the description of this barrier from density functional theory [47, 48] and LEED fine structure analysis [49].The data shown in Fig. 4.17 can be considered as a demonstration of electron momentum spectroscopy (EMS) – well established within the bulk in terms of $(e, 2e)$ high energy electron scattering [50] – in the selvedge of metal surfaces. A more detailed discussion on this topic can be found in the contribution of Lemell and Burgdörfer in Chap. 1 of this book.

The energy loss spectra displayed in Fig. 4.18 are recorded for 3 keV He° atoms scattered from a Cu(111) surface under $\Phi_{\text{in}} = 2.35°$. Inspection of the spectra coincident with the emission of no (full squares) and no electron (open circles) indicates a small but defined energy shift. This shift is interpreted as the additional energy needed to excite an electron to vacuum energies. Since the electron yields are still small in this energy/velocity regime, it is also evident from this data that most of the projectile energy loss is dissipated in terms of collisions with an energy transfer in the sub-eV domain. Therefore the energy difference between the two peaks in the spectrum shown in Fig. 4.18 can be directly related to the mean excitation energy of electrons emitted to vacuum.

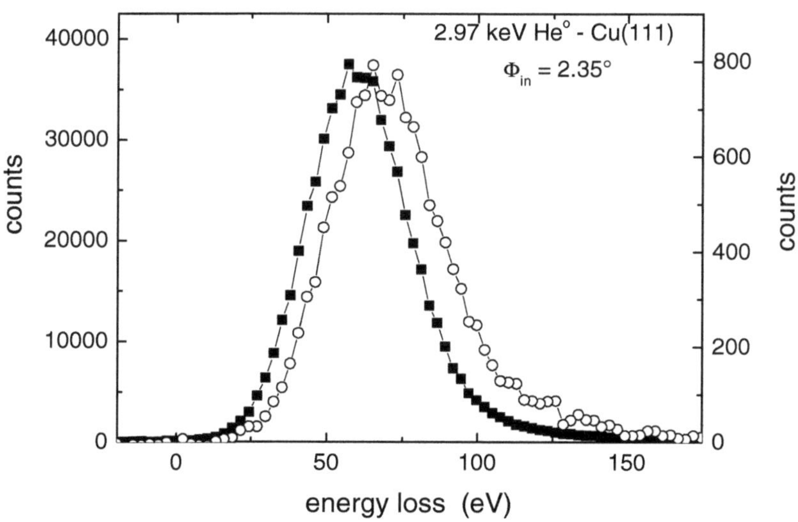

Fig. 4.18. Energy loss spectra coincident with emission of no electron (*full squares*) and one electron (*open circles*) for scattering of 3 keV He° atoms from Cu(111) under $\Phi_{\text{in}} = 2.35°$

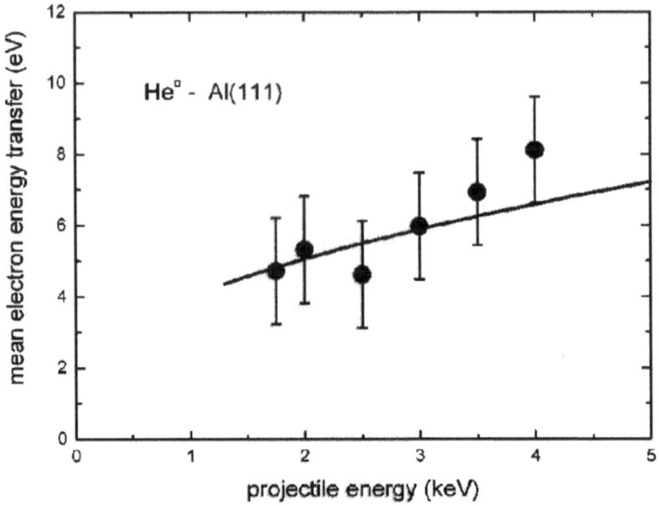

Fig. 4.19. Mean electron energy transfer ΔE_e as function of projectile energy for He atoms scattered from Al(111) under 1.9 deg. Solid curves: model calculations

In Fig. 4.19 we show the difference of the mean energy loss ΔE related to the emission of no and one electron for He atoms scattered from an Al(111) surface as function of projectile energy. Near the threshold for KE ($E_{th} \approx$ 1 keV) the energy transfer ΔE_e is close to the work function of the target ($W = 4.29$ eV) and increases with increasing projectile energy. The solid curve represents calculations based on the phase space for occupied and empty electronic metal states and their energies in the model of the shifted Fermi sphere [43]. The agreement with the data is fairly good. The energy transfer of about 5 eV is small compared to the overall energy loss of projectiles which amounts to about 50 eV at 2 keV. In the near threshold region total electron yields are 1 percent so that less than one per mille of the energy dissipated in the metal results in the emission of electrons.

In Fig. 4.20 we show the ratio $\gamma \Delta E_e / \Delta E$ as function of projectile energy. This ratio is equivalent to the fraction of energy loss leading to KE. The plot indicates that this ratio is generally small. From very small values near threshold it seems to saturate at about 0.1 at higher projectile energies/velocities. These results show that for atomic collisions with conduction electrons the majority of electronic excitations lead to a heating of the electron gas, i.e. the production of so called hot electrons, and only a minor fraction of electrons is ejected to vacuum.

In more detail, the electronic excitations can be understood in the framework of atom-electron collisions. In Figs. 4.21 and 4.22 experimental and simulated energy loss spectra are shown for 12 keV Ar atoms scattered from an Al(111) surface. Similar as for He° projectiles (cf. Figs. 4.18 and 4.19), a defined small energy shift of some eV between the spectra coincident with the

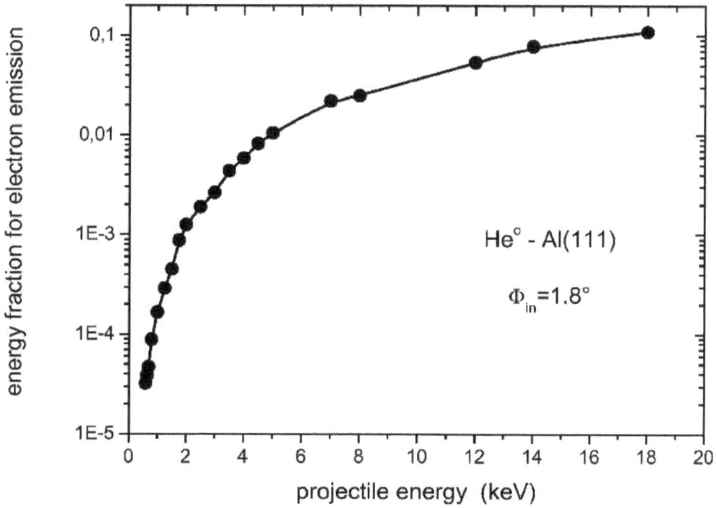

Fig. 4.20. Fraction of dissipated energy leading to KE as function of projectile energy for scattering of He° atoms from Al(111) under $\Phi_{\text{in}} = 1.8°$

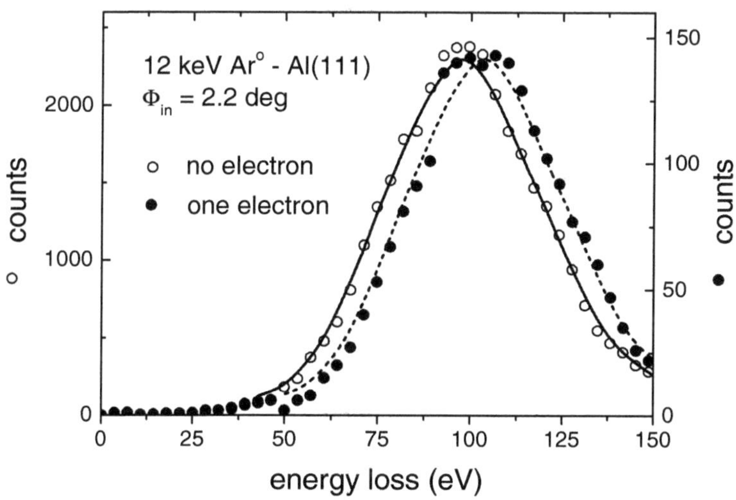

Fig. 4.21. Energy loss spectra for scattering of 12 keV Ar° atoms from Al(111) under $\Phi_{\text{in}} = 2.2°$ coincident with emission of no (*open circles*) and one electron (*full circles*). Curves are drawn to guide the eye

emission of no (open circles) and one electron (full circles) is observed. In computer simulations one can demonstrate that the origin of this shift is based on the clearly larger energy transfer related to KE compared to more probable weak excitations of the electron gas. In the projectile restframe one obtains for collisions of electrons with free Ar atoms from phase shifts [51] differen-

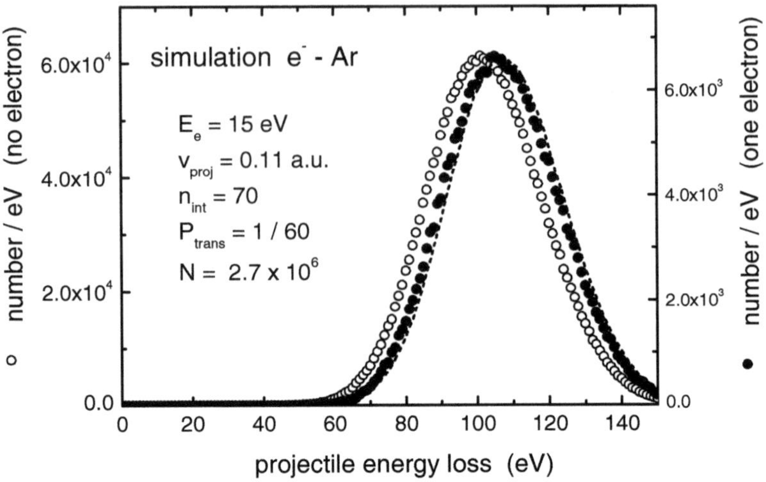

Fig. 4.22. Simulated energy loss spectra for collisions of Ar atoms with free electron gas using cross sections from electron-atom collisions. *Open circles*: events without electron in vacuum; *full circles*: events with one electron in vacuum

tial cross sections as function of scattering angle as shown in Fig. 4.23. The energy transfer from projectiles with $v_{\text{proj}} = 0.11$ a.u. (corresponds to an electron energy in the projectile rest frame of about 15 eV) shows a pronounced dependence on angle and exceeds only close to backscattering the energy for emission to vacuum, i.e. the kinetic energy (dashed curve in Fig. 4.23) is larger than $E_F + W = 14.9$ eV.

For a trajectory equivalent to 70 effective collisions, the probability for scattering under a specific scattering angle and the corresponding elastic energy transfer is taken from the data shown in Fig. 4.23. The energy loss in individual events for Fermi electrons is summed up over complete trajectories and plotted as total energy loss in Fig. 4.22 (open circles). Electron emission proceeds only for a sufficient energy transfer, where electron transport in bulk and through the interface is estimated by the assumption that about one out of 60 electrons reaches vacuum. The full circles show a spectrum coincident with the emission of one electron. The dashed curve represents the energy loss without emission of an electron plus maximum energy transfer. The simulation shows that the energy shift of the two spectra is close to the energy needed for the emission of a single electron. In view of the simple model the description of the energy loss spectra is fairly good.

The non-monotonic behaviour of the differential cross section with scattering angle in Fig. 4.23 is a pure quantum mechanical effect (relevant de Broglie wavelength is comparable with range of screened atomic potential). As a result one finds dominant forward but also backward scattering, where a substantial energy transfer is only achieved for backward scattering (cf. Fig. 4.23). It is this specific quantum mechanical feature which makes the description

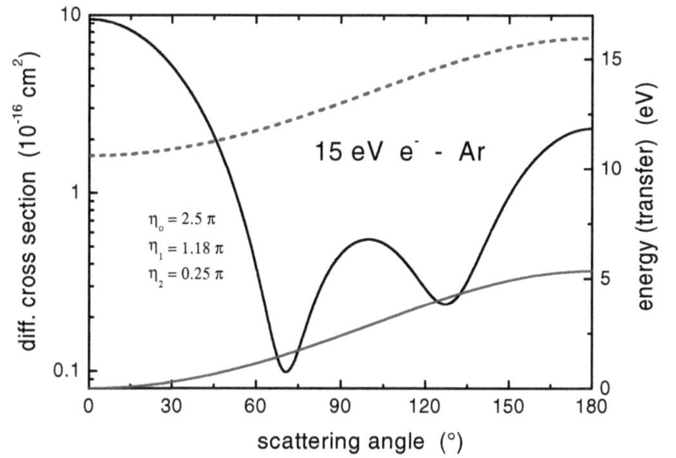

Fig. 4.23. Differential cross section (*black solid curve*) and energy transfer (*red/grey solid curve*) for collisions of 15 eV electron with Ar atom. *Dashed curve*: electron energy after collision

of threshold phenomena for KE from metals by the simple classical binary encounter model outlined above quite successful.

KE from Metal Surfaces Below the Classical Threshold

In the studies on threshold phenomena for KE induced by impact of light atoms on metal surfaces, the classical threshold velocity deduced for electronic properties of the bulk is found as lower limit for large angle impact. In general, the threshold velocities for KE during grazing collisions are larger than this limit. For impact of heavier noble gas atoms (Ne and Ar) small but significant fractions of emitted electrons are observed below v_{th} calculated from eq. (4.9) for the specific scattering conditions. As an example, we present in Fig. 4.24 a plot of the square root of total electron yields as function of projectile velocity for scattering of Ne° atoms (open circles) and Ar° atoms (full circles) from Al(111) under $\Phi_{\mathrm{in}} = 1.8°$ [53]. The dashed curve is the behaviour expected from the model outlined above. The plot indicates that electron emission takes place below the classical limit v_{th}.

This subthreshold emission is interpreted by the realistic band structure at the surface where enhanced momenta of Fermi electrons result in lower projectile velocities for KE. This shows the sensitivity of the data to the momentum distribution (Compton profile) of the electron gas in front of the surface and is different from e-2e-studies with high energy electrons where the momentum distributions in atoms or within the bulk of solids is probed [52]. A model calculation on this problem represents the solid curve in Fig. 4.24 which is in accord with the dependence found in the experiment [53]. An important input for this calculation is the differential cross section for elastic

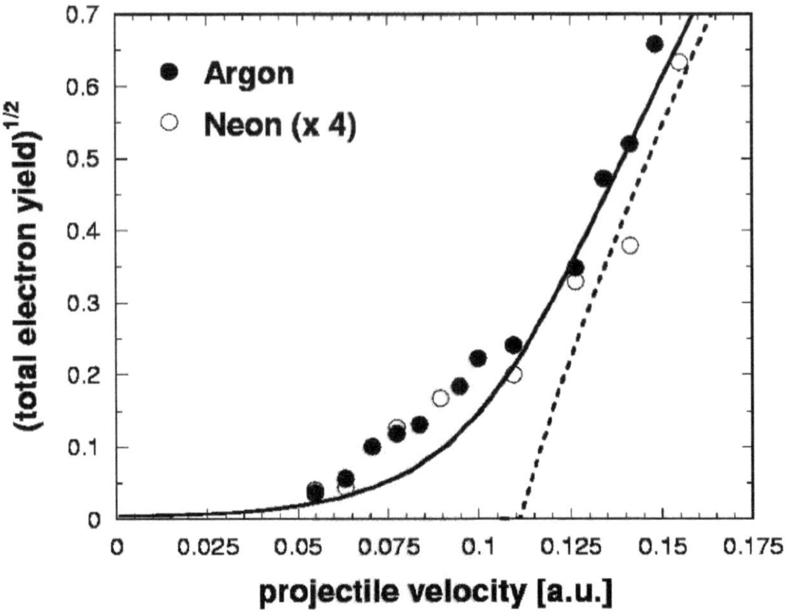

Fig. 4.24. Square root of total electron yield as function of projectile velocity for scattering of Ne (*open circles*) and Ar (*full circles*) atoms from Al(111) under $\Phi_{\text{in}} = 1.8°$

electron-atom scattering. These cross sections differ by a factor of about 4 for scattering from Ne° and Ar° atoms which is also observed for the measured total electron yields (yields for Ne are multiplied by factor of 4 in the plot shown in Fig. 4.24). For He atoms the corresponding cross sections are clearly smaller so that a subthreshold emission of electrons falls into the detection limit of the experimental method and is not observed in studies on the threshold behaviour of KE described in the previous section.

In Fig. 4.25 we show coincident energy loss spectra for 5 keV Ar° scattered from Al(111) under $\Phi_{\text{in}} = 2.2°$. The projectile velocity amounts to $v_p = 0.071$ a.u. which is below the classical threshold even for choosing bulk parameters. The total electron yield is still about 1 percent so that one can record spectra coincident with the emission of an electron. The solid curve in Fig. 4.25 represents a smoothing of data for the zero electron spectrum and the dashed curve is the same curve shifted by the work function of the Al(111) target ($W = 4.3$ eV). The data for the one electron spectrum reveals an energy shift less than the full work function, however, more than one half of this energy. This experimental finding points to an alternative interpretation of data where the emission of electrons below its classical threshold might proceed via the Auger decay of two spatially correlated excited conduction electrons ("hot electrons"). At present this process is still matter of debate.

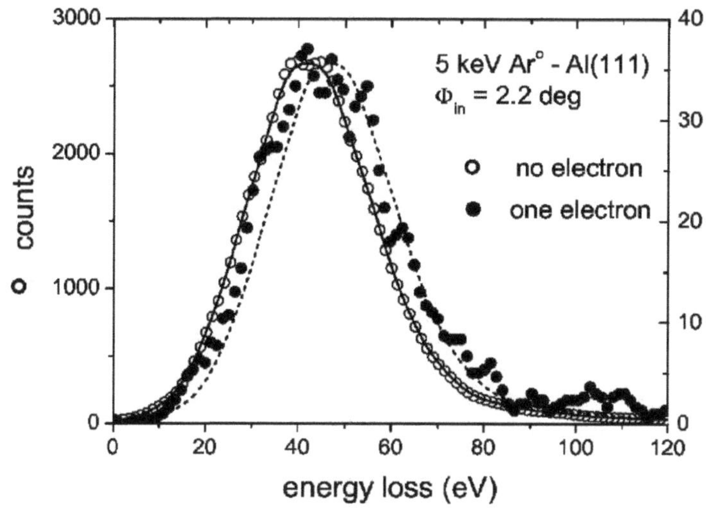

Fig. 4.25. Energy loss spectra for scattering of 5 keV Ar° atoms from Al(111) under $\Phi_{\text{in}} = 2.2°$ coincident with emission of no (*open circles*) and one electron (*full circles*). *Solid curve*: *Smoothed curve* through data for emission of no electron; *dashed curve*: *solid curve* shifted by workfunction of Al(111) ($W = 4.3$ eV) towards larger energy loss

KE from Metal Surfaces Via Electron Promotion

In studies on KE discussed so far the projectile energy and the angle of incidence were chosen in a way that the energy for the motion normal to the surface was sufficiently small to avoid collisions of atomic projectiles with atoms of the crystal lattice under smaller impact parameters, i.e. trajectories with distance of closest approach smaller than about 1.5 a.u. For He° atoms scattered from Al(111) this is met for normal energies smaller than about 20 eV. This feature is demonstrated in Fig. 4.26 for the mean energy transfer to emitted electrons (energy difference in spectra for emission of no and one electron) as function of projectile energy for different angles of incidence [54]. The solid curve represents calculations on the energy transfer using the binary encounter model outlined above. The data for the smallest angle is in fair agreement with the theoretical estimate, whereas for increasing angle of incidence deviations from the curve can be found at decreasing projectile energy. Inspection of data reveals that the deviation from the theoretical curve sets in at a normal energy close to 20 eV. Then projectiles interact at sufficiently small impact parameters for emission of electrons to vacuum via electron promotion of transient molecular orbitals.

Energy loss spectra for 5 keV Ne° atoms scattered from Al(111) under $\Phi_{\text{in}} = 5.5°$ are diplayed in Fig. 4.27. The energy for the normal motion of projectiles $E_z = E_o \sin^2\Phi_{\text{in}} = 46$ eV exceeds the energy limit for the onset of promotion processes of 20 eV. The spectra coincident with the emission of no

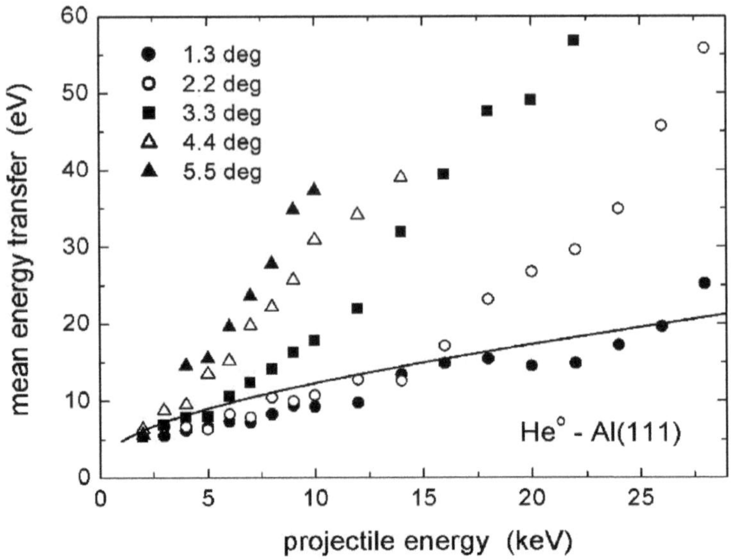

Fig. 4.26. Mean energy transfer for emission of one electron during scattering of He° atoms from Al(111) as function of projectile energy. $\Phi_{\text{in}} = 1.3°$ (*full circles*), 2.2° (*open circles*), 3.3° (*full squares*), 4.4° (*open triangles*), 5.5° (*full triangles*)

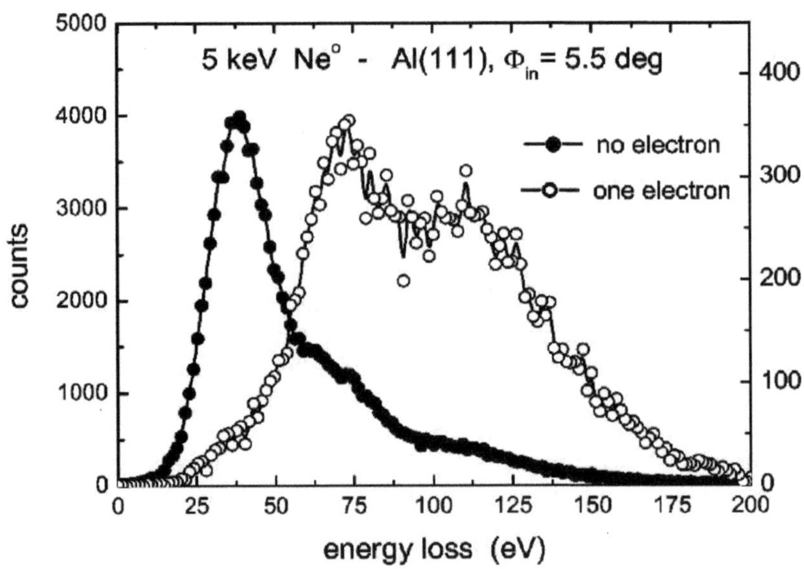

Fig. 4.27. Energy loss spectra for scattering of 5 keV Ne° atoms from Al(111) under $\Phi_{\text{in}} = 5.5°$ coincident with emission of no (*full circles*) and one electron (*open circles*) [54]

and one electron reveal a pronounced shift of about 35 eV which is attributed to electron emission via promotion of the $4f\sigma$ orbital for the Ne-Al system [55]. A fraction of events related to electron emission without promotion can be observed also in the spectra, and allows one to quantify the contributions owing to promotion in the electron emission process. We note that electron promotion takes place preferentially at smaller impact parameter. Then the transfer of projectile energy to the crystal lattice is enhanced and has to be taken into account in the analysis of data in terms of computer simulations [54]. Important issue of such studies is the quantitative information on the contributions of electron promotion compared to the overall emission of electrons.

Dependence of KE on Azimuthal Orientation of the Target Surface

Collisions of fast atoms with surfaces under a grazing angle of incidence take place in the regime of surface channeling. If projectiles are scattered in low index directions of the surface plane, scattering proceeds along strings of atoms in terms of axial surface channeling. In comparison to planar channeling for a "random" azimuthal orientation of the target surface, projectiles are scattered in a potential with cylindrical symmetry and have an enhanced probability to reach the subsurface region. Then projectiles interact with an electron gas of higher electron density than in the selvedge and interact with the solid in terms of longer trajectories. This leads to larger KE yields for axial surface channeling.

The effect can be demonstrated by recording the (uncompensated) target current to ground during azimuthal rotation of the target surface. As an example, we display in Fig. 4.28 the target current as function of the azimuthal angle for scattering of 16 keV He° atoms from an Al(111) surface, where the low index directions of the hexagonal structure of the (111) face are clearly identified [56]. In passing we note that this technique allows one during the preparation of the target surface via sputtering with, e.g., Ar^+ ions to monitor the quality of the target surface [13]. In most cases, the relative intensities of the peak structures of the target current increase with improvements in the quality of the surface.

An even more attractive application of this effect lies in the information on the surface geometrical structure inherent in the requirement of scattering along ordered strings of atoms in the surface plane. This allows one to apply concepts of triangulation in order to derive the surface structure in real space and in-situ. Such information is particularly valuable for the structure of ultrathin films during epitaxial growth on mono-crystalline substrates, where often complex arrangements of surface atoms in terms of super-structures are present. It turns out that this variant of ion-beam triangulation [57] is a powerful method to explore the surface structure with respect to the lateral arrangements of surface atoms. The technique is fairly complementary to the analysis based on detailed LEED studies [58] which is very sensitive to transversal arrangements of surface atoms.

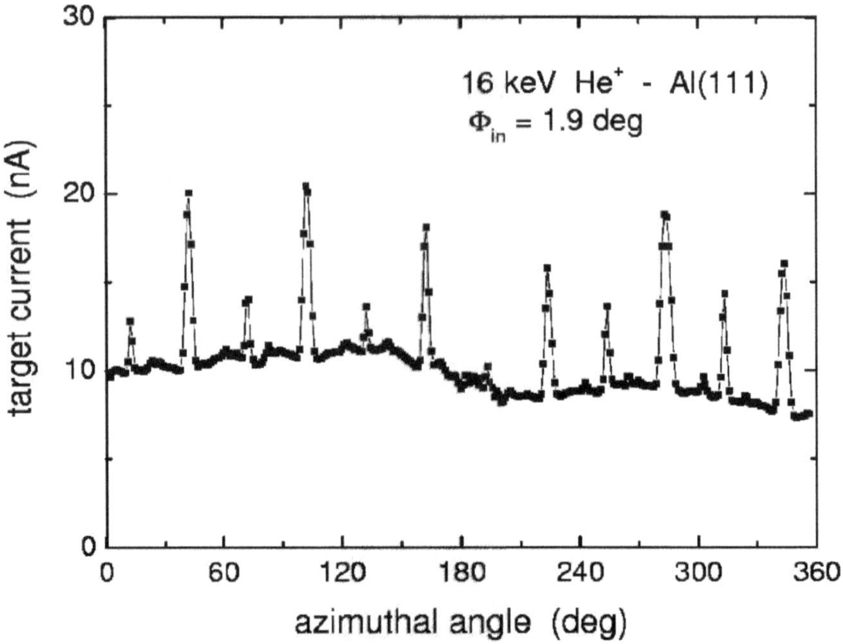

Fig. 4.28. Target current as function of azimuthal angle for scattering of 16 keV He^+ ions from Al(111) under $\Phi_{in} = 1.9°$ [56]

For applications of the ion beam triangulation method to the structure of ultrathin films it is of advantage to detect the electrons with high sensitivity using a SBD instead of the target current. Then only some 1000 atoms per second are scattered from the target, equivalent to an ion current in the sub-fA regime, and radiation damage of the surface structure can be excluded. In Fig. 4.29 we show pulse height spectra of a SBD biased to 25 kV for collisions of 29 keV H° atoms with a clean and flat Cu(001) surface under $\Phi_{in} = 1.5°$ which show a pronounced dependence on the azimuthal orientation of the target surface. The data reveals a pronounced increase of events related to the emission of a higher number of electrons for scattering along a low index direction owing to the penetration of projectiles into subsurface layers. As a result the number of events from scattering above the surface plane is reduced. Selection of these events by appropriate settings of discriminator levels allows one to monitor such events with high sensitivity to the surface layer. Since the total electron yield is sufficiently large here, the probability for emission of at least one electron is close to one and the data can be efficiently normalized to the overall SBD count rate [59].

Mn atoms expitaxially grown on a Cu(001) surface shows rich variety of complex superstructures depending on thickness and temperature of the ultrathin film [60]. In Fig. 4.30 we show triangulation curves using the normalized SBD signal for the emission of one to three electrons (cf. Fig. 4.29). For the

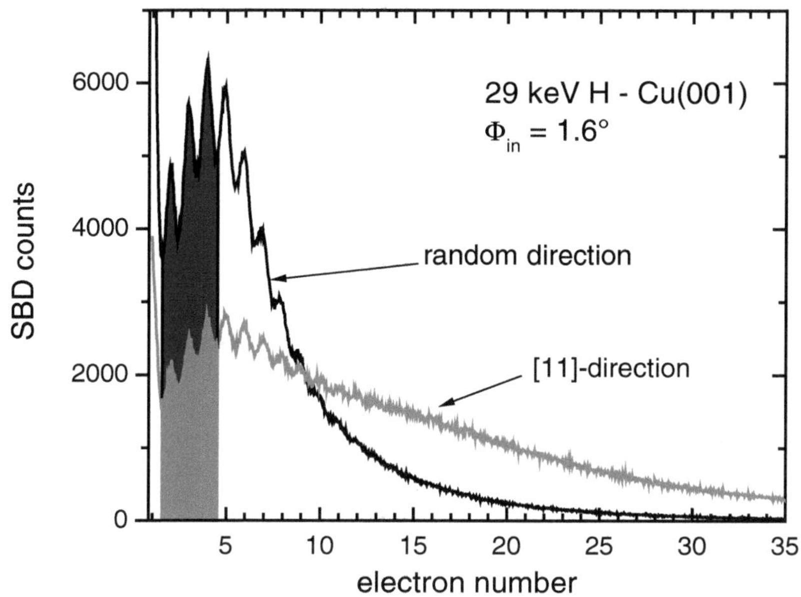

Fig. 4.29. Electron number spectra for scattering of 29 keV H° atoms from Cu(001) under $\Phi_{\rm in} = 1.5°$ along "random" azimuth (*black curve*) and [11] direction (*grey curve*)

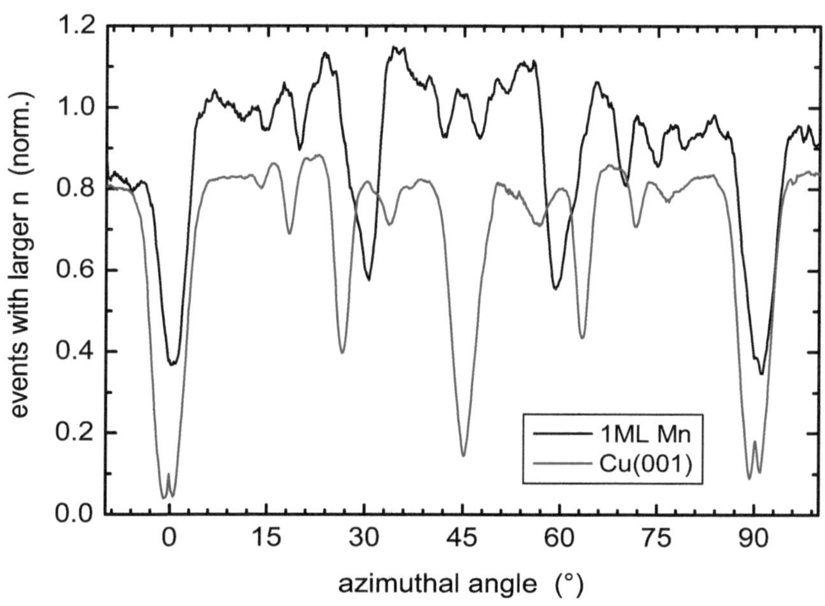

Fig. 4.30. SBD signal as function of azimuthal angle for scattering of 29 keV H° atoms from clean Cu(001) (*red/grey curve*) and from 1 ML Mn/Cu(001) grown at $T = 160$ K (*black curve*)

clean Cu(001) surface one finds the prominent low index directions of a simple fcc lattice. Growth of 1 ML Mn on the Cu(001) substrate leads to a different dependence of the signal on azimuthal angle. In particular, the prominent dip for scattering along [11] on the clean substrate at 45° completely disappears and most dips show specific shifts. This behaviour demonstrates the high sensitivity of the method to the structure at the surface. Curves as displayed in Fig. 4.30 can be analyzed in terms of computer simulations of trajectories where the penetration into the subsurface layer holds as criterion for the enhanced number of emitted electrons. Then the fractions of subsurface trajectories are directly related to the dips in the triangulations curves. The power of the method lies in direct and very sensitive tests of proposed structures for the film structure. For the case presented in Fig. 4.30 a (8×2) structure was proposed from the analysis of LEED data. The triangulation curve is, however, in conflict with such a structure, whereas the simulations are in good agreement with the experiment, if a (10x2) structure is assumed. A substantial potential of ion beam triangulation performed via the detection of KE in studies on structures of ultrathin films can be stated. The method is particularly sensitive to lateral positions of atoms in the topmost layer of a clean surface or ultrathin film.

Dependence of KE for Grazing Ion Surface Scattering on Surface Morphology

The total electron yield depends on the morphology of the target surface. For a "rough" surface violent binary encounter with atoms forming defect structures at the surface will occur and result in an enhanced probability for the emission of electrons. This effect plays an important role for grazing surface scattering, where projectiles probe the surface over a considerable length scale. Recently is has been shown that the variation of KE with surface morphology can be used to monitor epitaxial growth of ultrathin films [61].

In Fig. 4.31 we show results from studies on growth at room temperature of a Co film on a Cu(001) surface. In the experiments 25 keV H^+ ions (dashed curve), He^+ ions (solid curves), and N^+ ions (dotted curve) are scattered from the film surface under $\Phi_{\text{in}} = 1.6°$. The upper curves represent the intensities of electrons emitted with an energy of 145 eV, the lower curve is the intensity of specularly reflected He^+ ions. The data demonstrate that the variation of the surface morphology leads to growth oscillations of the intensity of emitted electrons which allow one to monitor and to study the growth of ultrathin films. In this respect the data is similar to oscillations observed via RHEED or the intensity of specularly reflected ions. For scattering of He^+ ions the oscillations of the electron intensities are quite prominent. A detailed understanding of the processes underlying the observations in Fig. 4.31 has not been worked out so far. However, it is obvious that whenever the surface structure shows a minimum of defects, the yields for KE are reduced accordingly. In passing we note that such growth oscillations are also detected for the overall

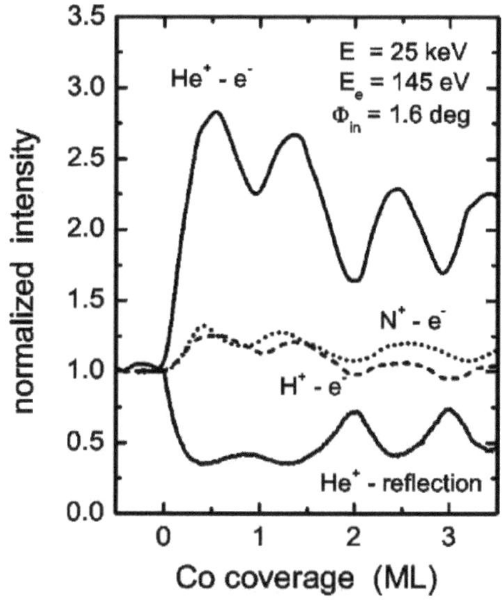

Fig. 4.31. Normalized intensity of electrons with energy of 145 eV as function of Co coverage of Cu(001) surface for impact of 25 keV H^+ (*dashed curve*), He^+ (*upper solid curve*), and N^+ ions (*dotted curve*) under $\Phi_{in} = 1.6°$. *Lower solid curve*: intensity of specularly reflected He^+ ions

electron emission observed with a SBD or measurement of the target current. The latter method provides a simple tool to monitor growth of ultrathin films, since only an ion gun and a current meter is needed.

Diffraction of Electrons Emitted During Grazing Ion Surface Scattering

In recent years considerable interest was devoted to studies on the role of plasmons for the emission of electrons during impact of singly and multiply charged ions on metal surfaces [62–65]. For a discussion we refer to detailed treatments of this topic in this book. A key issue of the problem is the observation of structures in electron spectra after impact of slow ions on metal surfaces. Since these spectral features appear at electron energies which correspond to the decay of surface and bulk plasmons, collective excitations of the electron gas, the experiments were interpreted in terms of such a process. Different mechanisms were proposed to overcome for excitation via impact of atomic projectiles the missing momentum transfer based on the dispersion relation for plasmon excitation [6]. Energy shifts of the electron energy observed for a variation of the observation angle were explained by effects of momentum matching with the crystal lattice [66].

An important development in the explanation of the observed effects, in particular the spectral features observed for impact of H^+ ions on an Al

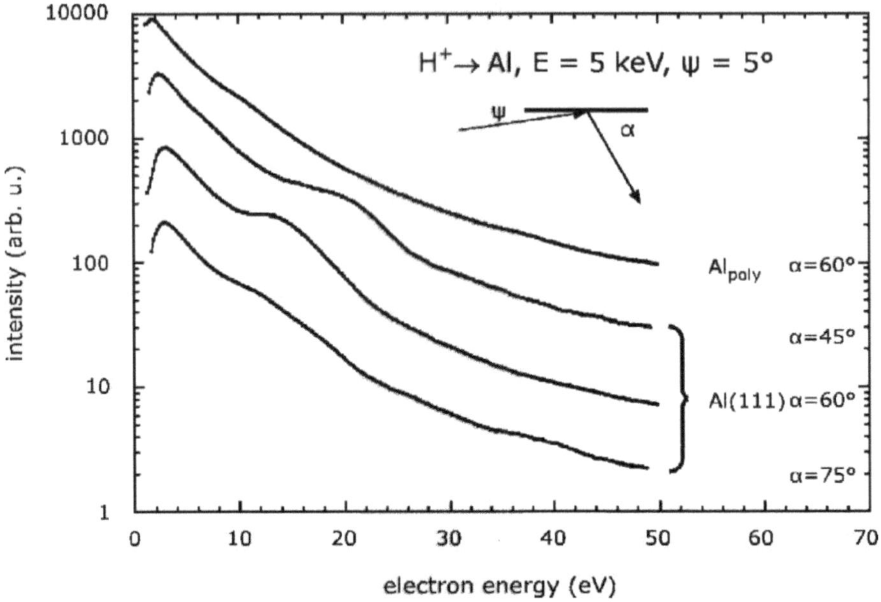

Fig. 4.32. Electron spectra for scattering of 5 keV H^+ ions scattered from Al(111) and polycrystalline Al under $\Phi_{\text{in}} = \Psi = 5°$ for different angles of observation α. For definition of angles see inset

surface, was a study by Eder et al. shown in Fig. 4.32 [65]. The spectra for impact of 5 keV H^+ ions on an Al(111) and on a polycrystalline Al surface under an angle of incidence $\Phi_{\text{in}} = \Psi = 5°$ are recorded for different angle of observation (see inset of Fig. 4.32). Striking feature are peaked structures for the monocrystalline Al(111) target which are completely missing in the spectrum for the polycrystalline sample (see also contribution of Aumayr and Winter in Chap. 3).

This observation makes the interpretation of the peaked structures in the spectra for the monocrystalline target in terms of plasmon excitation and subsequent decay rather unlikely. The authors concluded that possibly electron diffraction effects – well established for low energy electron scattering from surfaces (LEED) – is the origin of the experimental finding. Niehaus and coworkers [62] reanalyzed former electron spectra [66] in terms of a consistent interpretation of spectra based on electron diffraction at the topmost layer of surface atoms.

A direct demonstration of electron diffraction induced by impact of atomic projectiles has been recently presented by Bernhard et al. [67]. In case electron diffraction plays a noticeable role for KE of electrons, then diffraction pattern in terms of intensity spots in the angular distributions of scattered electron should be found (similar to spots observed in LEED). In Fig. 4.33 we show an intensity plot for electrons emitted in direction normal to the target surface

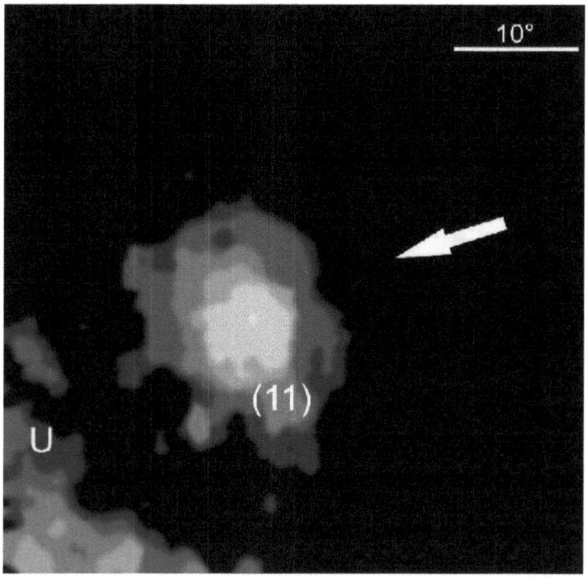

Fig. 4.33. Difference of intensity distributions for electrons recorded for energy cut offs of 52 eV and 57 eV with a SPALEED system for scattering of 25 keV H^+ ions from Cu(001) under $\Phi_{in} = 1.6°$ along ⟨100⟩. Arrow denotes direction of incident beam

for scattering of 25 keV H^+ ions from Cu(001) under $\Phi_{in} = 1.6°$. The arrow denotes the direction of the incident projectile beam along the ⟨100⟩ direction of the (001) face of the target surface and its tip the direction normal to the surface. The data shown in Fig. 4.33 are recorded with a scanning LEED system (Spot Profile Analysis LEED, SPALEED [68]) operated in a dispersive mode (difference of data sets recorded with pass energies of 52 eV and 57 eV). The intensity pattern shows a prominent peak for scattering at 10° in forward direction with respect to the surface normal. Using a modified Ewald sphere construction scheme of the reciprocal lattice, the diffraction peak can be ascribed to the (11) spot. In Fig. 4.34 we show results from similar studies performed with 30 keV H^+ and 25 keV He^+ ions for different azimuthal directions of the incident projectile beam and different final electron energies. The spot positions are in good agreement with the predictions from the Ewald construction which provides unequivocal evidence for the presence of electron diffraction.

An important consequence of the presence of diffraction effects on the spatial distribution of emitted electron is the modification of angular and energy distributions. This is evident from spectral features in electron spectra which have been (re)analyzed in term of electron diffraction effects [62] and from spectra obtained with 25 keV H^+ ions shown in Fig. 4.35. The energy spectra in Fig. 4.35 are recorded along a direction tilted by 12° with respect

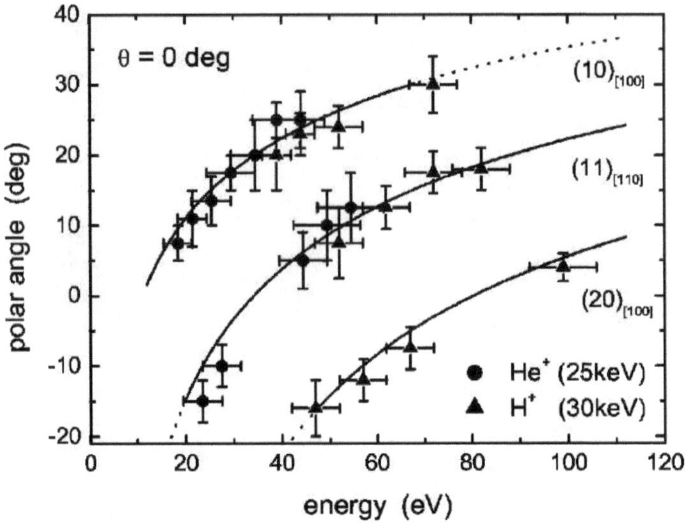

Fig. 4.34. Polar angle for position of intensity spot as function of electron energy (selected with SPALEED) for scattering of 30 keV H^+ (*full triangles*) and 25 keV He^+ ions (*full circles*) from Cu(001) along ⟨100⟩ and ⟨110⟩ under $\Phi_{in} = 1.6°$. *Solid curves*: Spot position calculated from Ewald construction scheme

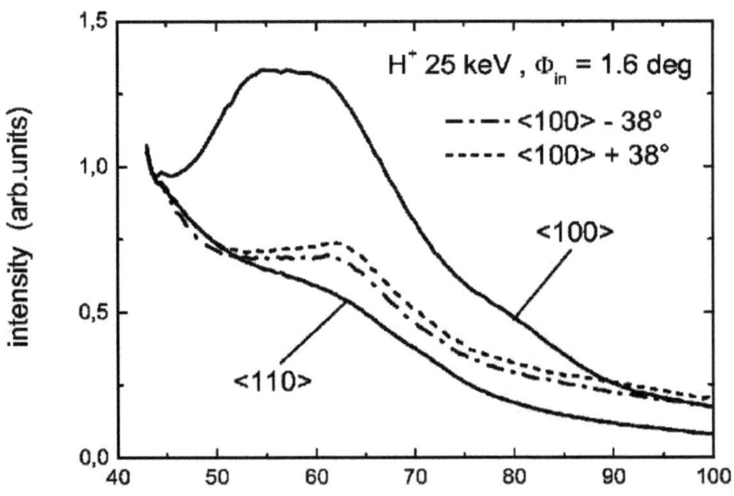

Fig. 4.35. Electron spectra for scattering of 25 keV H^+ ions scattered from Cu(001) under $\Phi_{in} = 1.6°$ for different azimuthal orientations of the target surface

to the surface normal, this is for about 60 eV electrons the direction of the (11) spot for scattering along ⟨100⟩. The strong variation of the spectrum with the azimuthal orientation of the target is a good demonstration, how strong diffraction effects can affect energy spectra of electrons emitted from the surface of a monocrystalline target. Therefore a thorough inspection of the scattering geometry has to be performed in order to possibly exclude those effects on the shape of energy spectra and to avoid misinterpretations of data.

4.4 Conclusions

We have demonstrated that the coincident combination of translation energy spectroscopy with the number of emitted electrons allows one to obtain detailed information on the electronic excitation and emission mechanisms for grazing impact of atomic projectiles on solid surfaces. The application of this method to the scattering of hydrogen atoms from a LiF surface (wide-badgap insulator) has led to a basic understanding of the interaction processes for electron emission, internal excitations of the target (surface excitons), and formation of negative ions. Basic feature is the (transient) formation of negative ions in local capture events of electrons from anion sites as dominant precursor for the electronic excitations. Furthermore, the wide band gap will affect transport of electrons within the bulk and at the surface of ionic crystals.

This interaction mechanism is clearly different from the transfer of projectile energy in direct atom-electron encounters which is the dominant mechanism for electron emission from metal surfaces near threshold. The efficient mechanism of formation of transient negative ions in atomic collisions with insulator surfaces provides a straightforward explanation for higher total electron yields observed for insulator compared to metal targets.

Acknowledgements

The contributions to the work presented here from Dr. S. Lederer, D. Blauth, and T. Bernhard (HU Berlin) and the fruitful collaboration with Profs. F. Aumayr and HP. Winter (TU Vienna) are gratefully acknowledged. The studies were supported by the Deutsche Forschungsgemeinschaft (DFG) under contract Wi1336.

References

1. D. Hasselkamp, *Particle Induced Electron Emission II* (Springer, Heidelberg 1992), Springer Tracts Mod. Phys., Vol. **123**, 1.
2. R.A. Baragiola, in "Low Energy Ion-Surface Interactions", ed. J. W. Rabalais, Wiley, New York 1994.
3. J. Schou, Scan. Micr. **2**, 607 (1988).
4. M. Kaminski, *Atomic and Ionic Impact Phenomena on Metal Surfaces,* Springer, Berlin, 1965.

5. W.O. Hofer, Scan. Microsc. Suppl. **4**, 265 (1990).
6. M. Rösler and W. Brauer, in *Particle Induced Electron Emission I,* ed. G. Höhler, Springer, Heidelberg 1992, Vol. **123**, p. 1.
7. M. Rösler, in "Ionization of Solids by Heavy Particles", ed. R.A. Baragiola, Plenum Press, New York, **193**, p. 27.
8. R.A. Baragiola, E.V. Alonso, and A. Oliva Florio, Phys. Rev. B **19**, 121 (1979).
9. G. Lakits, F. Aumayr, M. Heim, and HP. Winter, Phys. Rev. A **42**, 5780 (1990).
10. M. Ardenne, *Tabellen zur angewandten Physik,* Deutscher Verlag Wiss., Berlin 1973, Band I, p. 97.
11. K.H. Krebs, Ann. Phys. **10**, 2617 (1962).
12. D.S. Gemmell, Rev. Mod. Phys. **46**, 129 (1974).
13. H. Winter, Phys. Rep. **367**, 387 (2002).
14. M.W. Thompson and H.J. Pabst, Radiat. Eff. **37**, 105 (1978).
15. R. Sizmann and C. Varelas, Festkörperprobleme **17**, 261 (1977).
16. H. Niehus, W. Heiland, and E. Taglauer, Surf. Sci. Rep. **17**, 213 (1993).
17. K. Kimura, M. Hasegawa, and M. Mannami, Phys. Rev. B**36**, 7 (1987).
18. H.B. Gilbody, Adv. At. Mol. Opt. Phys. **32**, 149 (1994).
19. C. Rau and R. Sizmann, Phys. Lett. 43A, 317 (1973).
20. H.J. Andrä, Phys. Lett. 54A, 315 (1975).
21. H.G. Berry, Rep. Prog. Phys. **40**, 156 (1977).
22. C. Lemell, J. Stöckl, J. Burgdörfer, G. Betz, HP. Winter, and F. Aumayr, Phys. Rev. Lett. **81**, 1965 (1998).
23. P. Roncin, J. Vilette, J.P. Atanas, and H. Khemliche, Phys. Rev. Lett. **83**, 864 (1999).
24. S. Lederer, K. Maass, D. Blauth, H. Winter, HP. Winter, and F. Aumayr, Phys. Rev. B **67**, 121405 (R) (2003).
25. J. Stöckl, T. Suta, F. Ditroi, HP. Winter, and F. Aumayr, Phys. Rev. Lett. **93**, 263201 (2004).
26. A. Mertens, K. Maass, S. Lederer, H. Winter, H. Eder, J. Stöckl, HP. Winter, F. Aumayr, J. Viefhaus, and U. Becker, Nucl. Instrum. Meth. B **182**, 23 (2001).
27. C. Auth and H. Winter, Phys. Lett. A**176**, 109 (1993).
28. S. Wüstenbecker, H. Ebbing, W.H. Schule, H. Baumeister, H.W. Becker, B. Cleff, C. Rolfs, H.P. Trautvetter, G.E. Mitchell, J.S. Schweitzer, and C.A. Peterson, Nucl. Instrum. Meth. A**279**, 448 (1989).
29. K. Kimura, G. Andou, and K. Nakajima, Phys. Rev. Lett. **81**, 5438 (1998).
30. V.A. Morozov, F.W. Meyer and P. Roncin, Phys. Scr., T**80**, **69**, 1999.
31. K.H. Krebs, Vacuum **33**, 7 (1983).
32. G. Lakits, F. Aumayr, and HP. Winter, Rev. Sci. Instrum. **60**, 3151 (1989).
33. F. Aumayr, G. Lakits, and HP. Winter, Appl. Surf. Sci. **63**, **17** (1991).
34. H. Kurz, F. Aumayr, C. Lemell, K. Töglhofer, and HP. Winter, Phys. Rev. A **48**, 2182 (1993).
35. A. Mertens and H. Winter, Phys. Rev. Lett. **85**, 2825 (2000).
36. H. Winter, S. Lederer, K. Maass, A. Mertens, F. Aumayr, and HP. Winter, J. Phys. B: At. Mol. Phys. **35**, 3315 (2002).
37. Y.N. Demkov, Sov. Phys. JETP **18**, 138 (1964).
38. C. Auth, A.G. Borisov, and H. Winter, Phys. Rev. Lett. **75**, 2292 (1995).
39. A.G. Borisov, V. Sidis, and H. Winter, Phys. Rev. Lett. **77**, 1893 (1996).
40. R.E. Johnson, *Atomic and Molecular Collisions*, Plenum press, New York 1982, p. 116.

41. A.G. Borisov and J.P. Gauyacq, Phys. Rev. B **62**, 4265 (2000).
42. S. Lederer, H. Winter, F. Aumayr, and HP. Winter, Europhys. J. D (2007), in press.
43. H. Winter and HP. Winter, Europhys. Lett. **62**, 739 (2003).
44. H. Winter, S. Lederer, and HP. Winter, Europhys. Lett. **75**, 964 (2006).
45. J.C. Riviere, in *Solid State Surface Science,* Vol. I, Marcel Dekker, New York 1969, p. 198.
46. A. Schüller, G. Adamov, S. Wethekam, K. Maass, A. Mertens, and H. Winter, Phys. Rev. A **69**, 050901 (2004).
47. Z.Y. Zhang, D.C. Lengreth, and J.P. Perdew, Phys. Rev. B **41**, 5674 (1990).
48. A. Kiejna, Phys. Rev. B **43**, 14695 (1991).
49. R. O. Jones and P. J. Jennings, Surf. Sci. Rep. **9**, 165 (1988).
50. R. Rioux, J. Chem. Edu. **76**, 156 (1999).
51. A. Dasgupta and A. Bathia, Phys. Rev. A **32**, 3335 (1985).
52. A. Kheifets, M. Vos, an E. Weigold, Z. Phys. Chem. **215**, 1323 (2001).
53. HP. Winter, S. Lederer, H. Winter, C. Lemell, and J. Burgdörfer, Phys. Rev. B **72**, 161402 (2005).
54. Y. Matulevich, S. Lederer, and H. Winter, Phys. Rev. B **71**, 033405 (2005).
55. J. Lörincik and Z. Sroubek, Nucl. Instr. Meth. B **164**, 633 (2000).
56. H. Winter, K. Maass, S. Lederer, HP. Winter, and F. Aumayr, Phys. Rev. B**69**, 054110 (2004).
57. R. Pfandzelter, T. Bernhard, and H. Winter, Phys. Rev. Lett. **90**, 036102 (2003).
58. K. Heinz, S. Müller, and P. Bayer, Surf. Sci. **352**, 942 (1996).
59. T. Bernhard, M. Baron, M. Gruyters, and H. Winter, Phys. Rev. Lett. **95**, 087601 (2005).
60. T. Flores, M. Hansen, and W. Wuttig, Surf. Sci. **279**, 251 (1992).
61. T. Bernhard and H. Winter, Phys. Rev. B **71**, 1(R) (2005).
62. A. Niehaus, P.A. Zeijlmans van Emmichoven, I.F. Urazgildin, and B. van Someren, Nucl. Instr. Meth. B **182**, 1 (2001).
63. R. Baragiola and C. Dukes, Phys. Rev. Lett. **76**, 2547 (1996).
64. N. Stolterfoht, D. Niemann, V. Hoffmann, M. Rösler, and R.A. Baragiola, Phys. Rev. A **61**, 052902 (2000).
65. H. Eder, F. Aumayr, P. Berlinger, H. Störi, and HP. Winter, Surf. Sci. **472**, 195 (2001).
66. B. van Someren, P.A. Zeijlmans van Emmichoven, I.F. Urazgildin, and A. Niehaus, Phys. Rev. A **61**, 032902 (2000).
67. T. Bernhard, Z.L. Fang, and H. Winter, Phys. Rev. A **69**, 060901(R) (2004).
68. M. Henzler, Surf. Rev. Lett. **4**, 489 (1997).

5

Spin Polarization of Electrons Emitted in the Neutralization of He^+ Ions in Solids

M. Alducin, J. I. Juaristi, R. Díez Muiño, M. Rösler, and P. M. Echenique

5.1 Introduction

In the last years, a large amount of experimental work has been devoted to study spin effects in the interaction of atomic particles with metal surfaces [1–22]. Apart of its fundamental interest there has been a growing use of different atom/ion–scattering techniques aimed to probe magnetic phenomena at surfaces and thin films. The singularity of ion-scattering experiments, as compared with methods based on electron beams or photons, lies on its extreme surface sensitivity: by a suitable choice of the projectile normal energy one can gain information on electronic properties specific to the region of the target topmost layer. This is typically achieved by using atoms with thermal energies [23] or grazing incidence ion-scattering [24, 25]. Information on magnetic properties is extracted by analyzing the *products* (electrons, photons) resulting from the projectile–surface interaction. Next, we provide a brief description of the most common of these techniques.

In *electron capture spectroscopy* (ECS), ions are scattered from a magnetized target and capture surface electrons into atomic states of the projectile. The captured electrons bring information on the spin polarization of the probed region. In early studies of Rau et al. [1, 3], the projectiles used were D^+ ions. During the interaction with the magnetic surface, the ions capture electrons from the substrate and neutralize. The formed D^0 atoms are next directed to a tritium target, inducing a nuclear reaction in which ^{4}He particles are emitted. The spin polarization of the captured electrons is derived from the analysis of these ^{4}He particles. In present ECS, electrons are captured into atomic excited states and the spin polarization is studied by analysing the polarized light emitted during the decay of the excited atom [5, 6, 10–12]. A further development of this technique is the multiple electron capture spectroscopy (MECS) [22], which uses the characteristic electron emission of hollow atoms [26] produced during the interaction of the highly charged projectiles with surfaces. The advantage of this new method is that the number

M. Alducin et al.: *Spin Polarization of Electrons Emitted in the Neutralization of He*$^+$ *Ions in Solids,* STMP **225**, 153–183 (2007)
DOI 10.1007/3-540-70789-1_5

of electrons removed from the surface well exceeds the initial charge state of the incoming ion. This feature allows a more efficient sampling of the surface electron density in contrast to the traditional singly ECS.

After the pioneering work conducted at Rice University [2, 7, 8], *spin polarized metastable-atom deexcitation spectroscopy* (SPMDS) has been established as a robust technique that can be adapted to explore different aspects related to surface and thin film magnetism: variations of the spin polarization of the surface electrons upon adsorbate and film deposition [8, 13, 18, 20], detection of out-of-plane magnetization [19], etc. In the SPMDS, metastable noble atoms (typically He^* in the 2^3S triplet state) with thermal kinetic energies deexcite to the ground state due to their interaction with the surface. Electrons emitted in this decaying process are then detected and analyzed. From a theoretical point of view the quantitative description of the deexcitation process is a difficult task since it may involve different charge-exchange mechanisms. A crucial point that determines which approximations are adequate to model a certain experiment is the distance at which the charge-exchange process is taking place. At sufficient large distances where the projectile-surface interaction is weak enough to neglect the induced perturbation, the charge-exchange process can be treated using the non-interacting projectile and target states. Under these conditions, the deexcitation mechanism is determined, in a naive picture, by the relation between the ionization potential of the metastable atom IP and the surface work function ϕ. If IP $< \phi$, the excited atomic electron tunnels into an unoccupied state of the target (resonant ionization RI), and next the He^+ ion is neutralized by an Auger process AN. In this process, an electron from the conduction band decays to the empty bound state of the ion, whereas another electron is excited and, if possible, ejected from the target. If IP $> \phi$, the RI step is forbidden and deexcitation occurs via an indirect Auger deexcitation process (AD) in which a continuum electron occupies the empty 1s state and the 2s electron is excited to the continuum.[1] However, the above picture is no longer valid for those experimental situations in which the neutralization occurs near the surface. The interaction is then so strong that the projectile and the target cannot be treated separately. The electronic states have to be calculated for the entire projectile-target system and the concept of a excited bound state associated only to the projectile, in resonance with the continuum states of the target is misleading. The mechanism taking place consists in that initially there is an empty bound state in the system and this state is filled by an electron from the continuum of the projectile-target system. The distinction between AN and AD processes is thus meaningless. This point will be discussed in more detail in Sect. 5.3.

Parallel to the first experiments performed by Onellion et al. [2] that established the basis for SPMDS, Kirschner et al. [4] used noble-gas ions to

[1] The different charge exchange processes that may take place under these circumstances (i.e., weak projectile-target perturbation) are schematically represented in Fig. 3.1

obtain a 'magnetic sputter depth profile' of the Fe(110) surface. Thenceforth, *spin-polarized electron emission induced by a variety of singly- and multi-charged ions* has been widely applied to investigate spin effects and/or magnetic properties [15–17, 21]. Under certain experimental conditions, both kinetic and potential emission[2] are observed [4, 15–17]. The coexistence of both may sometimes make difficult the analysis and interpretation of the measured data. Detailed descriptions of KE and PE can be found in the contribution by H. Winter for the former and the contribution by F. Aumayr and HP. Winter for the latter. The conditions and recent experimental advances achieved to separate KE of PE in the measured data are also discussed in those chapters.

In all these techniques based on electron emission analysis, low velocity He^+ ions and He* metastable atoms are among the most used projectiles [2, 7–9, 14, 18–21]. On the one hand, there is only potential emission and, on the other hand, the absence of resonant electron capture processes from the metal valence band to the 1s bound state of the incident particles simplify the charge-exchange process. Both features contribute to a better understanding of the experiments.

Spin effects in this kind of processes are studied by using magnetic surfaces or/and spin polarized incident projectiles. The measured magnitudes are typically the spin polarization of emitted electrons and the asymmetry parameter, which indicates how the total number of emitted electrons depends on the polarization of the incoming particle. Only in the case of magnetic surfaces, the asymmetry factor is a useful parameter. For pure paramagnetic substrates, it is zero.

Frequently, the information obtained in this kind of experiments is analyzed by paying attention uniquely to the characteristics of the Auger processes in front of the surface. In many cases, this represents an oversimplification of the experimental situation, if one wants to understand the measured spectra of emitted electrons. For instance, a positive asymmetry parameter is measured in the interaction of spin-polarized He atoms with magnetic surfaces [2, 8, 9, 18–20]. This means that the total number of emitted electrons is larger when the spin polarization of the projectile is parallel to the spin direction of the target majority electrons than when it is antiparallel. Thinking in terms of the Auger process, this can be understood by considering that the target minority-spin electrons predominate in the surface region. In such a case, one would expect that the spin polarization of emitted electrons were also of the minority kind: if there were more spin-minority electrons in the surface region where the Auger process takes place, it would be easier to excite (and emit) this kind of electrons. Nevertheless, experimental measurements show that the spin polarization of emitted electrons is always of the majority

[2] Induced electron emission is traditionally separated in kinetic electron emission (KE) and potential electron emission (PE) depending on whether the ejected electrons are excited by transferring kinetic or potential energy (i.e. by charge-exchange processes) from the incident projectile (see Chaps. 3 and 4).

kind [8]. This apparent contradiction shows that apart of the Auger process, further mechanisms, such as the creation of a cascade of secondary electrons due to electron–electron scattering processes at the surface, are necessarily involved.

Therefore, a proper characterization of the spectra of emitted electrons measured in this kind of experiments has to include all these basic ingredients: (i) the perturbation induced by the surface/projectile interaction, (ii) the description of the Auger neutralization (or deexcitation) process, which is the mechanism responsible for electron excitation, and (iii) the study of the transport mechanism, i.e., electron–electron scattering processes that modify the initial distribution of excited electrons and induce a cascade of secondaries. A theoretical analysis of all these processes for the case of magnetic surfaces represents a formidable task that, up to our knowledge, has not been fully addressed yet.

Spin effects have been also investigated in different paramagnetic surfaces. In this chapter it will be shown how also in this case the inclusion of the three mentioned ingredients turns out to be crucial in order to understand the spectra of emitted electrons.

There exist recent measurements of the spin polarization of emitted electrons, when spin-polarized He^+ ions are incident on paramagnetic surfaces [14, 21]. In these experiments energy distributions and energy resolved spin polarization of the electrons emitted during the neutralization process are measured for clean Al (001), Au (001), and Cu (001) surfaces. The incident ion energies are in the range 10–500 eV. In this energy regime, any contribution from kinetic electron emission can be disregarded and the electronic excitation can be strictly associated to the Auger neutralization of the He^+ projectile. The spin polarization of the total ejected electrons is ~30% in all the probed surfaces and is larger for those electrons emitted at the highest energies. Only very recently these measurements have been explained in quantitative terms [27].

An important characteristic of these experiments is related to the estimated effective distances at which the incident particles are neutralized. The neutralization distance, defined as the atom/surface separation at which the number of surviving He^+ ions is one half of the incident flux, can be estimated from the theoretical Auger neutralization rate values [28–30] and the experimental values of the perpendicular component of the incident ion velocity. In this way, it was concluded that neutralization should occur at ~2–3 a.u. (atomic units) from the topmost atomic layer, i.e., close to the jellium edge, which is located around 2 a.u. As explained above, at these distances the perturbation induced by the projectile is so strong that a simple description of the problem in which the projectile and the surface are treated separately is no longer justified. The ion/surface system has to be modeled as a whole. Under these conditions, the traditional picture that distinguishes between the AD and AN processes is hardly justified.

In this chapter we review recent theoretical studies [27, 31–35] of the electron emission induced by a He^+ ion embedded in a free electron gas

(FEG). Within this approximation, the model incorporates all the stages of the emission process (steps (i), (ii), and (iii)) and allows to analyze separately their contribution to the spin polarization of the electron spectra. This model has been successfully applied to explain in quantitative terms the main experimental findings of [14, 21]. The key point to understand the success of a model based on a bulk description of the problem is the experimental neutralization distances (2–3 a.u. from the topmost atomic layer). At these distances the relative electron density variations are small, and a local model based on a uniform background density should provide a meaningful description of the problem.

The outline of this chapter is as follows. Section 5.2 is devoted to a detailed analysis of the screening of a He^+ ion in the embedding medium. Since the He^+ ion constitutes a spin-polarized object, special attention will be paid to the spin polarization of the induced density screening cloud. This study is important at least for two reasons. First, previous studies have shown that for small atom surface separations the strong perturbation created in the target by the projectile has to be taken into account in a proper calculation of the Auger rates [29, 30, 36]. Second, as in this case the perturbation is spin-dependent, its influence on the polarization of emitted electrons must be analyzed. In this respect, some studies have tried to explain the spin polarization of the emission, uniquely in terms of this local "magnetization" of the target induced by the projectile [14].

The calculation of the Auger neutralization rates for excitation of electrons with spin parallel and antiparallel to that bound to the incoming ion are described in Sect. 5.3. In these processes, the Coulomb interaction between the two electrons involved induce the decay of one of these electrons to the unoccupied bound state of the projectile and the promotion of the other to an excited continuum state. As a consequence, the Auger rates depend on the relative spin orientation of the two participating electrons. Since the electron to be captured has always spin antiparallel to that bound to the projectile, the spin of the two electrons participating in the Auger process is the same (opposite) for excitation of electrons with spin antiparallel (parallel) to that bound to the projectile. Hence, this fact will also contribute to the spin polarization of the emission.

As explained above, a proper characterization of the emitted electron spectra needs to include the modification of the initial distribution of excited electrons due to electron–electron scattering events. Within the bulk model reviewed here, this effect is studied in terms of the Boltzmann transport equation. Section 5.4 is devoted to a detailed analysis of this approach to calculate the electron transport. Attention is paid to the description of the main ingredients of the model, and to how the spin polarization of the emission is modified by this effect. The connection between the bulk calculations and the experiments in paramagnetic surfaces will be presented in Sect. 5.5. Finally, the limitations and lines for future improvements are discussed in Sect. 5.6.

The section ends up with some comments on the possible extensions to ferromagnetic surfaces and the new difficulties that may arise in this case.

Atomic units (a.u.) will be used unless it is otherwise stated.

5.2 Spin-Dependent Screening of a He$^+$ Ion in a Free Electron Gas

The study of the screening of a point impurity in an electron gas plays a central role in the characterization of the interaction of probe-particles with metal targets. The first approaches to this problem were based on linear response theory, and studied the effect of improving the description of the dielectric function [37, 38]. Nevertheless, it was soon realized that a static real charge represents a strong perturbation for the system, and that nonlinear approaches are necessary in order to properly characterize the rearrangement of the electron density it induces [39, 40]. In this respect, density functional theory (DFT) within the local density approximation (LDA) has been successfully used to describe the nonlinear screening of heavy impurities in metals [41–43].

For spin-dependent properties, the local spin density (LSD) approximation is needed [44]. The LSD approximation includes electronic exchange and correlation effects through approximate functionals, keeping the simplicity of a one-particle equation with a local potential. The starting point are Kohn–Sham (KS) equations [45]:

$$\left\{-\frac{1}{2}\nabla^2 + v^j_{\text{eff}}(\mathbf{r})\right\}\varphi^j_i(\mathbf{r}) = \varepsilon^j_i\varphi^j_i(\mathbf{r}) \;, \tag{5.1}$$

where $\varphi^j_i(\mathbf{r})$ and ε^j_i are the KS wave functions and eigenvalues respectively. The index j runs over the two spin components $\uparrow$ and $\downarrow$. KS equations are used to obtain in a self-consistent manner the electron density of the system $n(\mathbf{r})$:

$$n(\mathbf{r}) = \sum_{j=\uparrow,\downarrow}\sum_{i\in occ.}\left|\varphi^j_i(\mathbf{r})\right|^2 , \tag{5.2}$$

that can be separated in its continuum ($n_c(\mathbf{r})$) and bound ($n_b(\mathbf{r})$) components. The electron density for just spin-up (spin-down) electrons $n^\uparrow(\mathbf{r})$ ($n^\downarrow(\mathbf{r})$) can be defined in a similar way by limiting the sum over occupied states to the required spin component.

For the specific case of a static He$^+$ ion embedded in a paramagnetic FEG, the effective KS potential $v^j_{\text{eff}}(\mathbf{r})$ is

$$v^j_{\text{eff}}(\mathbf{r}) = -\frac{2}{r} + \int\frac{d\mathbf{r}'[n(\mathbf{r}') - n_0]}{|\mathbf{r}-\mathbf{r}'|} + v^j_{\text{xc}}[n(\mathbf{r}),\zeta(\mathbf{r})] \;. \tag{5.3}$$

The first term is the external Coulomb potential created by the He core, the second term is the electrostatic potential made by the induced density, and the third term is the change in the exchange-correlation potential $\mu^j_{\text{xc}}[n(\mathbf{r}), \zeta(\mathbf{r})]$ caused by the impurity,

$$v^j_{\text{xc}}[n(\mathbf{r}), \zeta(\mathbf{r})] = \mu^j_{\text{xc}}[n(\mathbf{r}), \zeta(\mathbf{r})] - \mu^j_{\text{xc}}[n_0] \tag{5.4}$$

The results shown in this review are calculated using the LSD parametrization of [44]. The potential $v^j_{\text{xc}}[n(\mathbf{r}), \zeta(\mathbf{r})]$ is different for each spin component j. It is a functional of both the local density $n(\mathbf{r})$ and the local spin polarization $\zeta(\mathbf{r})$, defined as

$$\zeta(\mathbf{r}) = \frac{[n^{\uparrow}(\mathbf{r}) - n^{\downarrow}(\mathbf{r})]}{[n^{\uparrow}(\mathbf{r}) + n^{\downarrow}(\mathbf{r})]} \,. \tag{5.5}$$

The electron density of the unperturbed FEG is denoted by n_0, and the customary one-electron radius r_s is defined from $1/n_0 = (4/3)\pi r_s^3$.

The KS wave functions $\varphi^j_i(\mathbf{r})$ are calculated numerically after expansion in the spherical harmonics basis set, and the set of KS equations are solved self-consistently using an iterative procedure. The He^+ ion, in which there is only one bound electron, is modeled by populating just one of the two bound KS 1*s*-states of the system (there is one for each spin orientation). By this procedure, the spin of the electron bound to the projectile is fixed. As a convention, let us denote in the following the spin orientation of the electron bound to the He^+ ion as up ($\uparrow$) orientation, i.e., $n^{\downarrow}_b(\mathbf{r}) = 0$ whereas $n^{\uparrow}_b(\mathbf{r})$ integrates to one electron. Therefore, $n^{\uparrow}(\mathbf{r})$ is different from $n^{\downarrow}(\mathbf{r})$ producing a local spin polarization $\zeta(\mathbf{r})$ different from zero. As a consequence, the KS effective potential $v^j_{\text{eff}}(\mathbf{r})$ is different for each spin orientation due to the spin dependent exchange-correlation terms $v^j_{\text{xc}}[n(\mathbf{r}), \zeta(\mathbf{r})]$. This implies that also the continuum part of the screening charge, $\Delta n^j_c(\mathbf{r}) = n^j_c(\mathbf{r}) - n_0/2$, is different for the two spins. Note that $\Delta n_c(\mathbf{r}) = \Delta n^{\uparrow}_c(\mathbf{r}) + \Delta n^{\downarrow}_c(\mathbf{r})$ integrates to the unit charge, providing complete screening of the He^+ ion. The important point is that this screening cloud is not spin compensated. Let us remark that the origin of this local magnetization of the FEG around the He^+ ion is the spin-dependent perturbation that the He^+ ion represents, and that this spin-dependent perturbation is introduced in the formalism by populating uniquely one of the KS bound states.

An important approximation made in this model refers to the description of the He^+ ion. The population of just one of he bound KS states implies that the KS wave functions are being used in an approximate way as monoelectronic wave functions. This approach has been widely used and successfully applied to the study of core-electron photoemission [46–48], calculations of core-hole Auger widths [49, 50], the neutralization via Auger processes of multicharged ions in metals [51–53], and the charge state dependent energy loss [54–56] and induced kinetic electron emission when slow multicharged ions are traveling through metals [57, 58]. Though DFT is only strictly valid for the ground state of each symmetry, and hence, there is not a rigorous theoretical justification to treat excited states with empty core holes within this scheme, it has been claimed that as much as the core-hole can be treated as

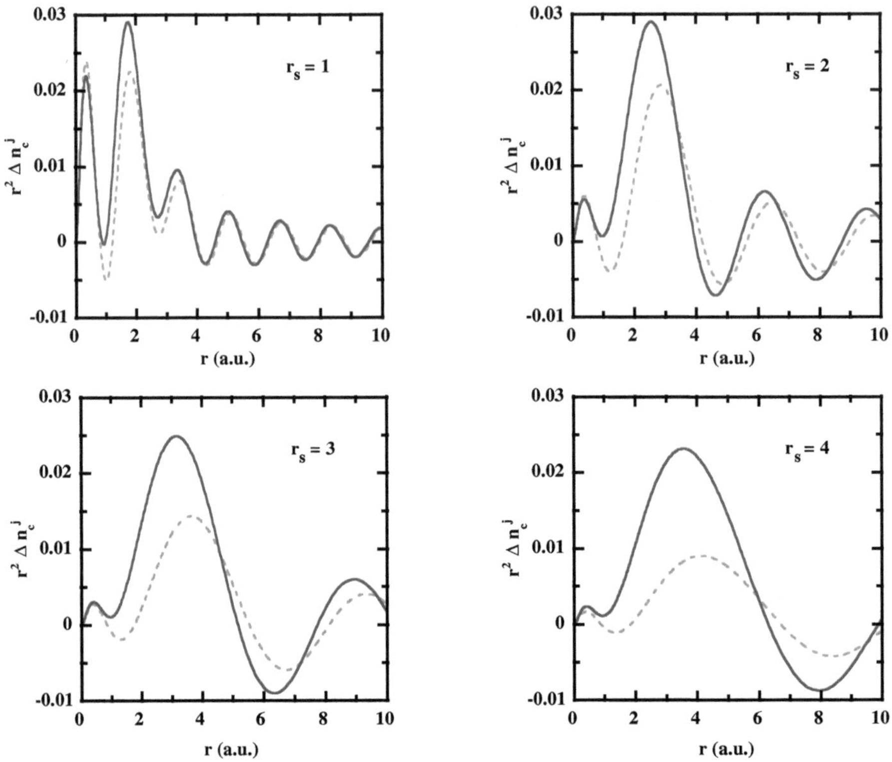

Fig. 5.1. Radial electron density $r^2 \Delta n_c^j(r)$ induced in the continuum by a He^+ ion embedded in a free electron gas. Different panels correspond to different medium electron densities. The He^+ ion with a spin-up electron is at the origin $r = 0$. The induced density with spin parallel to the bound electron $\Delta n_c^{\uparrow}$ is shown by *solid lines*, and that with spin antiparallel $\Delta n_c^{\downarrow}$ by *dashed lines*

an external potential the method should be physically sound [46]. He^+ inside an electron gas at metallic densities ($r_s = 1.5 - 6$) is a case in which this approach is justified.

Figure 5.1 shows the radial electron density induced in the continuum $r^2 \Delta n_c^j(r)$ by the spin-polarized He^+ ion embedded in a paramagnetic FEG. Different panels correspond to different metal-electron densities. The He^+ ion with a spin-up bound electron in its 1s state is at the origin, $r = 0$. In order to illustrate the effect of the spin-dependent perturbation in the medium, the induced electron density with spin parallel ($\Delta n_c^{\uparrow}(r)$) and antiparallel ($\Delta n_c^{\downarrow}(r)$) to that of the bound electron are shown separately. Clearly from this figure, the piling-up of electrons around the He^+ ion is a spin-dependent phenomenon, particularly for intermediate and low metallic densities ($r_s \geq 2$). We observe that close to the bound electron, the screening is preferable due to electrons of parallel spin ($\Delta n_c^{\uparrow}(r)$). This behavior is simply a manifestation of the Coulomb interaction and the Pauli principle, in other words, of *exchange*. Since the 1s state and the continuum states are well separated in energy and space, we focus on the effective electron–electron interaction to explain this effect. Thinking in terms of the mean-field formalism, one can treat short-range electron correlations by introducing local-field corrections to the bare Coulomb interaction between electrons. In a paramagnetic FEG, these local-field corrections depend only on the relative spin of electrons. Therefore, one should distinguish between two different kinds of effective electron–electron interactions: $V_{\uparrow\uparrow}$, the interaction between electrons of parallel spin, and $V_{\uparrow\downarrow}$, the interaction between electrons of opposite spin. For parallel spin electrons, both the Pauli principle and Coulomb interaction contribute to the creation of a exchange-correlation hole, reducing the effectiveness of the short-range part of the Coulomb interaction. For antiparallel spin electrons, however, only Coulomb correlations contribute to the creation of the hole, and therefore, the interaction is not so much reduced. As a result, the exchange is typically characterized by a weaker repulsive interaction between electrons of parallel spin, i.e., $0< V_{\uparrow\uparrow} < V_{\uparrow\downarrow}$. This allows us to understand the effect observed in Fig. 5.1. In DFT the correction to the bare Coulomb interaction is incorporated by the $v^j_{xc}(\mathbf{r})$ potential. In this respect, Gunnarsson et al. already showed the validity of the LSD to reproduce the Hund's first rule for an open-shell atom [44]. For instance, in case of a metastable He* atom this means that the lowest excited state is the triplet 2^3S instead of the singlet 2^1S. Here, our results indicate that the metal electrons participating in the screening of the spin-polarized He^+ ion also follow a kind of Hund's first rule that favours the alignment of electron spins, provided the Pauli principle is not violated.

In this respect, it is informative to focus in the integrated induced density for each spin direction ($j =\uparrow, \downarrow$),

$$Q^j_c = 4\pi \int_0^\infty dr\, r^2\, \Delta n^j_c(r)\ . \tag{5.6}$$

Note that $Q^\uparrow_c + Q^\downarrow_c = 1$. The values of $Q^\uparrow_c$ for different values of r_s are given in Table 5.1. At high densities, ($r_s = 1$), the screening is almost equally shared by the spin-up and spin-down electrons. As the density decreases, the screening charge is more spin polarized, and it is almost completely dominated by $Q^\uparrow_c$ for small densities such as $r_s = 5$.

Table 5.1. Contribution of $Q^\uparrow_c$ to the screening charge around a He^+ ion for different values of the medium electron density. The electron bound to the He^+ ion is spin-up

r_s	1	2	3	4	5
$Q^\uparrow_c$	0.57	0.61	0.66	0.75	0.90

The increasing importance of the spin polarization of the induced density, as the FEG density decreases, is a consequence of the dependence of the exchange-correlation energy on the FEG density [59]. At high metallic densities, exchange-correlation effects are small because the contribution of the kinetic energy term to the total energy of the system dominates over the interaction terms. As the electron density decreases, the potential energy becomes comparable to the kinetic one, and the exchange effects that favour the alignment of electron spins are important.

Another quantity of interest that gives information on the characteristics of the screening is the density of states in momentum space induced by the impurity in the continuum $\Delta\rho(k)$. The density of states induced in the continuum for each spin state ($j = \uparrow, \downarrow$) is calculated as follows [53, 60]:

$$\Delta\rho^{j}(k) = \frac{1}{\pi}\sum_{l}(2l+1)\,\frac{d}{dk}\delta_{l}^{j}(k) \tag{5.7}$$

where k is the electron momentum ($\epsilon = k^2/2$), l is the angular momentum in a partial-wave expansion, and $\delta_l^j(k)$ are the phase shifts of the KS radial wave functions. Figure 5.2 shows $\Delta\rho^{\uparrow}(k)$ for different electron gas densities. For high densities, the spin-up and spin-down bands show quite similar dependence on k. In the region of unoccupied states ($k > k_F$), the difference between $\Delta\rho^{\uparrow}(k)$ and $\Delta\rho^{\downarrow}(k)$ is small and keeps an almost constant value. As it is expected, the spin dependence of $\Delta\rho(k)$ is stronger for low densities. The k-structure of both spin bands are different. A resonance structure peak appears at low densities and is dominated by the spin-up component. This resonance is originated by the coupling of the He^{+} atomic states and the FEG continuum. When the resonance has a well defined partial wave symmetry it can be associated with a single electronic state in the gas-phase atom. However, in the case of He^{+}, there is not a clear connection between the resonance and one single atomic state: The resonance is basically composed of both s ($l = 0$) and p ($l = 1$) scattering states and it cannot be assigned to a single partial-wave symmetry [60]. This shows that the excited states are modified due to the interaction of the projectile with the metal electrons. Such a perturbation is expected to be also important in the surface region where the electron density is in the range of values analyzed here. Therefore, the use of a simplified atomic approach to describe the excited states of the He^{+} ion in close proximity to the target surface is unrealistic and misleading. Notice also that the quantitative difference between $\Delta\rho^{\uparrow}(k)$ and $\Delta\rho^{\downarrow}(k)$ is remarkably strong in the region of occupied states at low densities, in agreement with the high polarization of the induced density. In the unoccupied region the difference between both spin-bands is smaller.

The density of states are compared with those obtained within LDA, i.e., with a non-spin dependent local density approximation for the exchange correlation potential. The total density of states is almost unaffected by the spin-polarized nature of the perturbation at high densities, even though

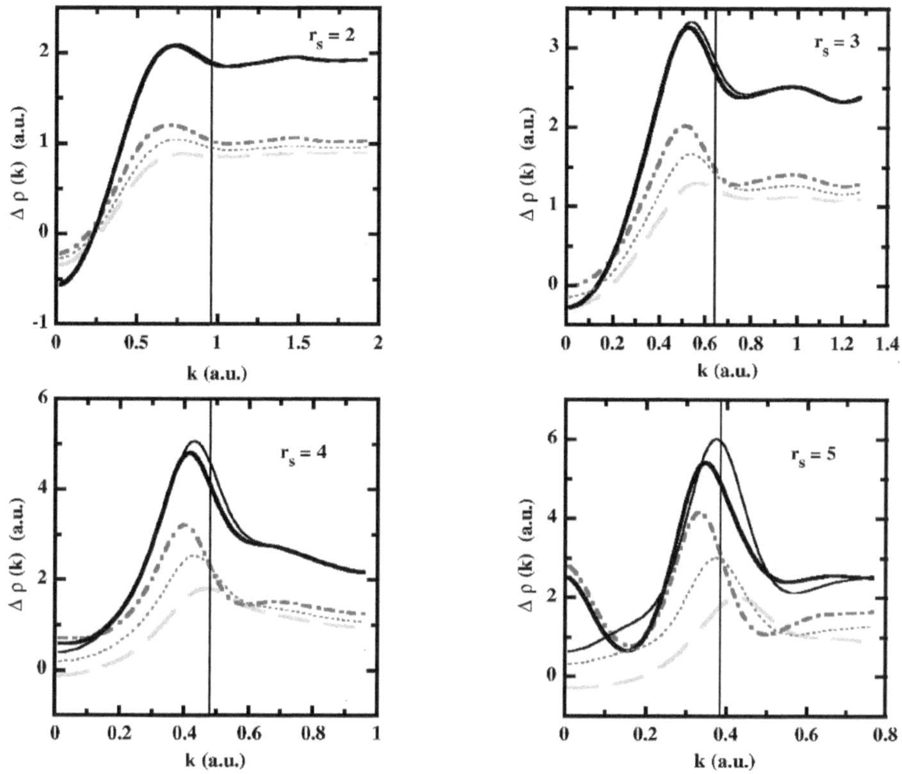

Fig. 5.2. Density of states induced in the continuum of a free electron gas by a He^+ ion as a function of the electron momentum k. The corresponding value of the Fermi wavevector k_F is indicated by a *vertical line* in each panel. *Thick lines* correspond to the results obtained within the LSD approximation: total $\Delta\rho(k)$ are shown by thick *solid lines*, $\Delta\rho^{\uparrow}(k)$ by *dash-dotted lines*, $\Delta\rho^{\downarrow}(k)$ by long *dashed lines*. *Thin lines* represent the results obtained by the LDA prescription: total $\Delta\rho(k)$ with thin *solid lines* and partial $\Delta\rho^{\uparrow}(k) = \Delta\rho^{\downarrow}(k)$ by *dotted lines*

the screening is not equally shared between both spins. For lower densities, the modification of $\Delta\rho(k)$ when including spin-effects is slight, and it is mostly observed in the small shift of the resonance peak to lower k values when the local spin polarization is taken into account.

As a main result of this section, we want to stress the fact that the He^+ ion induces a strong rearrangement of charge in the electron gas, and that the induced screening charge is spin unbalanced. More precisely, the screening is preferably due to electrons with spin parallel to that bound to the projectile, and the effect is more important at low densities. In connection to surface experiments, this fact should be important for processes that take place in the vicinity of the metal target (nearby the jellium edge) where the density is reduced from the bulk value but it is still significative.

5.3 Auger Neutralization of He^+ in Metals

Since the pioneering work of Hagstrum [61], a large amount of theoretical investigations has been devoted to the study of the Auger processes that take place in the neutralization of He^+ ions interacting with surfaces and to the calculation of the corresponding neutralization-deexcitation rates [28, 30, 62–68]. These works analyze different aspects of the Auger process for the case of unpolarized projectiles and nonmagnetic surfaces. As a consequence, a deep knowledge has been achieved about the value of these rates at large atom-surface separations, but difficulties arise at typical physisorption distances due to the strong perturbation represented by the projectile. This problem was analyzed in great detail in [30] for unpolarized projectiles, i.e., without including any spin-dependent effect. Spin-dependence in the Auger neutralization rates was included in [69] for the case of a spin-polarized metastable He^* atom in front of the surface. In this work, the influence in the Auger rate of the exchange process due to the indistinguishability of electrons with identical spin was analyzed. Nevertheless, the Auger rate was calculated without including the spin-dependent perturbation induced by the projectile that it has been analyzed in previous section. More recently, the deexcitation of a metastable 2^3S He^* in front of a Na surface [36] has been studied by including both the spin-dependent perturbation induced by the projectile, and the exchange term in the calculation of the Auger rate. The results show a strong spin polarization of the Auger rates. The applicability of this model is limited to projectile distances well above the jellium edge. It cannot be used to explain experiments with neutralization distances close to the surface, such as those of [14, 21].

In this section, we present a study of the Auger neutralization of a He^+ ion embedded in a FEG. The Auger process is due to the Coulomb interaction between two electrons in the excited states of the ion-target system. One of the electrons decays to the empty 1s bound state of the He^+ ion, whereas the second electron is promoted to the continuum. In ion–surface interaction problems it is customary to distinguish between the Auger neutralization and Auger deexcitation. This distinction has sense in situations in which the complete ion-target system can be described as two separated subsystems. In this way, one distinguishes between electronic states of the metal surface and electronic states of the projectile. Consequently, when the projectile is charged and the two involved electrons are initially in the target the resulting process is called Auger neutralization (AN), and when one of the two electrons is initially in the projectile, Auger deexcitation (AD).[3] The separation of the total system in two subsystems is reasonable and meaningful at large ion-metal separations, when the projectile-surface interaction is weak. Nevertheless, this is not true at close distances from the surface, when the projectile wave functions

[3] The theoretical description of the AN at surfaces, using the dielectric formalism can be found in Chap. 6.

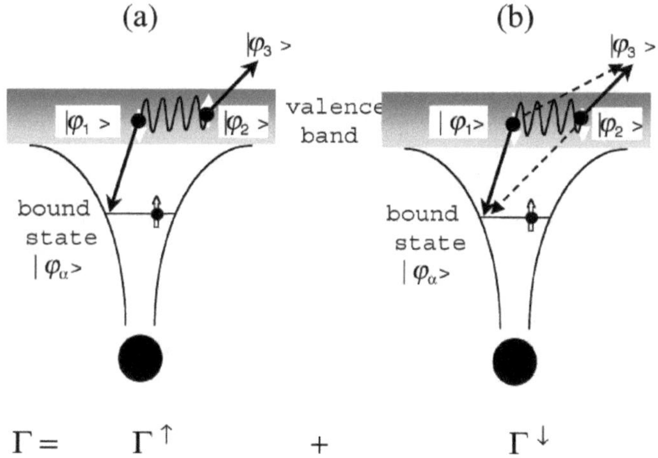

Fig. 5.3. Schematic representation of the Auger capture process: an electron from the valence band decays to the unoccupied 1s state of the He^+ ion, whereas a second electron is excited to an unoccupied continuum state. The Auger process can be viewed as the sum of two channels: (**a**) $\Gamma^{\uparrow}$, in which the excited electron has a spin parallel to that of the electron bound to the He^+ ion and (**b**) $\Gamma^{\downarrow}$, in which the spin of the excited electron is antiparallel. The indistinguishability of electrons gives rise in the latter to the two processes indicated by *solid* and by *dashed arrows*

start to overlap with the metal ones. In such cases, the electronic states belong to the entire projectile-target system. Therefore, it has no meaning to distinguish between AN and AD. One may consider that, actually, both processes are being considered by the bulk model described here, without making any distinction between them.

The Auger process is schematically shown in Fig. 5.3. Two different channels contribute to the total Auger capture (AC) probability, Γ: (a) capture of a spin-down electron and emission of a spin-up electron, $\Gamma^{\uparrow}$ and (b) capture and emission of spin-down electrons, $\Gamma^{\downarrow}$.

The AC probability can be calculated in first-order perturbation theory. The probability of AC per unit time involving the excitation of an electron of spin orientation parallel to the bound electron ($\Gamma^{\uparrow}$) is [69, 70]:

$$\Gamma^{\uparrow} = 2\pi \sum_{\varphi_1^{\downarrow} \in occ.} \sum_{\varphi_2^{\uparrow} \in occ.} \sum_{\varphi_3^{\uparrow} \notin occ.} \left| \int d\mathbf{r} d\mathbf{r}' \, [\varphi_{\alpha}^{\downarrow}(\mathbf{r})]^* [\varphi_3^{\uparrow}(\mathbf{r}')]^* v(\mathbf{r},\mathbf{r}') \varphi_2^{\uparrow}(\mathbf{r}') \varphi_1^{\downarrow}(\mathbf{r}) \right|^2 \times \delta\left(\varepsilon_1^{\downarrow} + \varepsilon_2^{\uparrow} - \varepsilon_{\alpha}^{\downarrow} - \varepsilon_3^{\uparrow}\right) , \tag{5.8}$$

where $v(\mathbf{r},\mathbf{r}') = 1/|\mathbf{r}-\mathbf{r}'|$ is the Coulomb potential, responsible for the decay. The wave functions of the electrons involved in the transition are approximated by the KS wave functions $\varphi_1^\downarrow(\mathbf{r})$ and $\varphi_2^\uparrow(\mathbf{r})$ of the He$^+$/FEG system, with KS eigenvalues $\varepsilon_1^\downarrow$ and $\varepsilon_2^\uparrow$ respectively. The wave function of the captured electron in the final state is approximated by the KS wave function of the unoccupied bound state $\varphi_\alpha^\downarrow(\mathbf{r})$ with eigenvalue $\varepsilon_\alpha^\downarrow$. The wave function of the unoccupied state in the continuum is also approximated by the KS wave function $\varphi_3^\uparrow(\mathbf{r})$ with eigenvalue $\varepsilon_3^\uparrow$.

The Auger probability per unit time when the spin orientation of the excited electron is antiparallel to the bound electron $\Gamma^\downarrow$ needs special care. In this case, the captured electron and the excited electron have the same spin orientation and are thus indistinguishable particles. Hence, the spatial part of the wave function of the two-electron states involved in the transition must be antisymmetric, and $\Gamma^\downarrow$ can be written as [69, 70],

$$\begin{aligned}\Gamma^\downarrow = 2\pi &\sum_{\varphi_1^\downarrow\in occ.}\ \sum_{\varphi_2^\downarrow\in occ.}\ \sum_{\varphi_3^\downarrow\notin occ.} \\ &\frac{1}{2}\Bigg|\int d\mathbf{r}d\mathbf{r}'[\varphi_\alpha^\downarrow(\mathbf{r})]^*\left[\varphi_3^\downarrow(\mathbf{r}')\right]^* v(\mathbf{r},\mathbf{r}')\varphi_2^\downarrow(\mathbf{r}')\varphi_1^\downarrow(\mathbf{r}) \\ &-\int d\mathbf{r}d\mathbf{r}'[\varphi_\alpha^\downarrow(\mathbf{r})]^*\left[\varphi_3^\downarrow(\mathbf{r}')\right]^* v(\mathbf{r},\mathbf{r}')\varphi_1^\downarrow(\mathbf{r}')\varphi_2^\downarrow(\mathbf{r})\Bigg|^2 \\ &\times\delta\left(\varepsilon_1^\downarrow+\varepsilon_2^\downarrow-\varepsilon_\alpha^\downarrow-\varepsilon_3^\downarrow\right)\,. \end{aligned} \tag{5.9}$$

The first double-integral describes the process indicated in Fig. 5.3 (b) by solid arrows. The last integral corresponds to the indistinguishable process represented by dashed arrows. Expanding the squared term, $\Gamma^\downarrow$ can be also written as,

$$\begin{aligned}\Gamma^\downarrow &= \Gamma_0^\downarrow - \Gamma_{int}^\downarrow \\ &= 2\pi \sum_{\varphi_1^\downarrow\in occ.}\ \sum_{\varphi_2^\downarrow\in occ.}\ \sum_{\varphi_3^\downarrow\notin occ.} \\ &\quad\left|\int d\mathbf{r}d\mathbf{r}'\ [\varphi_\alpha^\downarrow(\mathbf{r})]^*\left[\varphi_3^\downarrow(\mathbf{r}')\right]^* v(\mathbf{r},\mathbf{r}')\varphi_2^\downarrow(\mathbf{r}')\varphi_1^\downarrow(\mathbf{r})\right|^2 \delta\left(\varepsilon_1^\downarrow+\varepsilon_2^\downarrow-\varepsilon_\alpha^\downarrow-\varepsilon_3^\downarrow\right) \\ &\quad -2\pi \sum_{\varphi_1^\downarrow\in occ.}\ \sum_{\varphi_2^\downarrow\in occ.}\ \sum_{\varphi_3^\downarrow\notin occ.} \\ &\quad \mathrm{Re}\left\{\int d\mathbf{r}d\mathbf{r}'[\varphi_\alpha^\downarrow(\mathbf{r})]^*\left[\varphi_3^\downarrow(\mathbf{r}')\right]^* v(\mathbf{r},\mathbf{r}')\varphi_2^\downarrow(\mathbf{r}')\varphi_1^\downarrow(\mathbf{r})\right. \\ &\quad\times\left.\int d\mathbf{r}d\mathbf{r}'\varphi_\alpha^\downarrow(\mathbf{r})\varphi_3^\downarrow(\mathbf{r}')v(\mathbf{r},\mathbf{r}')[\varphi_1^\downarrow(\mathbf{r}')]^*\left[\varphi_2^\downarrow(\mathbf{r})\right]^*\delta\left(\varepsilon_1^\downarrow+\varepsilon_2^\downarrow-\varepsilon_\alpha^\downarrow-\varepsilon_3^\downarrow\right)\right\}, \end{aligned} \tag{5.10}$$

where $\Gamma_0^{\downarrow}$ corresponds to the first term, the one equivalent to the expression of $\Gamma^{\uparrow}$, and $\Gamma_{\mathrm{int}}^{\downarrow}$ is the second term preceded by the minus sign. The indistinguishability of the electrons gives rise to this interference term $\Gamma_{int}^{\downarrow}$, absent in the AC rate $\Gamma^{\uparrow}$. This term comes from the fact that the Auger rate is ruled by the Coulomb interaction between electrons, $v(\mathbf{r},\mathbf{r}') = 1/|\mathbf{r}-\mathbf{r}'|$, which is a two-body operator. Therefore, it depends on the two-body density $n(\mathbf{r},\mathbf{r}')$, i.e., the probability of finding a pair of electrons at the positions $\mathbf{r}$ and $\mathbf{r}'$. In $\Gamma^{\downarrow}$, two spin-down electrons participate: the one that is excited and the one that is captured and neutralizes the He^+ ion. Due to the exchange hole that surrounds an electron in the conduction band, the probability of finding two electrons with the same spin close to each other (e.g. the two spin-down electrons in the calculation of $\Gamma^{\downarrow}$) is reduced compared to the case in which the two electrons have different spin. In other words, for small values of $|\mathbf{r}-\mathbf{r}'|$, one has $n_{\downarrow\downarrow}(r,r') < n_{\downarrow\uparrow}(r,r')$. This fact is incorporated in the interference term.

Figure 5.4 represents the value of the Auger capture rate as a function of r_s. The contributions to the total rate coming exclusively from the excitation of electrons with spin-up ($\Gamma^{\uparrow}$) and spin-down ($\Gamma^{\downarrow}$) are shown separately. These results are compared with those obtained without including the interference term, i.e., the partial rate $\Gamma_0^{\downarrow}$ and the corresponding total rate $\Gamma_0 = \Gamma^{\uparrow} + \Gamma_0^{\downarrow}$.

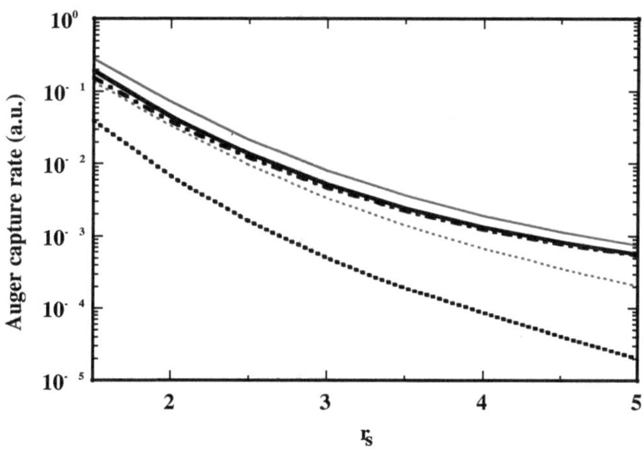

Fig. 5.4. r_s dependence of the Auger capture rate undergone by a He^+ ion. The *thick dash-dotted* (*thick dotted*) lines represent the AC rate in which a spin-up (spin-down) electrons is excited: $\Gamma^{\uparrow}(\Gamma^{\downarrow})$. The sum of both contributions is shown by a thick *solid line*. For comparison, we also show the results obtained when the interference term is not included. In this case, the total AC rate Γ_0 is indicated by a *thin solid line* and the partial rate $\Gamma_0^{\downarrow}$ by a *thin dotted line*

The large difference between $\Gamma^{\uparrow}$ and $\Gamma^{\downarrow}$ arises from the contribution of two ingredients. On the one hand, due to the spin-dependent screening explained in the previous section, it is easier to excite spin-up electrons, because the probability of finding them around the He^{+} ion is larger ($\Delta n_c^{\uparrow} > \Delta n_c^{\downarrow}$ close to the ion). The contribution of this effect can be obtained from the comparison of $\Gamma^{\uparrow}$ and $\Gamma_0^{\downarrow}$. In this case, the difference between the values obtained comes uniquely from the different wave functions entering the matrix elements of $\Gamma^{\uparrow}$ and $\Gamma_0^{\downarrow}$. As can be deduced from the results of the previous section, this effect is important at low electron densities for which the screening is strongly spin-dependent. At high densities the effect is much reduced. On the other hand, a further reduction of the value of $\Gamma^{\downarrow}$ is due to the interference term $\Gamma_{int}^{\downarrow}$, which accounts for the indistinguishability of electrons. A comparison between $\Gamma^{\downarrow}$ and $\Gamma_0^{\downarrow}$ shows that the contribution of this effect to the spin dependence of the AC rate is important over all the range of electron densities considered.

In order to quantify the spin dependence of the Auger process and the influence of the metal electron density on it, we define the spin polarization of the excitation ξ_{AC} by the following expression:

$$\xi_{AC} = \frac{\Gamma^{\uparrow} - \Gamma^{\downarrow}}{\Gamma^{\uparrow} + \Gamma^{\downarrow}} . \tag{5.11}$$

This quantity is related to the average spin polarization of the electrons excited during the Auger capture process. The dependence of ξ_{AC} on r_s is represented in Fig. 5.5. The curves shown correspond to three different calculations. Thick solid line represents the results obtained using LSD to calculate the screening

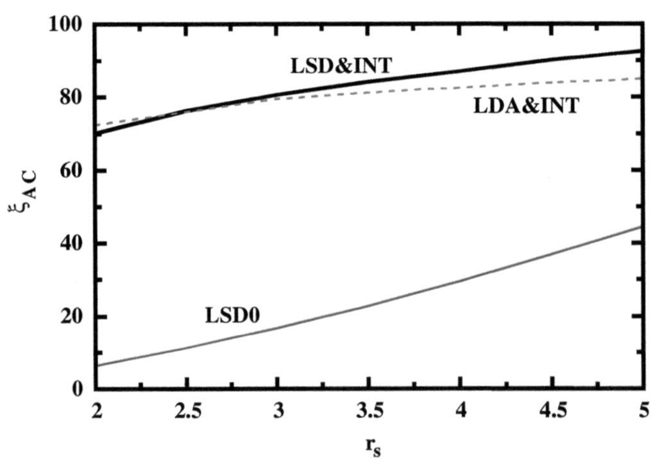

Fig. 5.5. Spin polarization of the Auger capture rate ξ_{AC}, defined in (5.11), as a function of the parameter r_s. The thick *solid line* indicates the results obtained within the LSD and including the interference term, thin *dashed line* the results obtained within the LDA and including the interference term, and thin *solid line* the results obtained within the LSD without including the interference term

of the He^+ ion and including the interference term in the calculation of $\Gamma^{\downarrow}$ (we will denote it by LSD&INT). Dashed line is obtained including the interference term but calculating the screening within the LDA (LDA&INT). In this case, the spin dependence of the screening is neglected and all the polarization comes from the interference term. Finally, thin solid line shows the result of neglecting the interference term in $\Gamma^{\downarrow}$ when the screening is calculated with the LSD prescription (LSD0). Hence, this curve shows the contribution to ξ_{AC} coming exclusively from the spin dependence of the screening.

According to the LSD&INT calculation, the spin polarization of the excited Auger electrons is very large (70%–90% in the range r_s = 2–5 a.u.). A comparative analysis of the different curves allows us to deduce the relative importance of these two effects in the spin polarization of the Auger process. By comparing LSD&INT to LSD0, we observe that the interference term plays the predominant role over all the range of densities. The effect due to the spin-dependent character of the screening is always smaller, although it gains relative importance at low densities (compare LSD&INT to LDA&INT).

It is noteworthy to stress that the spin polarization of the excited electrons is very high (70%–90% as a function of the FEG density) and that the main reason is the indistinguishability of electrons (interference term in $\Gamma^{\downarrow}$). The large polarization of the Auger excited electrons is not related to the bulk nature of the treatment: surface calculations of the Auger rates that include both the spin-dependent perturbation induced by the ion and the interference term in the calculation of the Auger rates give similar high values of the spin polarization [36]. However, the theoretical spin polarization is much higher than the one measured in the emitted electrons [14, 21]. As it is shown in next section, the role played by electron–electron scattering processes is crucial to interpret this kind of experiments even for paramagnetic surfaces. The usual tendency to compare directly the properties of the Auger excited electrons with the measured emitted electrons can be misleading, since gives an incomplete account of the full emission process. Furthermore, note that any theoretical model aimed to describe the spin dependence of the Auger process should include the interference term in the calculation of $\Gamma^{\downarrow}$. The existence of such a term is based on the fundamental indistinguishability of electrons. We stress this point because, paradoxically, theoretical calculations that neglect it may give a better agreement with the measured spin polarization [14, 31]. Actually, this is observed in the bulk LSD0 calculation discussed above. The LSD0 results are close to the experimental data in the r_s = 3–4 range, which is a range of reasonable densities in the surface region (see Fig. 5.5). However, since there is not any justification to neglect the interference term [27, 36, 69], this kind of approaches are underestimating the spin polarization of the excitation induced by the Auger process. Therefore, they cannot provide a correct interpretation of the experimental data.

In summary, the spin polarization of the excited electrons in the Auger neutralization of He^+ ions inside a metal and at its surface is very high due to the interference term that appears in the excitation of spin-down electrons.

Moreover, this polarization increases at low densities, which is expected to be important in the surface region. Therefore, the measured lower polarization of emitted electrons must be associated to different mechanisms not related to the Auger process itself. In the following, we will show that the scattering and creation of a cascade of secondary electrons is the main responsible for the drop of spin polarization.

5.4 Transport of Excited Electrons in a Paramagnetic Electron Gas

The scattering of the Auger-excited electrons and the creation of the cascade of secondary electrons in a metal can be properly accounted for by means of the Boltzmann transport equation [71, 72]. A detailed discussion of this issue can also be found in the contribution by W. S. M. Werner to this book. In the present chapter, we will concentrate on the description of the Boltzmann transport equation for a homogeneous excitation. This is a more simple approximation that applies to the Auger induced electron emission we are concern here.

The starting point is to obtain the source function $S^\sigma(E)$, i.e., the probability per unit time for excitation of an electron in the Auger process with a given spin σ, as a function of its final energy E:

$$S^\sigma(E) = \frac{d\Gamma^\sigma(E)}{dE} \,. \tag{5.12}$$

This quantity is obtained in a straightforward way, from the expression for the total rate without performing the integral over final energies of the excited electron.

The final distribution of excited electrons $N^\sigma(E)$ is different from $S^\sigma(E)$ due to electron–electron scattering processes. Within the Boltzmann transport equation and using the homogeneous excitation approach $N^\sigma(E)$ is obtained as follows:

$$\begin{aligned} \frac{v(E)}{l(E)} N^\sigma(E) &= S^\sigma(E) \\ &+ \sum_{\sigma'} \int_E^{E_{\max}} dE' F(E,\sigma,E',\sigma') N(E',\sigma'), \qquad \sigma = \uparrow, \downarrow \end{aligned} \tag{5.13}$$

where $E_{\max}$ is the maximum energy of the electron excited in the Auger process, $l(E)$ is the electronic mean free path which is spin independent in a paramagnetic FEG, $v(E)$ is the electron velocity, and $F(E,\sigma,E',\sigma')$ is the probability that an electron in the state (E', σ') is scattered and an electron is produced in the state (E, σ). This can happen in two ways. On the one hand, the electron at (E', σ') can scatter to the state (E, σ), by creating an excitation in the medium. The probability for this process will be denoted $W^{sc}(E, E')$. On the other hand, the electron at (E', σ') can scatter to any other energy with the subsequent excitation of an electron from the conduction band of the metal to the state (E, σ). The probability for this process will be denoted $W^{xc}(E, E')$. These two processes are schematically shown in Fig. 5.6. Note that, as shown in this figure, in the first process ($W^{sc}(E, E')$) the electron with final energy E has the same spin as the initial electron with energy E'. However, in the second process ($W^{xc}(E, E')$) the electron that is excited from the conduction band to the state with energy E can have any of the two spin directions with equal probability.

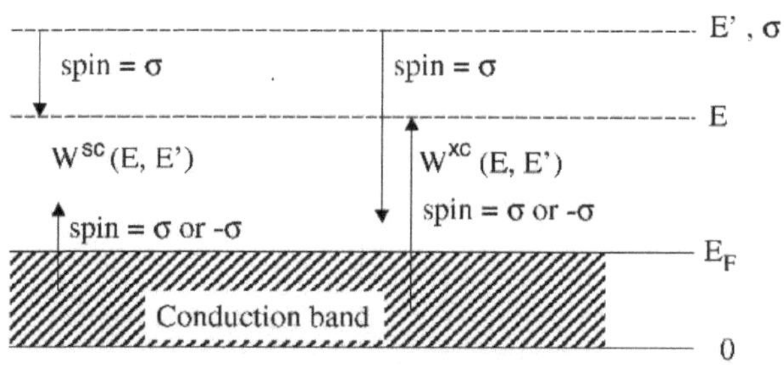

Fig. 5.6. Schematic representation of the electron–electron scattering processes for excited electrons that contribute to the creation of the cascade of secondaries

In a more rigorous treatment, when the two electrons that are scattered have the same spin, the two processes discussed above are indistinguishable. Therefore, the two processes could not be treated separately, and an interference term would appear. Nevertheless, it has been shown that this term may be only important at electron energies within 1 eV from the Fermi energy [73–75]. Since in the problem of electron emission, one is interested in electron energies higher than the work function of the metal, one is clearly in the situation in which this exchange effect can be definitely neglected.

In this way, and by defining $W^{xc}(E, E')$ as the probability that either a spin-up or a spin-down electron is excited from the conduction band to the energy state E, when another electron decays from the energy state E', one gets the following expressions for $F(E, \sigma, E', \sigma')$:

$$
\begin{aligned}
F(E, \sigma, E', \sigma) &= W^{sc}(E, E') + \frac{1}{2} W^{xc}(E, E') \, , \\
F(E, \sigma, E', -\sigma) &= \frac{1}{2} W^{xc}(E, E') \, .
\end{aligned}
\tag{5.14}
$$

As a result, the Boltzmann transport equations (one for each spin state) that one must solve self-consistently take the following form:

$$\frac{v(E)}{l(E)} N^\sigma(E) = S^\sigma(E) + \int_E^{E_{\max}} dE'(W^{sc}(E,E') + \frac{1}{2} W^{xc}(E,E'))N^\sigma(E')$$
$$+\frac{1}{2}\int_E^{E_{\max}} dE' W^{xc}(E,E')N^{-\sigma}(E'), \qquad \sigma = \uparrow, \downarrow \qquad (5.15)$$

In case of a paramagnetic FEG the electronic mean free path and the transition probabilities can be properly calculated in terms of the Lindhard dielectric function. Explicit expressions for these quantities can be found in [72]. More precisely, since the excitation function that results from the Auger process $S^\sigma(E)$ is isotropic, the quantities $W^{sc}(E,E')$ and $W^{xc}(E,E')$ entering (5.15) are obtained by taking the angular average of the expressions in [72].

The self-consistent solution of (5.15) is obtained by an iterative procedure. This also allows one to obtain the energy distribution of electrons after different number of scattering events. For instance, substituting $N^\sigma(E')$ by $(l(E')/v(E'))\ S^\sigma(E')$ in the right hand side of (5.15) one obtains $N_1^\sigma(E)$, the energy distribution of electrons after, at most, a single scattering event. In the next iteration, $N_1^\sigma(E')$ is introduced in the right hand side, obtaining $N_2^\sigma(E)$, the energy distribution of electrons after, at most, two scattering events. The procedure is continued until the self-consistency is achieved.

Finally, in order to fully describe the emission process, one must obtain from the distribution of inner excited electrons $N^\sigma(E)$ the energy distribution of the finally emitted electrons. In order to describe this escape process we use here the standard model of a planar surface barrier and free electrons inside the metal. The surface barrier height W is defined by the Fermi energy E_F and the work function Φ: $W = E_F + \Phi$. Now, using the conservation laws of energy and parallel momentum at this potential barrier, one can easily obtain the current density of emitted electrons $j^\sigma(E)$ [72]. For an isotropic distribution of inner excited electrons $N^\sigma(E)$, like the one in which we are interested here, $j^\sigma(E)$ takes the following form:

$$j^\sigma(E) = \pi\left(1 - \frac{W}{E+W}\right) v(E+W)N^\sigma(E+W)\ , \qquad (5.16)$$

where the energy E is measured from the vacuum level. An idea of the electrons emitted after n-scattering events is provided by using $N_n^\sigma(E)$ in (5.16).

5.5 Comparison with Experiments and Discussion

In this section we show the results of the complete calculation of the electron emission induced in the neutralization of a He^+ ion in a metal, in which all the different mechanisms discussed above are considered. These theoretical results are compared with the available experimental data in the neutralization of He^+ ions in front of metal surfaces.

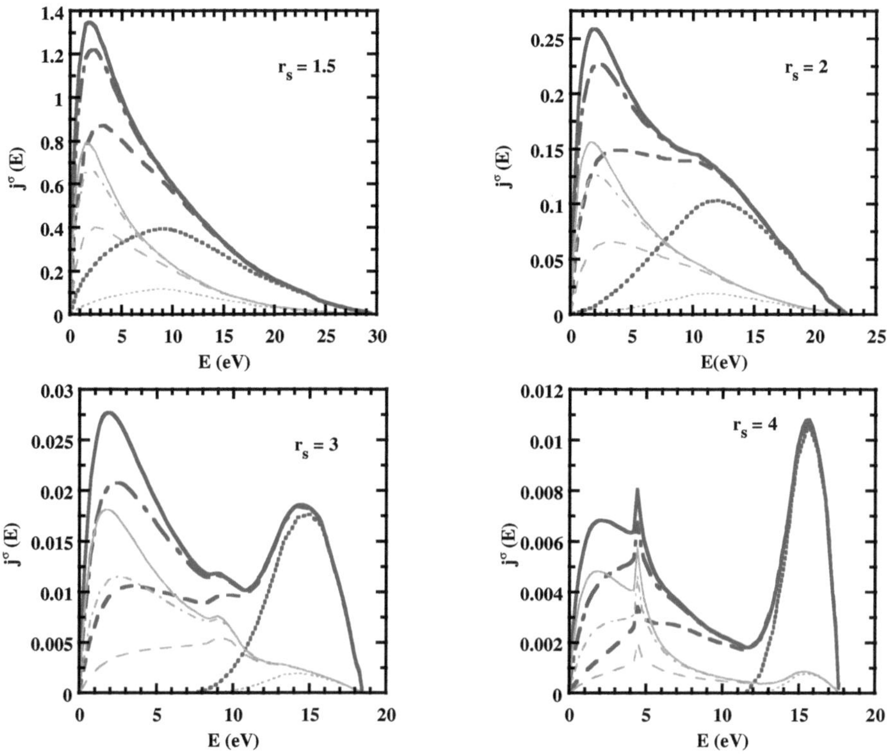

Fig. 5.7. Energy distribution of emitted electrons in the neutralization of a spin-polarized He^+ ion embedded in a free electron gas. Electron energy is measured from the vacuum level. Thick (*thin*) lines correspond to electrons with spin parallel (antiparallel) to that of the electron bound to the He^+ ion. *Dotted lines* correspond to the distribution of directly emitted electron, *dashed lines* to a single scattering approach, *dash-dotted lines* to the twofold scattering approximation, and *solid lines* to the complete self-consistent transport calculation

The calculated distributions of emitted electrons $j^\sigma(E)$ are represented in Fig. 5.7 for different values of the embedding medium electron density. The value of the work function is $\Phi = 4.25$ eV. Comparison of the results obtained for directly emitted electrons (i.e., solving (5.15) with $W^{xc}(E) = W^{sc}(E) = 0$) with those obtained after one-, two-scattering events and the self-consistent solution illustrates the role of the transport mechanism. The inclusion of transport has very different consequences in the spectra depending on the energy range considered. At low energies, the number of spin-up and spin-down electrons increases with the number of considered scattering events. Information on the neutralization process is therefore masked by the cascade of secondary electrons excited in electron–electron scattering processes. At high emission energies the number of emitted spin-up and spin-down electrons hardly changes with the transport process. These high energy electrons, are directly emitted after excitation without almost suffering further scattering events, and they provide direct information on the neutralization process.

The change in the spin polarization of the emitted electrons due to the transport process can be analyzed by means of the quantity:

$$P(E) = \frac{\int_E^{E_{\max}} dE' \, [\, j^{\uparrow}(E') - j^{\downarrow}(E') \,]}{\int_E^{E_{\max}} dE' \, [\, j^{\uparrow}(E') + j^{\downarrow}(E') \,]} \tag{5.17}$$

which can be directly compared to experimental results. $P(E)$ is the polarization of electrons emitted with energy higher than E. The results for different densities of the embedding medium are shown in Fig. 5.8. In the high energy range, for which the transport process plays a negligible role, the spin polarization is high, reflecting the high efficiency of the Auger process for exciting spin-up electrons. For $E = 0$, $P(E = 0)$ is the spin polarization of the total electron yield. Due to the transport process, $P(0)$ is much reduced as compared to the spin polarization of initially excited electrons. For the sake of comparison, the experimental data of [21] obtained in the neutralization of a He^+ ion in front of the Al (001) surface are also shown in the different panels of Fig. 5.8. The overall agreement between the model results (in the range of r_s = 2–3) and the experimental data is remarkable. This shows that the theoretical model captures well the main features of the problem.

Considering that the calculation is performed for a He^+ ion embedded in a FEG, it may seem surprising that the results are applicable to a surface problem. The main difficulty arises from the fact that a metal surface is a region of varying density, whereas in the bulk calculation the background density is uniform. Nevertheless, there are good reasons supporting the validity of the theoretical approach to reproduce the experimental data of [14, 21]. In these experiments the perpendicular energy of the projectile is high enough so that it probes a region close to the topmost layer of atoms. Based on different calculations of the distance dependent Auger neutralization rates for this system, it has been estimated that neutralization should occur at 2–3 a.u. from the topmost atomic layer. This is very close to the jellium edge. At these distances, the relative electron density variations are small, and a local model based on a uniform density background is reasonable.

Since the distance from the surface at which the ions are neutralized cannot be exactly known, there exists some uncertainty about the value of the r_s parameter to be used within the bulk model. In this respect, it is important that the spin polarization of emitted electrons does not depend very strongly on r_s. This can be observed in Fig. 5.8 where similar good overall agreement with the experimental data is obtained for different values of r_s ($r_s = 2 - 3$). This leads to conclude that the knowledge of the exact distance of neutralization and the precise value of the corresponding electron density probed by the projectile is not so critical to explain the measured spin polarization, provided that the Auger neutralization takes place at short distances from the surface where the electron density is not too low. Finally, it is worthy to

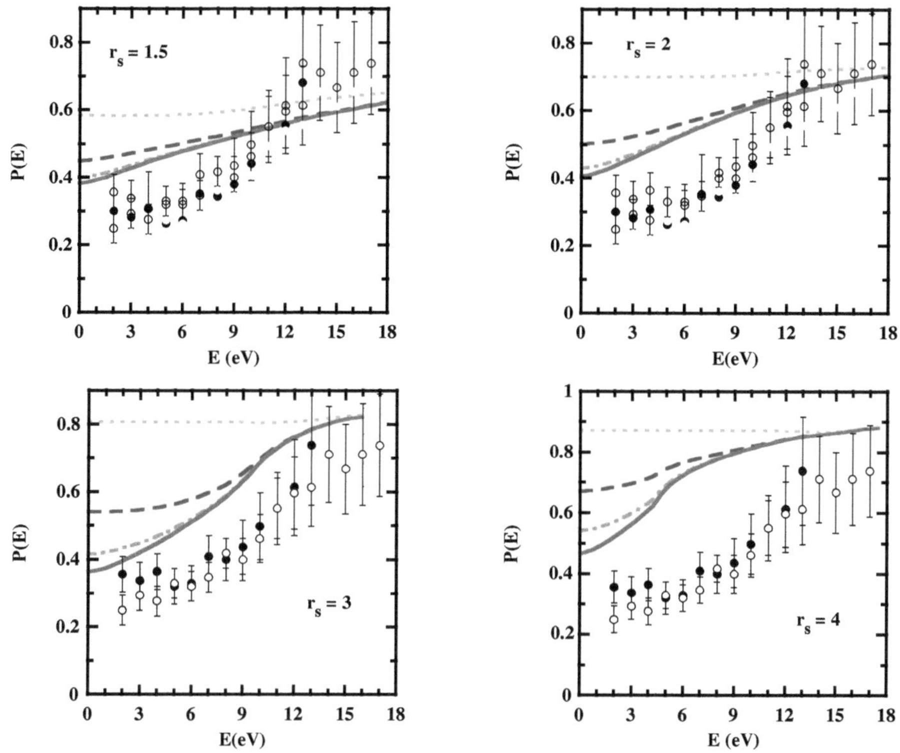

Fig. 5.8. Polarization of the electrons emitted with energy higher than E as defined by (5.17) in the neutralization of a He^+ ion embedded in a free electron gas. Results obtained for directly emitted electrons (*dotted lines*), for a single scattering approach (*dashed lines*), for a twofold scattering approximation (*dash-dotted lines*), and for the self-consistent transport calculation (*solid lines*) are shown. Experimental data of [21] for the polarization of the electrons emitted in the neutralization of 15 eV (*filled circles*) and 500 eV (*open circles*) He^+ ions incident on Al (001) are shown in all the panels

mention that although the density parameter for bulk aluminum is $r_s = 2.07$, one may expect that in the surface region the appropriate value of the density should be lower. In this respect, it is reasonably that in view of Fig. 5.8 one can invoke a better agreement with the experimental data when using $r_s = 3$.

Nevertheless, the dependence of the calculated spin polarization of the total electron yield on the electron density is weak. This feature explains the similar experimental results obtained for this quantity at different surfaces such as Al (001), Au (001), and Cu (001) [21]. The measured spin polarization of the total yield is around 30% in all three cases. Figure 5.9 shows the calculated spin polarization of the total yield ($P(E = 0)$ in (5.17)) for several values of the density parameter r_s. Experimental uncertainty for this quantity is shown as well. The calculations give values consistent with experiments in

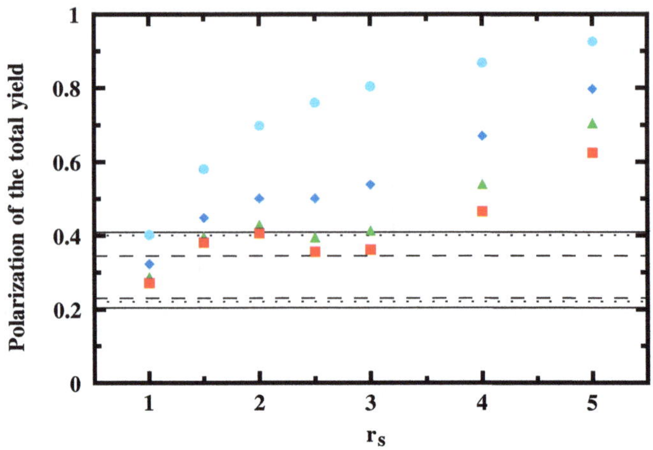

Fig. 5.9. Polarization of the total yield of emitted electrons ($P(0)$ in (5.17)) as a function of r_s. Circles correspond to the directly emitted electrons, diamonds to a single scattering approach, triangles to the twofold scattering approximation, and squares to the complete transport calculation. The lines limit, according to the error bars of [21], the experimental uncertainty for He^+ incident on Al (001) (*solid lines*), Cu (001) (*dotted lines*), and Au (001) (*dashed lines*)

the range $r_s = 1$–4. Note that, if the transport process were neglected, the value of the spin polarization would be much larger in general (circles).

More than in the value of the spin polarization of the total yield, a density effect can be observed in the dependence of the spin polarization on the outgoing electron energy. More precisely, the spin polarization at high energies is expected to decrease as the electron density increases. This is due to the lower spin polarization of excited electrons at high densities. In fact, this density effect can explain the difference in the spin polarization for high energy emitted electrons in the case of Au (001) and Cu (001) surfaces as compared to the Al (001) surface. In principle, Au and Cu are not really free electron metals, since the role played by *d* electrons cannot be neglected. Nevertheless, a free electron approximation for these metals can be performed by using effective density parameters obtained from the maximum of the experimental electron energy loss peaks. By this procedure, the values of the density parameters are $r_s = 1.83$ for Cu and $r_s = 1.49$ for Au [76]. Therefore, it is expected that even in the surface region (close to the topmost layer) the effective density probed by the projectile should be higher for these metals than for aluminum. In Figs. 5.10 and 5.11 the model calculations are compared with the experimental results obtained for the Cu (001) and Au (001) surfaces, respectively. The best overall agreement with experimental data is obtained in the two cases with r_s values lower than in the case of aluminum. Clearly, the $r_s = 3$ case overestimates the spin polarization at high energies in these surfaces, whereas it gives reasonable results for the Al (001) one. The r_s values that

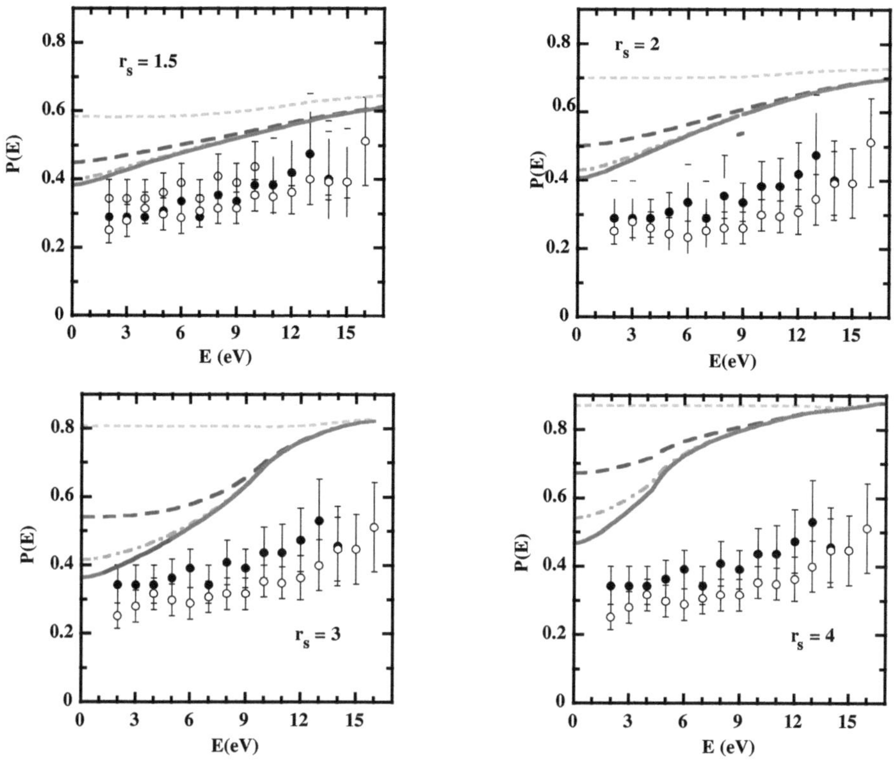

Fig. 5.10. Same as Fig. 5.8, but the experimental results from [21] are for the Cu (001) surface

give a better overall agreement in the Au (001) and Cu (001) surfaces are in the range $r_s = 1.5$–2. This range is consistent with the expected values of the density in the surface region where neutralization takes place.

A final difficulty with the applicability of the bulk model to the surface problem is related to the approach that has been used to estimate the effect due to the transport process. In this respect, calculations that take into account the surface explicitly have shown that the electron inelastic mean free paths keep their bulk value up to the position of the jellium edge [77]. This means that the use of the bulk values for the mean free paths and transition probabilities in (5.15) can be a reliable approximation. Nevertheless, a full self-consistent solution of the Boltzmann transport equation may overestimate the effectiveness of electron–electron scattering processes at the surface. The reason for this is that electrons excited in the surface region may undergo a limited number of scattering events before being emitted, though it is not completely clear the number of scattering events one should consider. In this respect, it is important that the results obtained within the twofold scattering approximation are very similar to those obtained from the

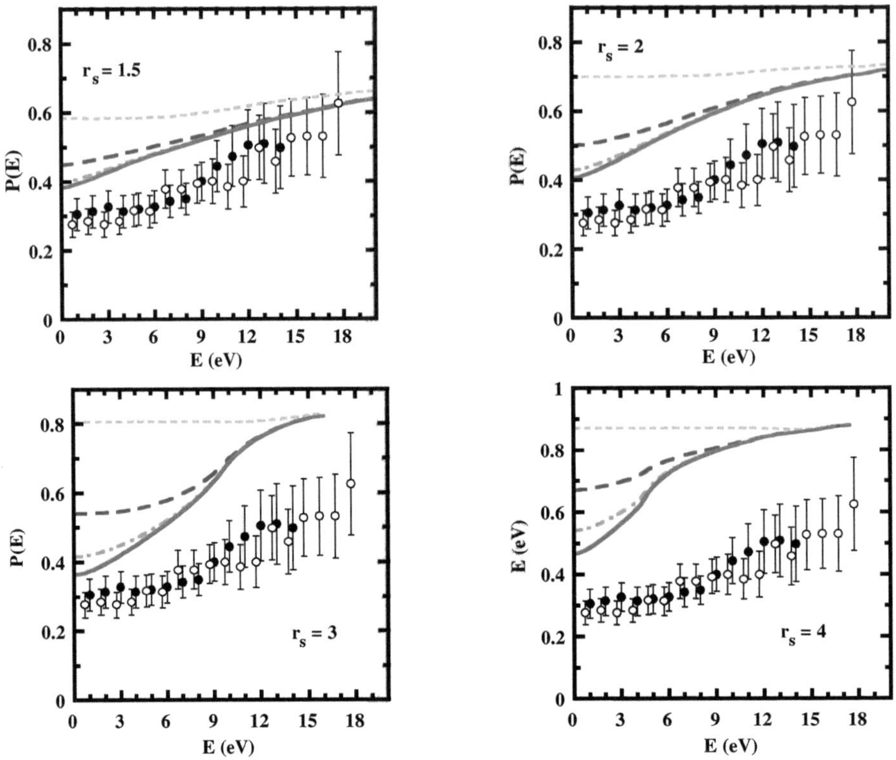

Fig. 5.11. Same as Fig. 5.8, but the experimental results from [21] are for the Au (001) surface

self-consistent solution, and show also good agreement with the experimental data (see Figs. 5.8 to 5.11). This fact, and its occurrence in all the range of densities that are most likely probed by the projectile when the neutralization takes place suggest that the efficiency of electron–electron scattering is not being overestimated.

In summary, the bulk model – that includes: screening, electron excitation, and electron transport – allows to understand the measured spin polarization of the electron emission and its energy dependence in the neutralization of He^+ ions in front of paramagnetic metal surfaces. The picture that arises is that of a strong spin polarization of the primary excited electrons in the Auger process, and a subsequent reduction of it to values of ~30% due to electron–electron scattering processes at the surface. These results depend weakly on the target density and are expected to be valid for those surfaces with typical electron densities $r_s < 4$, provided the Auger process takes place close enough to the topmost layer of atoms. This last condition is important to justify the local model. In the case of neutralization at large projectile-surface distances the usual picture based on the separation between target and projectile electronic levels, which invokes the distinction between Auger deexcitation and Auger neutralization is probably more suitable than the local model. Additionally, when the neutralization of the ion and the subsequent electronic excitation takes place at large distances from the surface, a reduction of the importance of the transport process is expected, at least for the Auger deexcitation. Therefore, in these cases one would expect a higher degree of spin polarization in the emission than that found in the experiments of [14, 21]. Finally, let us mention that even in the cases in which the transport process plays a relevant role, the strong spin polarization of the excitation is reflected in that of the electrons emitted at high energies.

5.6 Summary and Further Developments

We have reviewed some theoretical aspects of the spin polarization of electrons emitted in the Auger neutralization of He^+ ions embedded in a paramagnetic electron gas. All the processes that play an important role such as the spin-dependent screening of the ion, the Auger capture process, and the creation of a cascade of secondary electrons have been described in detail. This model has allowed to explain the measurements performed in ion–surface scattering experiments. Nevertheless, a bulk model may have some limitations when it is applied to a surface problem.

The most obvious limitation is the use of uniform electronic densities to describe situations where the surface is a region of varying density. This does not represent an important shortcoming for the calculation of the electronic excitation in the Auger process, when the neutralization takes place in regions around the jellium edge, where density non uniformities are small. In this respect, we refer the reader to [66], where it is proven the local nature of the Auger process for He^+ ions close to a metal surface.

The basic ingredient of the transport description given in Sect. 5.4 is the isotropic and homogeneous excitation function. In this special case it is possible to replace the correct transport equations (half space problem with suitable boundary conditions) by a system of transport equations for an infinite medium combined with a simple model of the escape of excited electrons in order to describe the emission properties with sufficient accuracy. However, in reality, the excitation function is not homogeneous, i.e., it is distance-dependent. Therefore, an analysis of how the inhomogeneity in the initial excitation function affects the creation of secondaries would be helpful and desirable. For instance, it would be of interest to perform a Monte-Carlo based transport calculation of the creation of secondaries,[4] by locating the electronic excitation at the surface region and describing properly electron–electron inelastic scattering processes at the surface, and electron-target atom elastic scattering processes. This approach might in principle allow to use distance dependent

[4] See Chap. 1 for a description of classical-trajectory Monte-Carlo methods

electron excitation functions for the electrons excited in the Auger process. For the conditions of the experiments that have been discussed here, it would be reliable to use the bulk model for the Auger excitation functions, obtaining the distance dependence by using for each position the corresponding r_s value. The correspondence between the position and r_s can be obtained from the calculated electron density profiles for each surface of interest. This model should allow a direct comparison with the experimental data for both the energy distributions of emitted electrons and their spin polarization. Moreover, it would help to definitely establish the picture provided in this review for the description of this kind of experiments.

In principle, a similar picture to that presented here should be valid in the case of ferromagnetic surfaces. One of the main differences as compared to the paramagnetic case is the role of electron transport. As shown in Sect. 5.5, the creation of a cascade of secondaries in the paramagnetic case involves a reduction of the spin polarization of emitted electrons. On the contrary, in the case of ferromagnetic targets, the cascade involves an increase of the spin polarization of the majority kind at low energies [74, 75]. Additionally, the distribution of the electrons excited in the Auger process depends as well on the spin orientation of the electron bound to the incoming projectile relative to the direction of magnetization of the target. In the following, let us denote $He^+_\uparrow$ ($He^+_\downarrow$) the case in which the spin direction of the electron bound to the projectile is parallel (antiparallel) to the spin of majority spin electrons. Calculations performed for a spin-polarized He^+ ion embedded in a spin-polarized electron gas [33] show that, as in the case of a paramagnetic target, the spin polarization of the excitation tends to be parallel to that of the electron bound to the He^+ ion. This means, for instance, that for $He^+_\uparrow$ the spin polarization of the excitation is of the majority kind. In the case of $He^+_\downarrow$, as there are less minority electrons to be excited, the sign of the spin polarization depends on the energy of the excited electron. Moreover, the total Auger capture rate is larger for $He^+_\downarrow$ than for $He^+_\uparrow$, which implies that the former is more efficient exciting electrons than the latter.

Nevertheless, important problems arise if one tries to use this information to understand the measured data. In the surface region of ferromagnetic materials not only the electron density varies with distance, but also the spin polarization. In the paramagnetic case, it has been shown that the main results concerning the spin polarization of emitted electrons are not very sensitive to the exact value of r_s, in the range of densities that correspond to distances at which the He^+ ion is more likely neutralized. Unfortunately, this is not the case concerning the effect of the variation of the spin polarization of the electrons in the surface of ferromagnetic materials. For instance, the measured yield is larger for $He^+_\uparrow$ than for $He^+_\downarrow$ ions [2, 8, 9, 18–20]. This is related to the inversion of the spin polarization of the electrons at the surface of these materials [9, 78], i.e., the spin polarization at the surface region where the projectile is neutralized is of the minority kind. Additionally, the spin polarization of the yield is of the majority kind for both $He^+_\uparrow$ and $He^+_\downarrow$. This fact may be related to the creation of the cascade of secondaries, for which the sign of the bulk magnetization plays the major role. In conclusion, the two step picture presented for the paramagnetic case (excitation in the Auger process + creation of secondaries) should be also useful to understand the spin polarization of the electron emission in the neutralization of He^+ ions in front of ferromagnetic surfaces. Nevertheless, the problem is much more complicated due to the variation of the spin polarization of the electron density with position in front of the surface. Certainly, more work along these lines will be necessary in future in order to better interpret the measured data in ferromagnetic surfaces.

We acknowledge partial support by the Basque Departamento de Educación, Universidades e Investigación, the University of the Basque Country UPV/EHU (Grant No. 9/UPV 00206.215-13639/2001), and the Spanish MCyT (Grant No. FIS2004-06490-CO3-00). M. A. acknowledges financial support by the Gipuzkoako Foru Aldundia. Computational resources were provided by the SGI/IZO-SGIker at the UPV/EHU (supported by the Spanish Ministry of Education and Science and the European Social Fund).

References

1. C. Rau and R. Sizmann, Phys. Lett: **43A**, 317 (1973)
2. M. Onellion, M. W. Hart, F. B. Dunning, and G. K. Walters: Phys. Rev. Lett. **52**, 380 (1984)
3. C. Rau, C. Schneider, G. Xing, and K. Jamison, Phys. Rev. Lett: **57**, 3221 (1986)
4. J. Kirschner, K. Koike, and H. P. Oepen: Phys. Rev. Lett. **59**, 2099 (1987)
5. H. Winter, H. Hagedorn, R. Zimmy, H. Nienhaus, and J. Kirschner: Phys. Rev. Lett. **62**, 296 (1989)
6. A. Närmann, M. Schleberger, W. Heiland, C. Huber, and J. Kirschner: Surf. Sci. **251/251**, 248 (1991)
7. M. W. Hart, M. S. Hammond, F. B. Dunning, and G. K. Walters: Phys. Rev. B **39**, 5488 (1989)
8. M. S. Hammond, F. B. Dunning, G. K. Walters, and G. A. Prinz: Phys. Rev. B **45**, 3674 (1992)
9. M. Salvietti, R. Moroni, P. Ferro, M. Canepa, and L. Mattera: Phys. Rev. B **54**, 14758 (1996)
10. M. Schleberger, M. Dirska, J. Manske, and A. Närmann: Appl. Phys. Lett. **71**, 3156 (1997)
11. A. Närmann, M. Dirska, J. Manske, G. Lubinski, M. Schleberger, and R. Hoekstra: Surf. Sci. **398**, 84 (1998)
12. T. Igel, R. Pfandzelter, and H. Winter: Phys. Rev. B **59**, 3318 (1999).
13. F. Bisio, R. Moroni, M. Canepa, and L. Mattera: Phys. Rev. Lett. **83**, 4868 (1999)
14. D. L. Bixler, J. C. Lancaster, F. J. Kontur, P. Nordlander, G. K. Walters, and F. B. Dunning: Phys. Rev. B **60**, 9082 (1999)
15. R. Pfandzelter, T. Bernhard, and H. Winter: Phys. Rev. Lett. **86**, 4152 (2001)

16. R. Pfandzelter, M. Ostwald, and H. Winter: Phys. Rev. B **63**, 140406 (2001)
17. R. Pfandzelter, H. Winter, I. Urazgil'din, and M. Rösler: Phys. Rev. B **68**, 165415 (2003)
18. R. Moroni, E. Oliveri, and L. Mattera: Nucl. Instrum. Methods Phys. Res. B **203**, 29 (2003)
19. M. Kurahashi, T. Suzuki, X. Ju, and Y. Yamauchi: Phys. Rev. B **67**, 024407 (2003)
20. M. Kurahashi, T. Suzuki, X. Ju, and Y. Yamauchi: Phys. Rev. Lett. **91**, 267203 (2003)
21. J. C. Lancaster, F. J. Kontur, G. K. Walters, and F. B. Dunning: Phys. Rev. B **67**, 115413 (2003)
22. M. Unipan, A. Robin, R. Morgenstern, and R. Hoekstra: Phys. Rev. Lett. **96**, 177601 (2006)
23. H. Conrad, G. Ertl, J. Küppers, and S. W. Wang, K. Gérard, and H. Haberland: Phys. Rev. Lett. **42**, 1082 (1979)
24. D. S. Gemell: Rev. Mod. Phys. **46**, 129 (1974)
25. H. Winter: Phys. Reports **367**, 387 (2002)
26. A. Arnau, F. Aumayr, P. M. Echenique, M. Grether, W. Heiland, J. Limburg, R. Morgenstern, P. Roncin, S. Schippers, R. Schuch, N. Stolterfoht, P. Varga, T.J. M. Zouros and HP. Winter: Surf. Sci. Rep. **27**, 113 (1997)
27. M. Alducin, J. I. Juaristi, R. Díez Muiño, M. Rösler, and P. M. Echenique: Phys. Rev. A **72**, 024901 (2005)
28. N. P. Wang, E. A. García, R. Monreal, F. Flores, E. C. Goldberg, H. H. Brongersma, and P. Bauer: Phys. Rev A **64**, 012901 (2001)
29. N. Lorente, M. A. Cazalilla, J. P. Gauyacq, D. Teillet-Billy, and P. M. Echenique: Surf. Sci. **411**, L888 (1998)
30. M. A. Cazalilla, N. Lorente, R. Díez Muiño, J. P. Gauyacq, D. Teillet-Billy, and P. M. Echenique: Phys. Rev. B **58**, 13991 (1998)
31. M. Alducin, R. Díez Muiño, J. I. Juaristi, and A. Arnau: J. Electron Spectrosc. Relat. Phenom. **137**, 401 (2004)
32. M. Alducin, R. Díez Muiño, and J. I. Juaristi: Phys. Rev. A **70**, 012901 (2004)
33. M. Alducin, R. Díez Muiño, and J. I. Juaristi: Nucl. Instrum. Methods Phys. Res. B **230**, 431 (2005)
34. J. I. Juaristi, M. Alducin, R. Díez Muiño, and M. Rösler: Nucl. Instrum. Methods Phys. Res. B **232**, 73 (2005)
35. M. Alducin: Nucl. Instrum. Methods Phys. Res. B **232**, 8 (2005)
36. N. Bonini, G. P. Brivio, and M. I. Trioni: Phys. Rev. B **68**, 035408 (2003)
37. J. S. Langer and S. H. Vosko: J. Phys. Chem. Solids **12**, 196 (1960)
38. K. S. Singwi and M. P. Tosi: Phys. Rev. **181**, 784 (1969)
39. A. Sjölander and M. J. Stott: Phys. Rev. B **5**, 2109 (1972)
40. Z. D. Popović and M. J. Stott: Phys. Rev. Lett. **33**, 1164 (1974)
41. Z. D. Popović, M. J. Stott, J. P. Carbotte, and G. R. Piercy: Phys. Rev. B **13**, 590 (1976)
42. C. -O. Almbladh, U. von Barth, Z. D. Popović, and M. J. Stott: Phys. Rev. B **14**, 2250 (1976)
43. E. Zaremba, L. M. Sander, H. B. Shore, and J. H. Rose: J. Phys. F **7**, 1763 (1977)
44. O. Gunnarsson and B. I. Lundqvist: Phys. Rev. B **13**, 4274 (1976); O. Gunnarsson and B. I. Lundqvist, Phys. Rev. B **15**, 6006(E) (1977)

45. W. Kohn and L. J. Sham: Phys. Rev. **140**, A1133 (1965)
46. C. -O. Almbladh and U. von Barth: Phys. Rev. B **13**, 3307 (1976)
47. R. M. Nieminen and M. J. Puska: Phys. Rev. B **25**, 67 (1982)
48. T. T. Rantala: Phys. Rev. B **28**, 3182 (1983)
49. C. -O. Almbladh, A. L. Morales, and G. Grossmann: Phys. Rev. B **39**, 3489 (1989)
50. E. K. Chang and E. L. Shirley: Phys. Rev. B **66**, 035106 (2002)
51. A. Arnau, P. A. Zeijlmans van Emmichoven, J. I. Juaristi, and E. Zaremba: Nucl. Instrum. and Methods Phys. Res. B **100**, 279 (1995)
52. R. Díez Muiño, N. Stolterfoht, A. Arnau, A. Salin, and P. M. Echenique: Phys. Rev. Lett. **76**, 4636 (1996)
53. R. Díez Muiño, A. Salin, N. Stolterfoht, A. Arnau, and P. M. Echenique: Phys. Rev. A **57**, 1126 (1998)
54. J. I. Juaristi and A. Arnau: Nucl. Instrum. and Methods Phys. Res. B **115**, 173 (1996)
55. J. I. Juaristi, A. Arnau, P. M. Echenique, C. Auth, and H. Winter: Phys. Rev. Lett. **82**, 1048 (1999)
56. J. I. Juaristi, A. Arnau, P. M. Echenique, C. Auth, and H. Winter: Nucl. Instrum. and Methods Phys. Res. B **157**, 87 (1999)
57. J. I. Juaristi, R. Díez Muiño, A. Dubus, and M. Rösler: Phys. Rev. A **68**, 012902 (2003)
58. R. Vincent and J. I. Juaristi: Nucl. Instrum. and Methods Phys. Res. B **232**, 67 (2005)
59. P. Nozieres and D. Pines: *The Theory of Quantum Liquids*, (Persseus Books, Cambridge, Massachusetts, 1999)
60. R. Díez Muiño: Nucl. Instrum. Methods Phys. Res. B **203**, 8 (2003)
61. H. D. Hagstrum: Phys. Rev. **96**, 336 (1954)
62. F. M. Propst: Phys. Rev. **129**, 7 (1963)
63. R. K. Janev and N. N. Nedeljković: J. Phys. B **18**, 915 (1985)
64. M. Alducin, A. Arnau, and P. M. Echenique: Nucl. Instrum. Methods Phys. Res. B **67**, 157 (1992)
65. T. Fonden and A. Zwartkruis: Phys. Rev. B **48**, 15603 (1993)
66. N. Lorente, R. Monreal, and M. Alducin: Phys. Rev. A **49** , 4716 (1994)
67. M. Alducin: Phys. Rev. A **53**, 4222 (1996)
68. E. A. García, N. P. Wang, R. C. Monreal, and E. C. Goldberg: Phys. Rev. B **67**, 205426 (2003)
69. L. A. Salmi: Phys. Rev. B **46**, 4180 (1992)
70. P. J. Feibelman, E. J. McGuire, and K. C. Pandey: Phys. Rev. B **15**, 2202 (1977)
71. P. A. Wolff: Phys. Rev. **95**, 56 (1954)
72. M. Rösler and W. Brauer: *Particle Induced Electron Emission I*, Springer Tracts in Modern Physics Vol. 122 (Springer, New York, Berlin, Heidelberg, 1991), pp. 1–65
73. R. W. Rendell and D. R. Penn: Phys. Rev. Lett. **45**, 2057 (1980)
74. D. R. Penn, S. P. Apell, and S. M. Girvin: Phys. Rev. Lett. **55**, 518 (1985)
75. D. R. Penn, S. P. Apell, and S. M. Girvin: Phys. Rev. B **32**, 7753 (1985)
76. D. Isaacson: New York, University Document No. 02698 (1975), (National Auxiliary Publication Service, New York, 1975)
77. C. O. Reinhold, J. Burgdörfer, K. Kimura, and M. Mannami: Nucl. Instrum. Methods Phys. Res. B **115**, 233 (1996)
78. D. R. Penn and S. P. Apell: Phys. Rev. B **41**, 3303 (1990)

6

Electron Emission from Surfaces Mediated by Ion-Induced Plasmon Excitation

Raúl A. Baragiola and R. Carmina Monreal

We review theoretical and experimental research on electron emission due to plasmon excitation and decay in ion-solid interactions for low projectile velocities (less than the Fermi velocity). The main excitation mechanisms in this case are plasmon-assisted neutralization of ions with sufficiently high potential energy, and indirect excitation by fast secondary electrons.

6.1 Introduction – Basic Notions

Plasmons, quantized charge-density oscillations in solids [1, 2], were discovered in 1942 in the form of multiple discrete energy losses of electrons traversing thin films [3] and first explained theoretically by Pines and Bohm in 1952 [4]. Plasmons arise due to the overshoot of the screening response of the electronic system to a rapidly varying electric field such as that produced by fast charges moving in solids and in ion neutralization. These excitations are evidenced not only in the discrete energy losses of electrons in EELS (electron energy loss spectroscopy), photoelectrons or Auger spectroscopy but also in the "shake-up" excitation produced by sudden changes in the occupation of localized electron states, e.g., in the photoionization and decay of inner-shells. In photoionization, the electrons respond by trying to screen the suddenly created hole but overshoot the static screening position, leading to an oscillation of the screening charge. The switching of the perturbation must be fast, so that the transient electric fields have Fourier components with frequencies overlapping with the plasmon frequencies at $1 - 5 \times 10^{15}$ Hz, in the optical (UV) range. These shake-up excitations produce energy losses in X-ray photoelectron spectroscopy, and can give useful information for surface analysis of solids [5].

6.1.1 Bulk Plasmons

A plasma oscillation is a collective oscillation of the electrons that occurs as a consequence of the long range of the Coulomb interaction in systems with

R.A. Baragiola and R.C. Monreal: *Electron Emission from Surfaces Mediated by Ion-Induced Plasmon Excitation,* STMP **225**, 185–211 (2007)
DOI 10.1007/3-540-70789-1_6

a large number of electrons, such as plasmas or metals. These oscillations are approximately normal modes of the system if they are weakly damped. A normal mode of a system is an oscillation that, once excited, can be self-sustained in the absence of an external driving field.

For the case of plasma oscillations, the constitutive relation between the electric field **E** and the displacement **D** is:

$$\mathbf{E} = \mathbf{D}/\varepsilon \tag{6.1}$$

where $\varepsilon = \varepsilon_1 + \varepsilon_2$ is the (complex) dielectric function ($1/\varepsilon$ is called the response function). Equation (6.1) implies that $\mathbf{E} \neq 0$ for $\mathbf{D} = 0$ if $\varepsilon = 0$. Therefore the oscillation is self-sustained if $\varepsilon_1 = 0$ and $\varepsilon_2 \ll 1$. In the simplest case of a free electron gas described by the Drude dielectric function:

$$\varepsilon(\omega) = 1 - \frac{\omega_P^2}{\omega^2} . \tag{6.2}$$

Plasma oscillations are produced at the plasma frequency given by

$$\omega_P = \sqrt{\frac{4\pi n e^2}{m}} . \tag{6.3}$$

where n is the electron density and e and m the electron charge and mass. In general, ε depends on the frequency ω and on the wave vector $\mathbf{k}$. Then, the oscillation frequency depends on $k = |\mathbf{k}|$ trough its dispersion relation:

$$\omega(k) \approx \omega_P + \frac{3}{10}\frac{k^2 v_F^2}{\omega_P} \tag{6.4}$$

Where v_F is the Fermi velocity. The oscillation is quantized (plasmon) with a quantum of energy $E_p = \hbar\omega_P$.

In reality, electrons do not move freely but are subject to forces due to the polarization cloud of correlated electrons. Hence, the term "quasi-particle" is preferred to electron, and an effective mass m^* is used in Eq. (6.3). Plasmon energies deviate strongly from the free-electron value in the case of metals with localized d-bands, such as silver. In other cases, such as Cu and Au, these polarization effects also damp the plasma oscillation so heavily that they cannot be considered to be normal modes.

6.1.2 Monopole and Multipole Surface Plasmons

A boundary introduces different surface or interface oscillation modes, first predicted by Ritchie in 1957 [6], which have been treated in various reviews, e.g., [2, 7, 8]. Surface normal modes are defined as the poles of the surface response function that is the equivalent to the bulk response function $1/\varepsilon$. In the simplest case of a planar surface separating a medium of dielectric constant $\varepsilon(\omega)$ from the vacuum, the surface response function is given by

$$g(\omega) = \frac{\varepsilon(\omega) - 1}{\varepsilon(\omega) + 1} \tag{6.5}$$

which, for the Drude dielectric function ε Eq. (6.2), has a pole that describes a surface plasmon oscillation at an energy $\hbar\omega_{SP} = E_P/\sqrt{2}$ where E_P is the energy of the volume plasmon (VP) given above.

When an atomic particle or a photon interacts with a perfect planar surfaces, the component of the momentum parallel to the surface, $\hbar q$, has to be conserved. This means that, in general, the surface response function depends on ω and q. Feibelman [9] showed that in the limit of small q the dispersion relation of the surface plasmon can be expressed as

$$\omega(q) = \omega_{SP}\left(1 - \frac{1}{2} q d(\omega_{SP})\right) \tag{6.6}$$

where $d(\omega_{SP})$ is the position of the centroid of the induced charge density, evaluated at the frequency of the surface plasmon. This quantity can be positive or negative depending on the surface model: a negative dispersion results if the electronic density spills out into the vacuum while the dispersion is positive for confined electrons (for infinite surface potential barrier).

The normal surface plasmons (nSP) just described are not the only normal modes of a surface. The surface response function of realistic surface can also present poles at other frequencies named multipole surface plasmons (mSP). These plasmons, which at $q = 0$ have energies close to that of the volume plasmon: $(0.80\text{–}0.85)E_p$, and a positive dispersion [2, 10], can be viewed as a volume plasmon of the low electron density surface region [11]. They are responsible for the large enhancement of photoelectron emission [12, 13] at the mSP energy and seen in EELS spectra of Al(111) [14, 15].

6.1.3 Plasmon Decay

The excited plasmon can subsequently decay by interband transitions assisted by scattering with impurities, grain boundaries, etc. after a short time (< 1 fs), which allows at most a few oscillations. This short lifetime τ introduces an energy broadening $\Gamma = \hbar/\tau$ of the plasmon loss peak seen in EELS. Interband transitions to states above the vacuum level can lead to electron escape into vacuum [16, 17]. Since the plasmon energy can be transferred to any electron in the valence band, the internal electron energy distribution, $N_p(E)$, is broad, as wide as the valence band if E_p is larger than the Fermi energy [18–20]. In metals, the discontinuity at the Fermi level causes a shoulder in $N_p(E)$, broadened by Γ. Since $N(E)$, the energy spectrum of all emitted electrons, contains also electrons excited by other mechanisms, visualization of the plasmon decay shoulder is enhanced in the derivative $dN(E)/dE$ [21], a standard method in Auger spectroscopy. An example of an electron energy spectrum from Al and its derivative is shown in Fig. 6.1 [15], for the case of incident electrons.

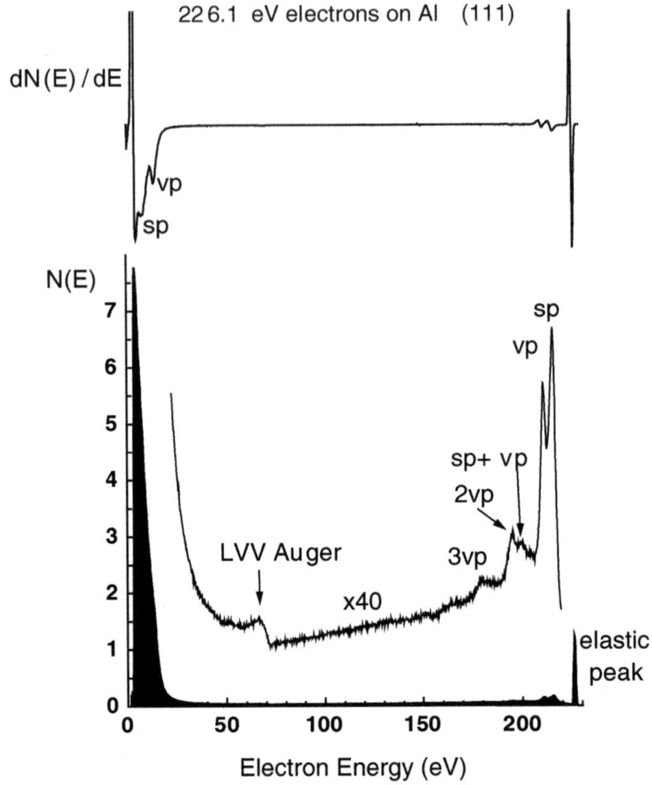

Fig. 6.1. Electron energy spectrum from clean Al(111) excited by 226.1 eV electrons at normal incidence, together with its derivative. The energy loss structure near the elastic peak is assigned to excitation of single and multiple surface plasmons (SP) and volume plasmons (VP). The derivative indicates the structure due to the decay of surface and volume plasmons. The central region of the derivative spectrum is magnified to show the structures due to multiple plasmon and $2p$ Auger decay. From [15]

6.2 Theory for Plasmon Excitation

6.2.1 Direct Excitation by Fast Charged Particles

Charged particles exchange energy and momentum with the electrons of a solid, and create a wake of electron density fluctuations behind them. If the charges are fast enough, the valence electrons cannot react adiabatically; e.g., electrons that rush to screen a fast positive ion find that it is gone by the time they reach its path, creating an excess of screening charge behind the ion that starts the plasmon oscillation. In addition to the energy loss into electron-hole pairs, the energy loss to plasmon excitations is a major mechanism for stopping of fast electrons and ions in solids, especially for light elements, where the number of core electrons is small. EELS experiments have been used to measure the dispersion relation of normal and multipole surface plasmons of Al and the alkaline metals. While the normal surface plasmons, being dipole active, can be easily excited, the detection of multipole surface plasmons is more subtle. This was first achieved in photoelectron yield experiments, where the excitation of the normal plasmon is forbidden by energy and momentum conservation, and in reflection EELS measurements performed in off-specular direction.

In the case of ion excitation, one cannot use the energy loss of the projectile, as in EELS [1], since it is obscured by multiple continuum energy losses of the ion to target nuclei. The alternative to study plasmons is to observe their decay in characteristic electrons or, far less likely, photons [22, 23].

Allowed electronic excitations in solids are typically described with a diagram of energy transfer vs. momentum transfers, such as that in Fig. 6.2 for the case of a free electron gas. The region of single-particle excitation is bound by $E/E_F = (k/k_F)[(k/k_F) + 2]$ and $E/E_F = (k/k_F)[(k/k_F) - 2]$ resulting from energy and momentum conservation. The plasmon line in the figure represents Eq. (6.14), and the width is omitted for clarity. There is a maximum or cut-off momentum $\hbar k_c$ for the plasmon, which is where the plasmon line enters the single-particle region. At such high momenta, the concept

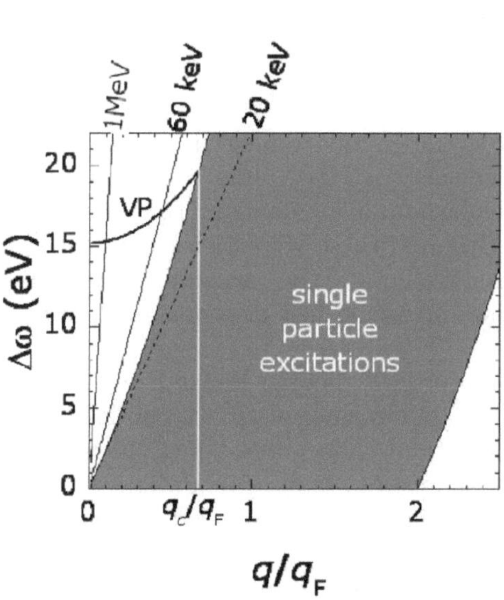

Fig. 6.2. Diagram showing the region of allowed energy transfer $\hbar\omega$ and momentum transfer q (in units of the Fermi momentum q_F) in an electron gas. Excitations can be to single-particles (shaded) or along the volume plasmon line (VP, shown for Al). Higher order processes are not included. The straight lines indicate maximum energy transfers for a given q, for protons at the indicated energy. q_c is the cutoff q (maximum plasmon momentum)

of a plasmon loses its meaning, since the wavelength becomes comparable to the average distance between electrons, and Γ becomes large due to strong coupling to single-particle excitations. The maximum energy transfer is given by $\Delta E_{\max} = k(2k_0 - k)/2M$, where k_0 and M are the momentum and mass of the projectile charge. Thus, for electrons, the maximum energy loss forms a parabola in $E - k$ space, where for ions, $k_0 \gg k$ and the maximum energy transfer is the straight line $\Delta E_{\max} = \hbar kv$, where $v = |\mathbf{v}|$ is the speed of the ion.

The rate of energy loss of fast charges in electron gas is given by:

$$W(k,\omega) = -\frac{(4\pi Z)^2}{k^2}\mathrm{Im}[-1/\varepsilon(k,\omega)]\delta(\omega - \mathbf{kv}) \tag{6.7}$$

where Z is the value of the charge and where $\mathrm{Im}(-1/\varepsilon) = \varepsilon_2/(\varepsilon_1^2 + \varepsilon_2^2)$ is called the energy loss function. The Coulomb divergence for small q (large distances) in Eq. (6.6) is suppressed by screening, described by $\mathrm{Im}[-1/\varepsilon(k,\omega)]$, and initially proportional to k^2. The condition for plasmon excitation is that $\varepsilon_1 \ll 1$ and $\varepsilon_2 \ll 1$; other criteria have been proposed [24, 25]. Equation (6.6) thus shows that the energy loss has a pronounced peak at the energy of the normal modes.

Kinetic plasmon excitation by ions has attracted several theoretical studies [19, 26–31]. Ritzau et al. [32] discovered that, for proton impact on Al and Mg, plasmons are excited below the threshold velocity v_{th} predicted by theory obtained from the condition $\hbar k_c v_{\mathrm{th}} = E_p(k_c)$. For a free-electron gas with Fermi velocity v_F, the cutoff momentum $\hbar k_c \approx E_p/v_F$, leading to $v_{\mathrm{th}} \approx 1.3 v_F$; that is a ratio of projectile energy to mass of ~40 keV/amu for Al [26] and 25 keV/amu for Mg. The energy of the volume plasmon increases with increasing v due to the positive plasmon dispersion and agrees with theory except near or below v_{th} [32, 33]. Although these velocities are higher than those considered in this chapter (and this book), the mechanism of Coulomb excitation still intervenes in processes involving fast electrons excited by the projectile, that will be seen later.

Similar to the case of ions moving inside the solid, discussed above, fast particles traveling outside but near a surface can lose energy by exciting surface modes, with the rate of energy loss being proportional to the imaginary part of the surface response function. Calculations of plasmon excitation and decay by fast particles incident upon a metal surface at glancing angles have been performed, for instance, in references [34–36]. These calculations show that the spectrum of emitted electrons have a pronounced peak at the energy of the surface plasmon minus the work function, as a consequence of the excitation of surface plasmons and their consequent decay.

6.2.2 Neutralization

Plasmon excitation can occur at even lower velocities ($v \ll v_{\mathrm{th}}$) by the rapid switch of the potential during electron capture by ions of sufficiently high

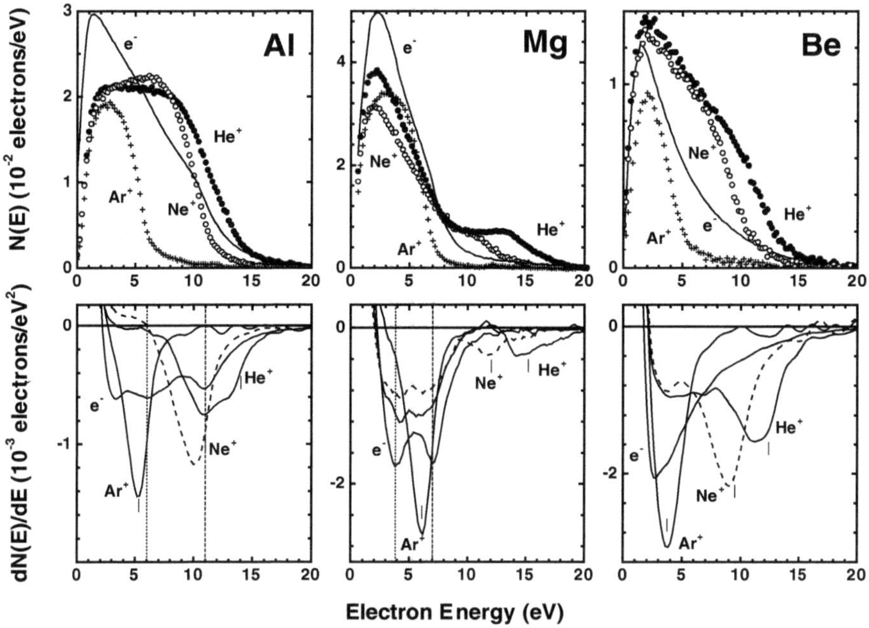

Fig. 6.3. Electron energy distributions and their derivatives for polycrystalline Al, Mg and Be induced by 106 eV He^+, Ne^+, and Ar^+ ions, and 1 keV electrons [15]. The *vertical dotted* and *dashed curves* represent the position of the monopole surface and volume plasmons, which are only evident for Al and Mg due to their small widths. The short *solid lines* are the positions of the high-energy edge of Auger neutralization assuming an image shift of 2 eV of the ionic levels

potential energy, as discovered in experiments using 50–4500 eV He^+ and Ne^+ ions on Al and Mg (Fig. 6.3) [37]. Several theories of potential plasmon excitation during neutralization have been published [38–44]. Electron emission from this process competes with the more studied Auger neutralization (AN) (also called Auger capture–AC), which can occur if the potential energy of the ion exceeds twice ϕ, the work function of the surface (Fig. 6.4). For low ϕ surfaces, a competing channel is resonance neutralization followed by Auger de-excitation (AD) [49]. The potential plasmon excitation mechanism is allowed if the potential energy released when the ion neutralizes near the surface, $E_{\rm n} = I' - \phi - E_h$ equals E_p. Here I' is the ionization potential of the ion I, shifted by the interaction with its image charge (~2 eV) [37] and E_h the energy of the final hole in the solid, measured from the Fermi level. With a work function of 4.3 eV for Al, slow He^+ ($I = 24.6$ eV) and Ne^+ ($I = 21.6$ eV) have enough energy to excite the volume plasmon of Al but Ar^+($I = 15.8$ eV) has not ($E_{pv}(q) > 15$ eV) [2, 8]. On the other hand, low momentum monopole surface plasmons (nSP) can be excited by the three ions ($E_{pv}(0) = 10.6$ eV). It is important to note that plasmon excitation is a resonance process, and

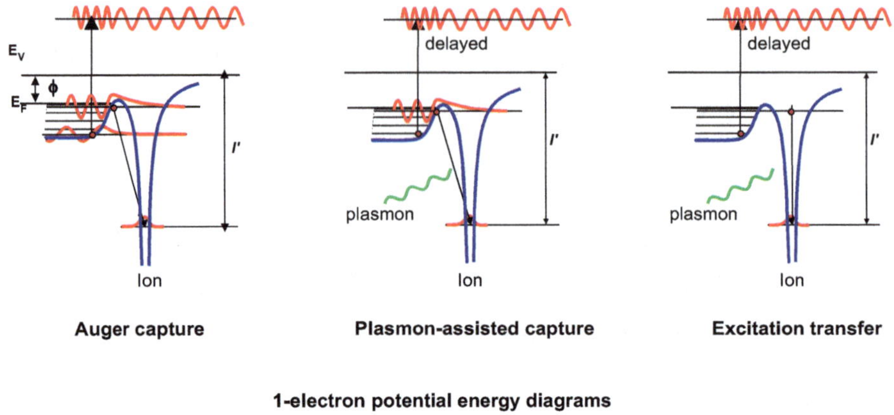

Fig. 6.4. One-electron energy diagrams appropriate for Auger capture, electron capture with plasmon excitation, and electronic excitation transfer from the projectile to a plasmon

hence it will not occur if the energy released in electron capture is far from the plasmon energy plus Γ.

Plasmons were proposed to be important in ion neutralization and deexcitation of metastable atoms at surfaces about two decades ago [38, 39]. Here we review a very general theory of plasmon assisted neutralization based on the dielectric formalism. The basic magnitude appearing in the theory is the dielectric susceptibility χ of the solid (related to the dielectric constant) whose imaginary part describes excitations. The key quantity is the transition rate $1/\tau$ given by Fermi's golden-rule,

$$\frac{1}{\tau} = 2\pi \sum_{i.j} \left| \left\langle f | \hat{V} | i \right\rangle \right|^2 \delta \left(E_f - E_i \right) \tag{6.8}$$

where $|i\rangle$ and $|f\rangle$ are the initial and final states of the ion (atom) plus the solid, with energies E_i and E_f, and $\hat{V}$ is the Coulomb interaction between the charge density in the atom and that induced in the metal.

The initial and final states are expressed as:

$$\begin{aligned} |i\rangle &= |0\rangle \otimes |\mathbf{k}\rangle \\ |f\rangle &= |n\rangle \otimes |a\rangle \end{aligned} \tag{6.9}$$

with $|0\rangle$ and $|n\rangle$ being the ground and excited states of the many-electron system respectively, $|\mathbf{k}\rangle$ is the state representing the neutralizing electron and $|a\rangle$ is the final atomic state. The interaction potential can be written down as the Coulomb interaction between operators representing the fluctuations of charge densities in the atom $\hat{\rho}_a$ and in the metal $\delta\hat{n}$ upon the neutralization event as

$$\hat{V} = \int d\mathbf{r}_1 \int d\mathbf{r}_2 \frac{\delta\hat{n}(\mathbf{r}_1)\hat{\rho}_a(\mathbf{r}_2)}{|\mathbf{r}_1 - \mathbf{r}_2|} \tag{6.10}$$

Then, from the Fourier transform in the coordinates parallel to the surface, we obtain the contribution to the matrix elements of all the metallic excitations that can be produced in the neutralizing event. Then, these are summed up yielding the imaginary part of the dielectric susceptibility defined in Chap. 1 to yield our final expression for the Auger neutralization rate:

$$\frac{1}{\tau}(z_a) = 2 \sum_{k<k_F} \int_0^\infty d\omega \int \frac{d\mathbf{q}}{4\pi^2} \int dz \int dz' \mathrm{Im}\left(-\chi\left(\mathbf{q},\omega;z,z'\right)\right) \\ \times \Phi\left(\mathbf{k},\mathbf{q},z\right) \Phi\left(\mathbf{k},\mathbf{q},z'\right) \delta\left(E_a - E_k + \omega\right) \tag{6.11}$$

In this equation, z_a is the distance of the atom to the surface, z, z' are electron coordinates along the surface normal, $\mathbf{k}$ is the wave vector of the neutralizing electron below the Fermi level of energy E_k, E_a is the energy of the atomic level, and ω and $\mathbf{q}$ are the energy and momentum parallel to the surface transferred to the metal. The potential Φ is given by:

$$\Phi(\mathbf{k},\mathbf{q},z) = \frac{2\pi}{q} \langle \varphi_a(\mathbf{r}' - z_a)| e^{i\mathbf{q}\cdot\boldsymbol{\rho}'} e^{-q|z-z'|} |\varphi_k(\mathbf{r}')\rangle \tag{6.12}$$

where $\varphi_a\left(\mathbf{r}' - z_a\right)$ is the wave function of the atomic state which depends on the electron coordinate $\mathbf{r}'$ and $\varphi_k\left(\mathbf{r}'\right)$ is that of the neutralizing electron.

The physics behind Eq. (6.11) is the following. $\Phi(\mathbf{k},\mathbf{q},z)\exp(i(E_a - E_k)t)$ is the effective potential that causes the transition from the metal state $\varphi_k \exp(-iE_k t)$ to the final atomic state $\varphi_a \exp(iE_a t)$. This effective potential oscillates in time with frequency $\omega = E_a - E_k$ inducing a corresponding fluctuation in electronic density:

$$\delta n(\mathbf{q},\omega,z) = \int dz' \chi(\mathbf{q},\omega,z,z') \Phi(\mathbf{k},\mathbf{q},z)$$

whose imaginary part is is related to energy losses. In fact the quantity $\mathrm{Im}(\int dz\ \delta n(\mathbf{q},\omega,z)\Phi(\mathbf{k},\mathbf{q},z))$ gives the excitation rate of the electron gas. Then, the total neutralization rate in Eq. (6.11) contains the contributions of all possible events that can neutralize the ion, which include excitation of single-particles (Auger processes) and plasmons.

The surface susceptibility $\chi(\mathbf{q},\omega;z,z')$ is a smooth function of the spatial coordinates z, z' that interpolates from zero very far from the surface to the bulk susceptibility of the solid. In addition to containing the excitation spectrum of the single-particle and collective modes (normal and multipole surface plasmons), χ also contains the coupling between these modes, which accounts for the decay of surface plasmons of $q \neq 0$ into electron-hole pairs and the accompanying width of the plasmon resonance (Landau-damping). On the other hand, volume plasmons cannot decay into electron-hole pairs below a critical frequency in this linear theory. The fact that this theory includes implicitly the decay of the surface modes is what allows the description of the electron emission that is observed experimentally.

The surface susceptibility $\chi(\mathbf{q}, \omega; z, z')$ can be obtained from the Time-Dependent Density Functional Theory (TDDFT). When the effective interaction is simply the Coulomb potential (neglecting exchange and correlation between electrons), the evaluation is said to be done in the Random Phase Approximation (RPA). The use of TDDFT or the RPA for calculating EELS produce good agreement with experiments [10]. The computation of the neutralization rate is much more involved because it is necessary to evaluate χ for as many values as required to perform the 8-dimensional integral. This is one of the reasons why the problem of calculating a realistic neutralization rates at surfaces has been long-standing and why there are so many different approximations to the matrix elements Eq. (6.11) and/or χ in the literature. One common approximation is the calculation of the matrix elements using the bare Coulomb potential, which is reasonable only for $\omega \gg \omega_P$ when screening can be ignored. This is clearly seen in the plot of Fig. 6.5 [45] of the neutralization rate per unit frequency $d(1/\tau)/d\omega$ for He^+ at $z_a = 5$ a.u. from a Lang-Kohn jellium surface of $r_S = 2.0$ a.u. representing Al. The calculation neglecting screening, represented by the line with squares in Fig. 6.5, gives a smooth function of ω, while that with screening, in which χ is evaluated in the TDDFT, raises sharply in the region of the collective surface modes of Al, which are obviously absent in the calculation neglecting screening. Therefore,

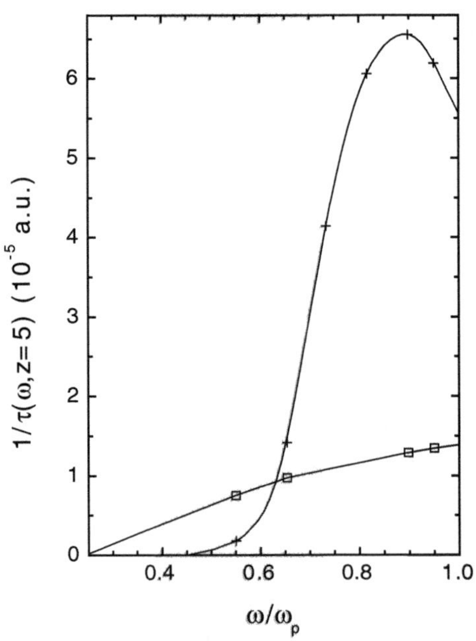

Fig. 6.5. Neutralization rate per unit energy transfer for He^+ at 5 au from the jellium edge of Al. The *squares* are the results of the unscreened calculation and the crosses those for the full calculation including all surface modes

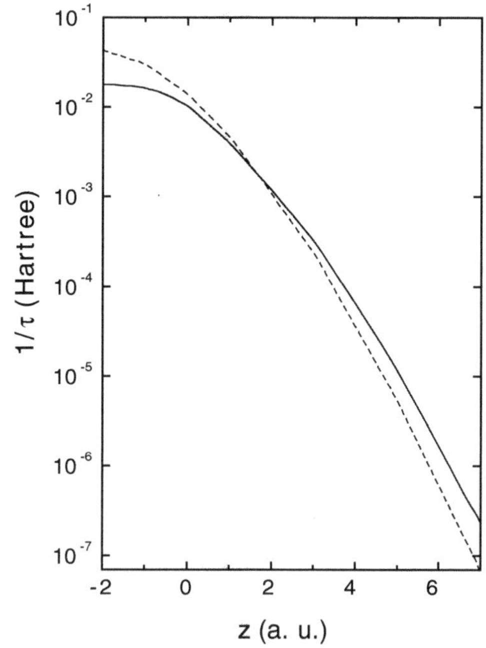

Fig. 6.6. Neutralization rate for He^+ vs distance from the jellium edge of Al. *Dashed line*: unscreened calculation; full line: full calculation including all surface modes. From Monreal and Flores [46]

the comparison between both calculations shows that the plasmon-assisted neutralization channel is indeed dominant at these energies. In the limit of small energy transfers, the use of an unscreened potential overestimates the transition rate, as expected. The overall effect on the total neutralization rate of He^+ on Al is not so large, however, as shown in Fig. 6.6 [46], because there is a compensation when integrating over ω, but such a compensation might not hold in other systems.

The same approach has been used by Cazalilla et al. [47]. These authors refine the theory by including the possible tunneling of metal electrons to the 2s-levels of He. They found that the neutralization rate is not much affected in the region of distances between ion and surface where neutralization occurs.

Other calculations of the plasmon assisted neutralization process [48] improve on the atomic wave functions of He but use an interaction potential with a spatial dependence only appropriate for very large distances between ion and solid. In addition, the approach only describe an undamped surface plasmon decoupled from electron-hole pair excitation and thus overestimates the plasmon neutralization channel.

The plasmon excitations seen in the neutralization of slow ($v \ll v_F$) rare gas ions at surfaces can also occur for fast ions. For Al and Mg, stationary protons cannot excite volume plasmons since $I' \approx 11.6$ eV is insufficient. For moving protons, the velocity distribution of valence electrons is Doppler shifted in the frame of the ion, increasing the maximum energy release accompanying neutralization by $\Delta E_s = mvv_F + {}^1\!/\!{}_2 mv^2$ [32]. The shift may then enable neutralization, effectively increasing the potential energy available for electron capture. Also contributing is the broadening $\delta E = v_n/a$ caused by the finite time the ion spends near the surface, where a is the width of the tail of electron density spilling outside the surface, and v_n the ion velocity normal to the surface. Although the argument is being used specifically for ion-surface collisions, it is actually the same occurring in plasmon-assisted electron capture *inside* the solid. In the atomic picture, the broadening occurs due to the finite collision time, and the transition probability peaks at the velocity $v \approx \hbar a/\Delta E$ (the Massey criterion), where ΔE is the adiabatic energy defect in an electron capture collision.

6.2.3 Electron Energy Distributions

As mentioned, potential excitation of plasmons competes with Auger electron emission due to Auger capture and with Auger deexcitation of excited atoms [49]. The characteristic energy distribution of the plasmon decay electrons allows their separation from the electrons originating from those Auger processes, from direct excitation in ion-atom and ion-electron collisions, and Auger decay of core excitations [49, 50].

The theoretical study of the spectrum of the electrons emitted in neutralization is much more involved than that of the neutralization mentioned above. It was performed by Monreal [51] for He^+ and Ne^+ on Al. The calculation is based on the fact that in the self-consistent field theory the rate at which metallic excitations are produced by any external potential Φ has the following golden-rule expression:

$$\frac{1}{\tau}(z_a) = 4\pi \int_{E_F}^{\infty} dE_A \int d\Omega \sqrt{2E_A} \sum_{k\langle k_F} \int d\mathbf{k}' f(\mathbf{k}') |\langle \mathbf{k}_A | \Phi_{SCF} | \mathbf{k}' \rangle|^2 \delta(E_A - E_{k'} - E_k + E_a) \quad (6.13)$$

Where $|\mathbf{k}_A\rangle$ and $|\mathbf{k}'\rangle$ are one-electron states representing the excited Auger electron of energy E_A and the hole left behind with energy $E_{k'}$ respectively, $d\Omega$ is the differential of solid angle and the total potential producing the excitation is just the sum of the "external potential" Φ of Eq. (6.12) plus the potential induced in the metal (see Eqs. (1.14)–(1.16) of Chap. 1), that is:

$$\Phi_{SCF}(\mathbf{k}, \mathbf{q}, z) = \Phi(\mathbf{k}, \mathbf{q}, z) + \frac{2\pi}{q} \int dz_1 \times \int dz_2 \chi(q, \omega, z_1, z_2)\, e^{-q|z-z_1|}\, \Phi(\mathbf{k}, \mathbf{q}, z_2) \quad (6.14)$$

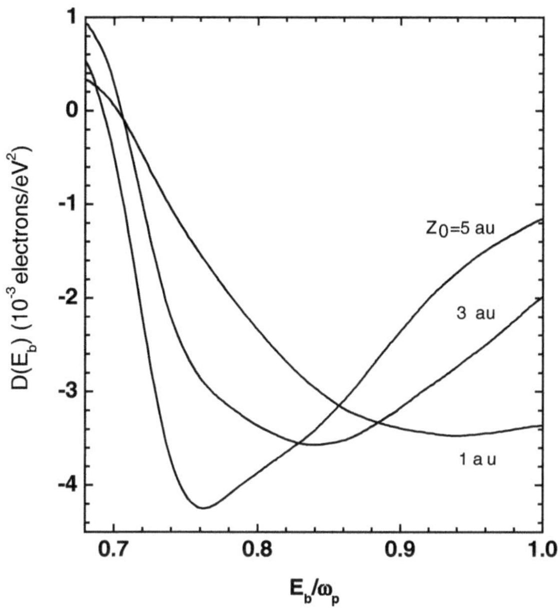

Fig. 6.7. Derivative of the energy distribution of electrons emitted by Ne^+ from Al calculated at different distance from the jellium edge indicated in the figure using a hydrogenic model of the atom. The energy scale is with respect to the Fermi level and normalized to the bulk plasmon energy, ω_P

In this way, the integrand of Eq. (6.13) gives the number of excited electrons with energy E_A within a solid angle $d\Omega$ that is calculated consistently with the neutralization rate. As we said above, surface plasmons are Landau-damped and can show up in electron emission spectra. The results of this calculation show that the dip in the derivative of the electron spectra does not depend on the ionization potential (Fig. 6.7) and that the minima are due to the excitation of surface plasmons, which are the only surface modes that the simple model surface used in the calculation can sustain, since multipole plasmons require a gradual surface barrier. The normal surface plasmons are excited mainly at the short distances where the ions are neutralized and have high parallel momentum q (typically $q \sim q_F/2$). Even though the calculations shown in Fig. 6.8 are in good agreement with experiments for Ne^+ on Al, there is the need of further theoretical studies using a more realistic surface barrier that allows for both monopole and multipole plasmons, in the same way as was done for the total neutralization rate. The possibility that the observed plasmons are mSP is suggesting by the idea that the density fluctuation of the mSP has dipole character [2] and could couple more easily with the surface dipole formed by the incoming ion and its image charge, that rapidly disappears upon neutralization (the image charge transfers to the ion and the incoming hole transfers to the solid).

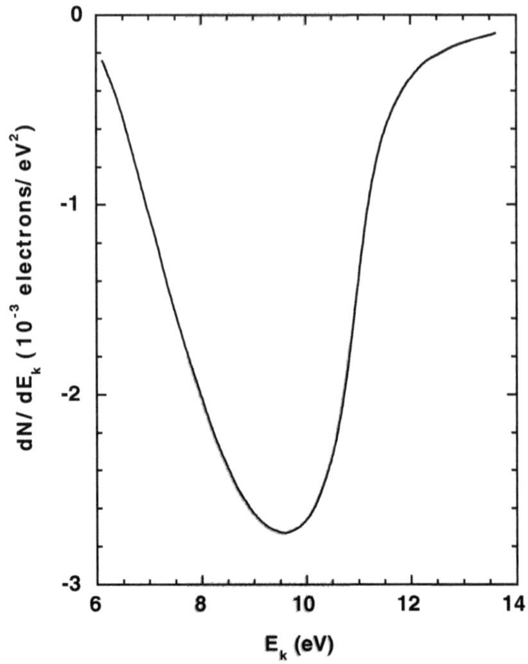

Fig. 6.8. Derivative of the energy distribution of electrons emitted by Ne^+ from Al calculated at 1 a.u. from the Al jellium edge, from the results of Fig. 6.7, using an additional Gaussian broadening of 1 eV to take into account of the experimental energy resolution. The energy scale is referred to the vacuum level

We note here that an additional mechanism for potential plasmon excitation is possible during the redistribution of charge that occurs during the deexcitation of an atom or ion at a surface, if it occurs fast enough (currently unknown). The resonant condition is achieved when the energy difference between the (broadened) atomic levels equals the plasmon energy [52]. This excitation transfer (Fig. 6.4) may contribute to the decay of the excited cloud of the "hollow atom" that forms when a multiply charged ion captures electrons from a surface.

It was argued theoretically that significant volume plasmon excitation could occur by *external* charges not too distant from the surface [53]. However, in core level photoemission of adsorbates, where both the hole and the emitted electron remain outside the surface, only low q SP are excited.

6.2.4 Secondary Processes

Mechanisms that can excite volume plasmons at velocities lower than v_{th} Eq. (6.3) were first discussed by Ritzau et al. [32]. These are: (a) the effect of finite width of plasmons that allows lower energy excitations with energies lower than $E_p(q = 0)$, (b) second-order electronic processes, (c) absorption of momentum by lattice atoms by an Umklapp process, (d) the increase in neutralization energy in potential excitation afforded by the shift in target electron energy in the projectile frame (discussed at the end of Sect. 2B), and (e) excitation by electrons faster than the projectile.

Ritzau et al. [32] considered processes (a)–(c) for protons on Al and Mg and concluded that they are unlikely. Nevertheless, the idea of absorption of momentum by lattice atoms was retaken by von Someren et al. [54], to explain prominent peaks in the energy distribution of electrons emitted in grazing collisions of 2–6 keV protons with Al (111) surfaces. The structure was attributed to plasmons, but this assignment cannot go over the objection that the electron energies are too large for plasmon decay; they extend to more than 20 eV for 6 keV protons. As discussed in Sect. 6.4.1 below, the explanation of these peaks is electron diffraction.

The most likely sub-threshold mechanism for protons is the indirect plasmon excitation by fast secondary electrons produced in binary proton-electron interactions [32]. The minimum energy that an electron must have to excite a volume plasmon is 23 eV for Al and 17 eV for Mg, measured from the Fermi level, when calculated for an electron gas model in the RPA. This mechanism cannot be easily verified by correlating, in a given electron energy spectrum, the number of plasmon decay electrons and the number of electrons ejected with sufficient energy to excite a plasmon. This is because the path of the fast electrons near the surface, which can be affected by the angle of ion incidence [55], is undetermined. In addition, the correlation may be misleading since the fast emitted electrons represent only a tiny fraction of the total number of fast electrons moving *inside* the solid, due to the small escape depths (~1–2 nm). In particular, most of the electrons that have excited plasmons (and thus lost energy) will appear in the low energy part of the energy spectrum. Rather, one may calculate the energy distribution of fast electrons *inside* the solid from that of *ejected* electrons plus a model for attenuation and escape [32] or do a full theoretical estimate that includes the calculation of the initial distribution of electrons ejected in binary collisions and a transport calculation [56]. The results of both methods differ, not surprisingly given the approximations needed in both cases.

Plasmon excitation by fast electrons from binary collisions requires a threshold velocity, because the maximum energy transfer from a proton to an electron at the Fermi level, $2mv(v + v_F)$ [57] must lead to a final electron energy exceeding the minimum required for plasmon excitation. The value of this threshold $v_{\rm th}$ is 0.74 (0.64) $\times$ 10^8 cm/s for Al (Mg), quite lower than $v_{\rm th}$ for direct excitation, and corresponds to 2.9 (2.1) keV/amu.

6.3 Experimental Techniques for Studying Plasmon Excitation and Decay

In the case of electron impact, plasmon excitation is typically studied by measuring the characteristic energy losses of the electrons reflected from surfaces or transmitted through thin films. The assumption is made that broad energy loss peaks, resembling resonances, are due to the excitation of plasmons, if there is agreement with some theoretically predicted value of the energy loss. Outside from the case of nearly-free electron metals, it is not normally unclear how to distinguish between plasmons and interband transitions. As an example, the literature abounds on the assumption that water has a plasmon peak in energy loss at around 21 eV, but this assignment is ruled out by studying the effect of incident electron energy [58].

The energy loss technique is not easily available for ions due to the masking effect of energy losses in momentum transfer collisions with target atoms. It may be possible to circumvent this problem by using channeling or by having glancing ion-surface collisions over regions of limited extent. Given this limitations, plasmon excitation is inferred from the emission of electrons and photons when the plasmon decay, a method available for other forms of excitations as well.

6.3.1 Energy Distribution of Plasmon-Decay Electrons

The first detailed analysis of energy distribution of electrons from plasmon decay is by Chung and Everhart [18] using electron impact on Aluminum. The analysis assumes that the energy of the plasmon is given to a valence electron (with momentum taken up by the lattice) with equal probability across the valence band. Thus, in a free-electron model, the energy distribution of electrons from plasmon decay is the density of states (DOS) displaced by the plasmon energy E_p, convoluted with the lifetime energy broadening of the plasmon. The energy distribution is then affected by electron-electron scattering (although this is not very important at small energy losses) and by the energy-dependent probability $P(E)$ of transmission through the surface barrier. $P(E)$ in turn depends on the angular distribution of electrons and diffraction effects, and thus is generally not known, although some approximations exist [49].

The plasmon yields are obtained from the derivative of the energy distribution dN/dE, by subtracting a background in the region 6–16 eV and integrating the plasmon dip [See also discussion in Ref. [59]]. The procedure to obtain absolute yields of plasmon decay electrons is discussed by Stolterfoht et al. [20].

A new development, the deconvolution of spectral features using the technique of factor analysis [60] hold the promise of circumventing the need to make approximations to extract the profile of plasmon decay electrons.

6.4 Experimental Results

6.4.1 Dependence on Type of Solid

Experiments with slow He^+ ions on Al and Mg, allow an unequivocal assignment of the spectra features to plasmon decay, rather than AN. First, the difference between the high-energy edge of the two distributions (Fig. 6.3) is $I' - E_p - \phi$ which indicates that a general way to separate the plasmon decay structure from that of AN is to use incident ions with very different I, such as He^+, Ne^+, and Ar^+. In addition, the width of the high-energy edge in AN depends on ion velocity normal to the surface [49, 61, 62], while that of plasmon decay is nearly fixed, determined by the plasmon lifetime plus a possible dispersion with q. From these arguments, it is clear that the prominent plasmon shoulder that appears for impact with He^+ and Ne^+ ions (and fast electrons), but not with Ar^+ ions (Fig. 6.3), cannot be attributed to AN involving structure in the density of valence states [63], since its energy is not correlated with I, and is not broadened by increasing the ion velocity. For slow He^+ ions on Al and Mg, plasmon decay is more important than AN; this is more clearly seen for Mg in Fig. 6.9, where the different groups of electrons are well separated in energy, and allow the determination of each component with a simple model [64]. For Be, only AN is seen in Fig. 6.3 (the structure shifts with the potential energy of the ion), probably due to the relatively large plasmon width.

A few experiments have investigated the effect of adsorption on the electron spectra. The sensitivity of the measured plasmon structure to slight cesiation or oxidation of the surface [15] is a very strong confirmation that the excitations are indeed *surface* plasmons. The fact that oxidation suppresses plasmon assisted neutralization, rather than just shift the plasmon energy, as measured with primary electrons [65] indicates that the role of oxygen is to suppress the creation of plasmons under ion impact.

Experiments of ion-induced electron emission with single crystal targets, under conditions of narrow detection angle produce energy distributions with pronounced diffraction peaks [66, 67], which have been mistakenly attributed to plasmon decay in keV proton collisions with Al(111) [68], as demonstrated in the comparison of electron spectra from single- and poly-crystalline targets [69] and in the absence of the peaks when using large solid angle of collection of electrons from Al(111) [15].

6.4.2 Dependence on Type of Projectile

The first study by Baragiola and Dukes [37] used 50 eV–4.5 keV He^+, $Ne^+$and Ar^+ ions on poly-crystalline Al and Mg and established the importance of the potential energy of the projectile, which must be larger than the plasmon energy. Eder et al. studied plasmon excitation by H^+, H_2^+, He^+, Ne^+, Ne^{2+}, Ar^+ and Ar^{2+} ions below 10 keV on clean poly- and mono-crystalline Al surfaces.

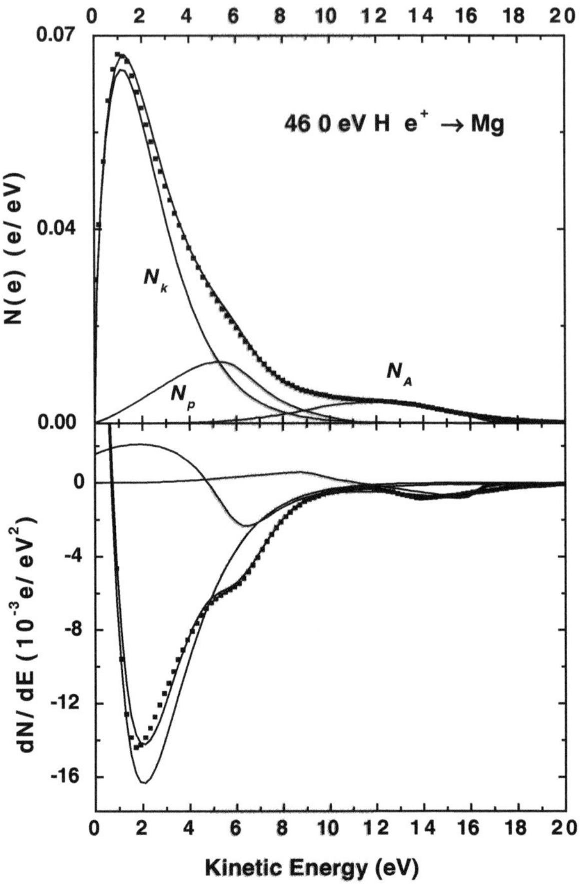

Fig. 6.9. *Top*: The energy spectrum of electrons ejected from Mg by 460 He^+ eV ions is reproduced by the sum of three contributions: collision cascade (Nk), plasmon decay (Np) and Auger neutralization (NAN). *Bottom*: Derivative of the experimental and the calculated spectra. From [64]

Only He^+, Ne^+, Ne^{2+}, and Ar^{2+} were energetically capable of exciting plasmons by neutralization on the surface. Other experiments have reiterated that plasmon excitation by keV heavy ions on Al is produced by Auger electrons from the decay of Al-2p holes excited by electron promotion in binary atomic collisions [60, 70].

Plasmon excitation has been studied as a function of the incident charge n of Ne ions [20, 71–73], with the result that the plasmon decay intensity increases with charge state n and saturates around $n = 5, 6$. In this case, plasmons are produced indirectly in the bulk by fast secondary electrons, and the plasmon decay energy corresponds to volume plasmons. Further studies using Ne^{4+} ions showed, in addition to the excitation of volume plasmons, that

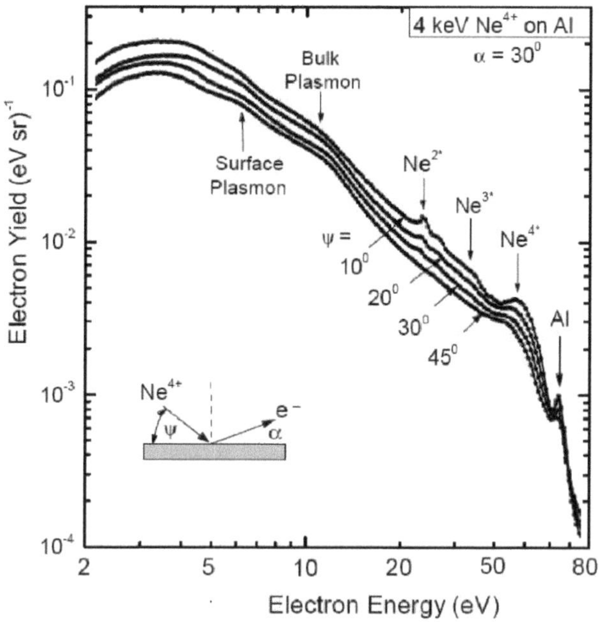

Fig. 6.10. Electron yields measured for 4 keV Ne^{4+} impact on Al vs. electron energy. The parameter is the incidence angle Ψ. The peak labeled Al is produced by L-Auger transitions filling a vacancy in the L-shell of Al. The peaks labeled Ne2*, Ne3*, Ne4* are due to L-Auger transitions in hollow Ne with, respectively, 2, 3 and 4 vacancies in the L-shell. From Stolterfoht et al. [74]

of monopole surface plasmons, again attributed to fast secondary electrons (Fig. 6.10) [74].

6.4.3 Dependence on Incident Energy

Experiments show that the intensity of the plasmon decay structure, where it is observed, is independent of impact energy at low energies. This is characteristic of a potential emission mechanism, enabled by the neutralization energy of the incoming projectile. In this case, the peak in the derivative spectra corresponds to an energy about 1 eV below the energy of the bulk plasmon, but too large to be a shifter (high momentum) monopole surface plasmon. For this reason, the plasmons have been assigned to multipole modes, or mSP (multipole surface plasmons). At energies larger than about 1 keV, the spectra of Al and Mg show, for all ions studied so far, a plasmon that appears at energies consistent with volume plasmons, such as observed by electron impact excitation. This independence of ion type shows that the excitation is not produced by neutralization, but by an indirect mechanism through fast secondary electrons produced during penetration. The transition between surface and bulk excitation is illustrated in experiments with He^+ and Ne^+projectiles on Al and Mg. One observes an apparent shift in the plasmon energy with projectile energy, as shown in Figs. 6.11 and 6.12 for He^+ on Mg [37]. A careful analysis of electron energy spectra [64, 75, 76] allowed separation of the contribution of SP and VP allowing to ascertain that the excitation of SP be nearly constant or slightly decrease with projectile energy, while that of VP increased from a threshold at ~1 keV. An independent method yields a similar threshold for Ne^+ on Al (Figs. 6.13 and Figs. 6.14)

Kinetic excitation of plasmons can occur by a variety of mechanisms, as discussed above: direct Coulomb mechanism, by projectile neutralization, enabled or enhanced by the Doppler effect or by energetic electrons traveling inside the solid. For Ar^+ ions on Al, the first two mechanisms have an onset of 1.6 MeV and 8 keV, respectively. Therefore, plasmon excitation between a threshold at ~1 keV [37, 77] and 8 keV is attributed to energetic secondary electrons, such as those resulting from the Auger decay of 2p vacancies [78] (e.g., 72.7 eV above E_F for Al-LVV Auger electrons).

6.4.4 Angular Dependences

Riccardi et al. [79] showed that for Ne^+ on Al, the plasmon intensity is independent of incidence angle at 1 keV, but increases rapidly with angle at 5 keV. These two contrasting behaviors are consistent with the dominance of potential electron emission(occurring outside the solid) at low energies and of kinetic electron emission (occurring mainly inside) at the higher impact energies [49]. For Ne^{4+} ions on Al, where plasmons are excited mainly by secondary electrons, Stolterfoht et al. [80] finds that the dependence of plasmon intensity on incidence angle has two components, one prominent at glancing incidence, attributed to Auger processes above the surface, and one dominant at larger angles of incidence, attributed to Auger electrons produced inside the solid by penetrating ions.

Regarding the angular distribution of electrons from plasmon decay, the only observations are for volume plasmons excited by secondary electrons [20, 59], which yield a cosine distribution, which is characteristic of a source of electrons *below* the surface.

6.5 Summary, Outlook, and Open Questions

In the last few years, results produced in different laboratories on mechanisms for plasmon excitation in ion-surface collisions consistently show that potential excitation can occur during apparently adiabatic conditions (slow ion motion) because the energy and high frequencies required for excitations are provided by the fast capture of a surface electron. This type of excitation can be tuned by choosing projectile ions of different potential energy. Moreover, by choosing sufficiently low perpendicular velocity to prevent penetration, slow ions become the only known way of exciting plasmons purely outside solids.

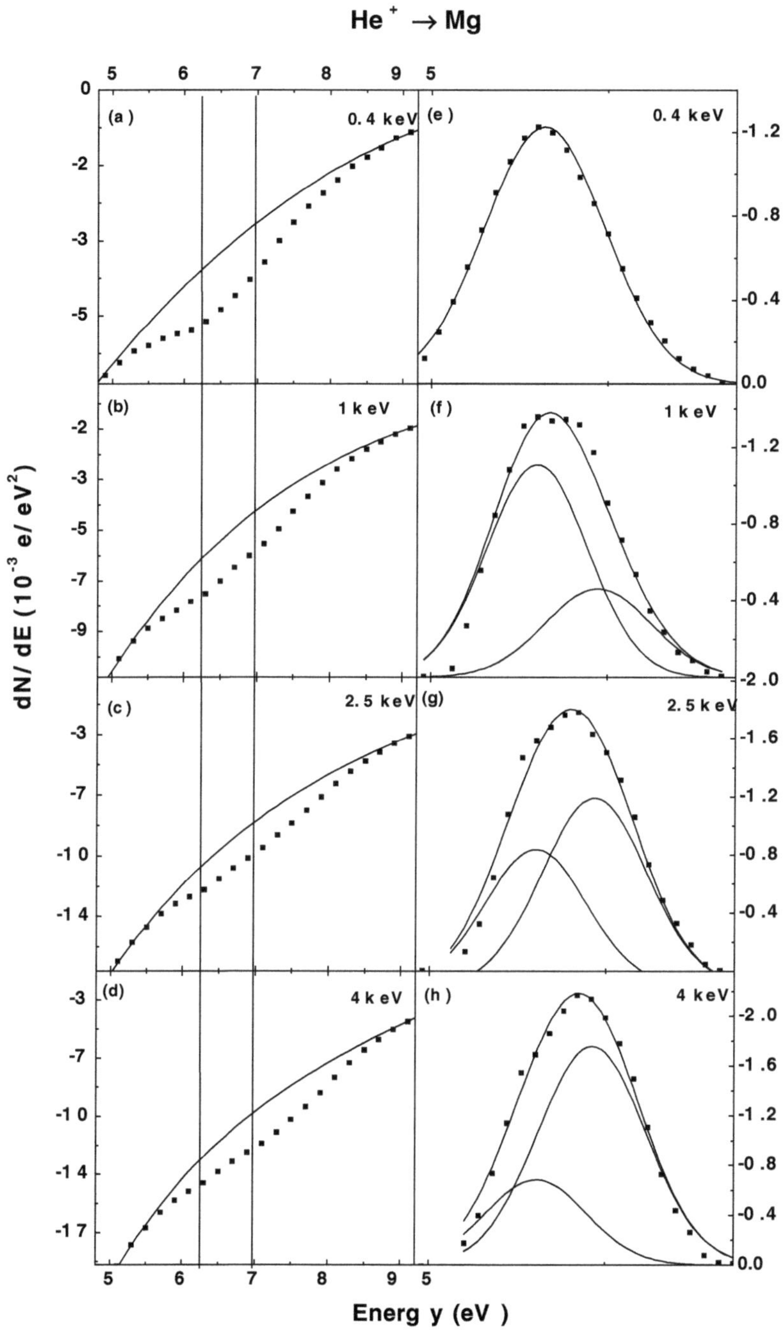

Fig. 6.11. Results of polynomial background subtraction of continuum secondary electrons in dN/dE in Al and fit of two components, surface plasmons excited at low energies, and volume plasmons that are increasingly excited with increasing He^+ energies. From Riccardi et al. [64]

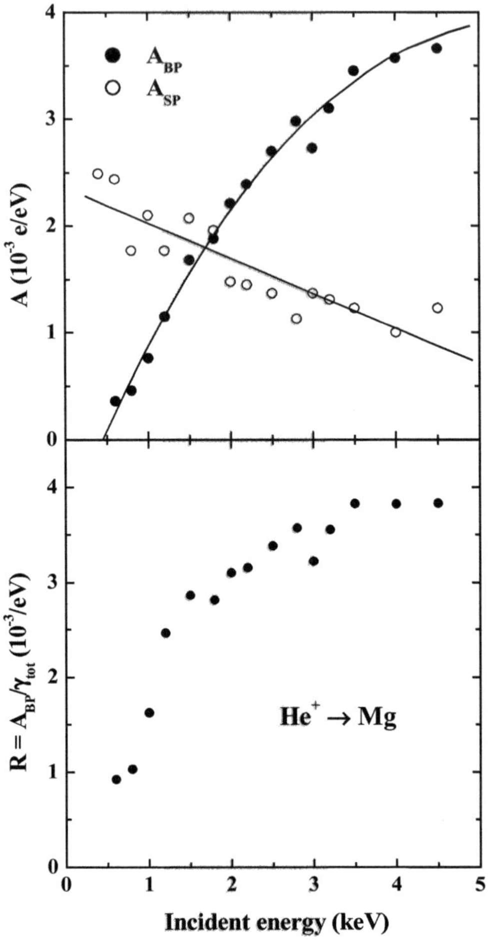

Fig. 6.12. Areas of the surface (A_{SP}) and volume plasmon (A_{BP}) Gaussians in Fig. 6.11 versus He^+ incident energy. From Riccardi et al. [64]

Plasmon and Auger neutralization are separable by their distinct energy distributions and also, in principle, by their different time dependence (Auger emission occurs promptly during neutralization, whereas plasmon decay is delayed by the plasmon lifetime.) The energy separation is very clear in Mg due to the relatively small difference between its plasmon energy and the work function, and serves to show that neutralization leads preferentially to plasmon shake-up rather than to an Auger electron. A further difference between Auger and plasmon processes is that the electron energy distribution in AN broadens strongly when changing the velocity of the ion perpendicular to the surface while that from plasmon decay is determined by the solid and depends only slightly on excitation conditions.

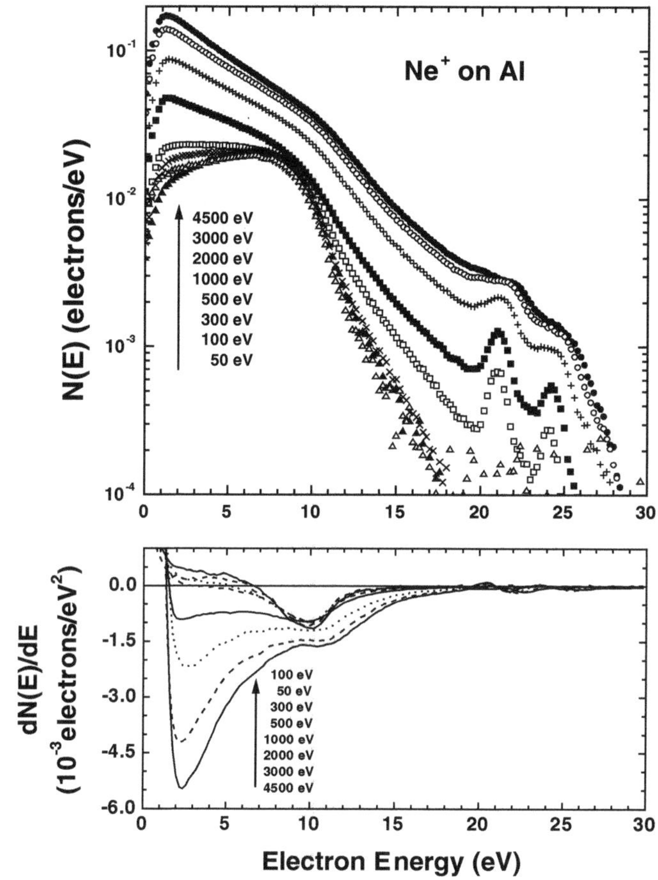

Fig. 6.13. *Top*: electron energy spectra dN/dE for 50–4500 eV Ne^+ ions on aluminum at 12° grazing incidence. *Bottom*: dN/dE. The structure above 20 eV is due to autoionization from backscattered, doubly excited Ne^{**}. From ref. [37]

Plasmons excited outside the surface by the potential mechanism are most likely surface modes, as judged by their sensitivity to surface conditions. They can be either multipole surface plasmons or high q monopole surface plasmons, a question that is still unresolved. Plasmon excitation is predicted to dominate neutralization when it is energetically allowed, and thus may be relevant in secondary ion mass spectrometry, and electron stimulated desorption of ions from surfaces.

In the impact regime where kinetic electron emission is important, plasmons can be produced additionally by fast secondary electrons inside the solid, e.g., Auger and binary electrons with energies above the threshold for plasmon excitation. Fast secondary electrons are responsible for the sub-threshold plasmon excitation by protons, for the kinetic plasmon excitation by keV Ne and

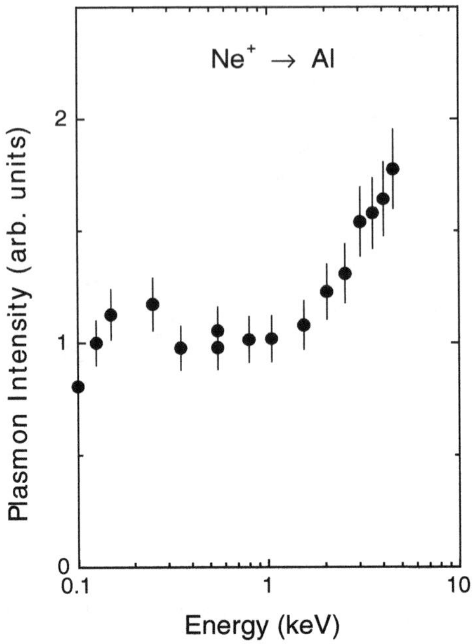

Fig. 6.14. Intensity of plasmon decay electrons for Ne^+ on polycrystalline Al at 12° grazing incidence. The intensity is derived from the strength in dN/dE in Fig. 6.13, after background subtraction. From Ref. [37]

Ar ions, and likely dominates plasmon excitation observed in the interaction of multiply charged ions with solids.

Acknowledgements

The writing of this chapter was assisted by a grant from the Spanish Comisión Interministerial de Ciencia y Tecnología, Proyecto FIS2005-02909.

References

1. H. Raether, *Excitation of plasmons and interband transitions by electrons*, (Springer-Verlag, Berlin, 1980)
2. A. Liebsch, *Electronic excitations at metal surfaces* (Plenum, New York, 1997)
3. G. Ruthemann, Naturwissenschaften **29b**, 648 (1941)
4. D. Pines and D. Bohm, Phys. Rev. **85**, 338 (1952)
5. T.L. Barr, *Modern ESCA* (CRC Press, Boca Raton, 1994) Chap. 6
6. R.H. Ritchie, Phys. Rev. **106**, 874 (1957)
7. H. Raether, *Surface Plasmons on Smooth and Rough Surfaces and on Gratings* (Springer-Verlag, Berlin, 1988)

8. M. Rocca, Surface Sci. Repts. **22**, 1 (1995)
9. P.J. Feibelman, Prog. Surf. Sci. **12**, 287 (1982)
10. K.-D. Tsuei, E.W. Plummer, A. Liebsch, K. Kempa, and P. Bakshi, Phys. Rev. Lett. **64**, 44 (1990)
11. J.J. Quinn, Sol. St. Comm. **84**, 139 (1992)
12. H.J. Levinson, E.W. Plummer, and P.J. Feibelman, Phys. Rev. Lett. **43**, 952 (1979)
13. S.R. Barman, P. Häberle, and K. Horn, Phys. Rev. B **58**, R4285 (1998)
14. G. Chiarello, V. Formoso, A. Santaniello, E. Colavita, and L. Papagno, Phys. Rev. B **62**, 12676 (2000)
15. R.A. Baragiola, C.A. Dukes and P. Riccardi, Nucl. Instr. Meth. Phys. Res. B **182**, 73 (2001)
16. N.B. Gornyi, L.M. Rakhovich, and S.T. Skirko, Sov. Phys. J. **20**, 15 (1967)
17. C. von Koch, Phys. Rev. Lett. **25**, 792 (1970)
18. M.S. Chung and T.E. Everhart, Phys. Rev. B **15**, 4699 (1977)
19. M. Rösler and W. Brauer, Springer Tracts Mod. Phys. **122**, 1 (1991)
20. N. Stolterfoht, D. Niemann, V. Hoffmann, M. Rösler, and R.A. Baragiola, Phys. Rev. A **61**, 052902 (2000)
21. L.H. Jenkins and M.F. Chung, Surf. Sci. **28**, 409 (1971)
22. Yu. A. Bandurin, L.S. Belykh, I.W. Mitropolsky, and S.S. Pop, Nucl. Instr. Meth. Phys. Res. B **58**, 448 (1991)
23. Yu. Bandurin, S. Lacombe, L. Guillemot, and V.A. Esaulov, Surf. Sci. **513**, L413 (2002)
24. U. Fano, Phys. Rev. **118**, 451 (1960)
25. H. Ehrenreich and H.R. Philipp, *Proc. Int. Conf. Phys. Semiconductors*, **367**, Edited by A.C. Strickland (Bartholomew Press, Dorking, U.K., 1962)
26. M. Rösler, Nucl. Instr. Meth. Phys. Res. B **69**, 150 (1992)
27. A.A. Lucas and M. Šunjic, Surf. Sci. **32**, 439 (1972)
28. M. Rösler, Appl. Phys. A **61**, 595 (1995)
29. R. Zimmy, Surf. Sci. **260**, 347 (1992)
30. F.J. García de Abajo and P.M. Echenique, Nucl. Instr. Meth. Phys. Res. B **79**, 15 (1993)
31. J.A. Gasspar, A.G. Eguiluz, and D.L. Mills, Phys. Rev. B **51**, 14604 (1995)
32. S.M. Ritzau, R.A. Baragiola, and R.C. Monreal, Phys. Rev. B **59**, 15506 (1999)
33. D. Hasselkamp and Scharmann, Surf. Sci. **119**, L388 (1982)
34. F.J. García de Abajo and P.M. Echenique, Nucl. Instr. Meth. Phys. Res. B **79**, 15 (1993)
35. F.J. García de Abajo, Nucl. Instr. Meth. Phys. Res. B **98**, 445 (1995)
36. Y.H. Song, Y.N. Wang and Z.L. Miskovic, Phys Rev A **68**, 022903 (2003)
37. R.A. Baragiola and C.A. Dukes, Phys. Rev. Lett. **76**, 2547 (1996)
38. P. Apell, J. Phys. B: At. Mol. Opt. Phys. **21**, 2665 (1988)
39. A.A. Almulhen and M.D. Girardeau, Surf. Sci. **210**, 138 (1989)
40. R. Monreal and N. Lorente, Phys. Rev. B **52**, 4760 (1995)
41. F.A. Gutierrez, Surf. Sci. **370**, 77 (1997)
42. M.A. Vicente Alvarez, V.H. Ponce, and E.C. Goldberg, Phys. Rev. B **27**, 14919 (1998)
43. F.A. Gutierrez, J. Jouin, S. Jequier, and M. Riquelme, Surf. Sci. **431**, 269 (1999)
44. H, Jouin, F.A. Gutierrez, C. Harel, S. Jequier, and J. Rangama, Nucl. Instr. Meth. Phys. Res. B **164**, 595 (2000)

45. N. Lorente and R. Monreal, Surf. Sci. **370**, 324 (1997)
46. R.C. Monreal and F. Flores, Adv. Chem. Phys. **45**, 175 (2004)
47. M.A. Cazalilla, N. Lorente, R. Diez Muiño, J.P. Gauyacq, D. Teillet-Billy, and P.M. Echenique, Phys. Rev. B **58**, 13991 (1998)
48. F.A. Gutierrez and H. Jouin, Phys. Rev. A **68**, 012903 (2003)
49. R.A. Baragiola, in *Low Energy Ion-Surface Interactions*, ed. by J.W. Rabalais (Wiley, 1994) Chap. 4.
50. S. Valeri, Surface Sci. Repts. **17**, 85 (1993)
51. R. Monreal, Surf. Sci. **388**, 231 (1997)
52. J.I. Gersten and N. Tzoar, Phys. Rev. B **9**, 4038 (1973)
53. A. Bergara, J.M. Pitarke, and R.H. Ritchie, Phys. Lett. **256**, 405 (1999)
54. B. Van Someren, P.A. Zeijlmans van Emmichoven, I.F. Urazgil'din, and A. Niehaus, Phys. Rev. A **61**, 32902 (2000)
55. E.A. Sánchez, J.E. Gayone, M.L. Martiarena, O. Grizzi, and R.A. Baragiola, Phys. Rev. B **61**, 14209 (2000)
56. M. Rösler, Nucl. Instr. Meth. B **164–165**, 873 (2000)
57. R. A Baragiola, E.V. Alonso, and A. Oliva-Florio, Phys. Rev. B **19**, 121 (1979)
58. C.D. Wilson, C.A. Dukes, and R.A. Baragiola, Phys. Rev. B **63**, 1101 (2001)
59. P. Riccardi, P. Barone, A. Bonanno, M. Camarca, A. Oliva, F. Xu, and R.A. Baragiola, Nucl. Instr. Meth. Phys. Res. B **182**, 64 (2001)
60. N. Bajales, S. Montoro, E.C. Goldberg, R.A. Baragiola, and J. Ferron, Surf. Sci. **579**, L97 (2005)
61. A. Sindona, R.A. Baragiola, S. Maletta, G. Falcone, A. Oliva, and P. Riccardi, Nucl. Instr. Meth. Phys. Res. **230**, 298 (2005)
62. A. Sindona, R.A. Baragiola, G. Falcone, A. Oliva, and P. Riccardi, Phys. Rev. A **71**, 052903 (2005)
63. A. Hitzke, J. Günster, J. Kolaczkiewicz, and V. Kempter, Surface Sci. **318**, 139 (1994)
64. P. Riccardi, A. Sindona, P. Barone, A. Bonanno, A. Oliva, and R.A. Baragiola, Nucl. Instr. Meth. Phys. Res. B **212**, 339 (2003)
65. M. Alducin, S.P. Apell, I. Zoric, and A. Arnau, Phys. Rev. B **64**, 125410 (2001)
66. J. Mischler, M. Negrè, and N. Benazeth, Rad. Eff. Def. Solids **70**, 117 (1983)
67. T. Bernhard, Z.L. Fang, and H. Winter, Phys. Rev. A **69**, 060901 (2004)
68. B. van Someren, P.A.Z. van Emmichoven, I.F. Urazgil'din, and A. Niehaus, Phys. Rev. A **61**, 032902 (2000)
69. H. Winter, H. Eder, F. Aumayr, J. Lörincik, and Z. Sroubek, Nucl. Instr. Meth. Phys. Res. B **182**, 15 (2001)
70. M. Commisso, M. Minniti, A. Sindona, A. Bonanno, A. Oliva, R.A. Baragiola, and P. Riccardi, Phys. Rev. B **72**, 165419 (2005)
71. D. Niemann, M. Grether, M. Rösler, and N. Stolterfoht, Nucl. Instr. Meth. Phys. Res. **146**, 70 (1998)
72. D. Niemann, M. Grether, M. Rösler, and N. Stolterfoht, Nucl. Instr. Meth. Phys. Res. B **161**, 90–95 (2000)
73. D. Niemann, M. Grether, M. Rösler, and N. Stolterfoht, Phys. Rev. Lett. **80**, 3328 (1998)
74. N. Stolterfoht, J.H. Bremer, V. Hoffmann, M. Rösler, and R.A. Baragiola, Nucl. Instr. Meth. Phys. Res. B **193**, 523–529 (2002)
75. P. Barone, R.A. Baragiola, A. Bonanno, M. Camarca, A. Oliva, P. Riccardi, and F. Xu, Surf. Sci. **480**, L420 (2001)

76. P. Barone, A. Sindona, R.A. Baragiola, A. Bonanno, A. Oliva, and P. Riccardi, Nucl. Instr. Meth. Phys. Res. B **209**, 68 (2003)
77. P. Riccardi, P. Barone, M. Camarca, A. Oliva, and R.A. Baragiola, Nucl. Instr. Meth. Phys. Res. B **164–165**, 886 (2000)
78. R.A. Baragiola, E.V. Alonso, and H. Raiti, Phys. Rev. A **25**, 1969 (1982)
79. P. Riccardi, P. Barone, A. Bonanno, A. Oliva, and R.A. Baragiola, Phys. Rev. Lett. **84**, 378 (2000)
80. N. Stolterfoht, J.H. Bremer, V. Hoffmann, M. Rösler, R.A. Baragiola, and I. de Gortari, Nucl. Instr. Meth. Phys. Res. B **182**, 89 (2001)

7

Slow Ion-Induced Electron Emission from Thin Insulating Films

P.A. Zeijlmans van Emmichoven and Y.T. Matulevich

7.1 Introduction

Thin insulating films are known to be excellent electron emitters. They are widely applied as cover material for conductive electrodes in devices where the performance critically depends on the secondary electron emission yield. Well known examples are Micro Channel Plates [1] and Plasma Display Panels [2]. In recent years, a lot of research has been devoted to find stable materials with the highest possible electron yield and to study the underlying mechanisms for electron emission. In spite of these research efforts, the emission mechanisms are not yet understood in detail and research continues.

In this chapter, we give an overview of the present understanding of slow ion-induced electron emission from thin insulating films. We mainly focus on those research results that are relevant for Plasma Display Panels (PDPs), thereby limiting ourselves to slow noble-gas ions and to specific materials. With "slow" we mean that the ions have kinetic energies of tens of an eV. In industries, research groups have carried out many experiments to find stable materials with the highest possible electron yield. They mainly focussed on metal oxides and in particular on MgO, widely used in present-day PDPs. The most important results will be presented in this chapter. At the same time, significant research activities took place at universities and in research laboratories to study the fundamental processes occurring in ion-insulator collisions. As opposed to metals, many of the phenomena are not yet very well understood. The poor conductivity of the insulators causes well-controlled experiments to be more complicated to carry out. In addition, the different band structure of insulators enables other ways for energy deposition, thereby opening other channels for neutralization of the incoming ions. A lot of the experiments have been carried out on surfaces of the alkali halides and in particular on LiF. The results that are of importance for understanding the electron emission under conditions as in PDPs will be presented in this chapter as well. Only very recently, in collaborations between industries and universities, well-controlled studies on MgO have been carried out. Although progress

P.A. Zeijlmans van Emmichoven and Y.T. Matulevich: *Slow Ion-Induced Electron Emission from Thin Insulating Films,* STMP **225**, 213–239 (2007)
DOI 10.1007/3-540-70789-1_7

has been made, it will also be shown that for revealing the details of the electron emission induced by ions further studies are required.

In Sect. 7.2, we start out with a brief introduction on the role of insulating films in PDPs. The relationship between firing voltage of a discharge cell and total electron yield of the protective insulating film will be discussed. In Sect. 7.3, some fundamental aspects of the interactions between an ion and an insulator will be discussed. The shift of the atomic energy levels of an ion in front of a surface will be considered as well as the channels via which a slow ion can be neutralized. The contents of this section comprises mostly the results of research activities of slow ions interacting with the alkali halides. Based on these results, a simple diagram will be constructed for the anticipated neutralization of ions interacting with MgO. In Sect. 7.4, a brief overview will be given of the experimental results of the electron emission from MgO thin films: results from experiments performed in discharge cells as well as results from ion-beam experiments. In discharge-cell experiments, the results are obtained under conditions close to those in actual PDPs. Experiments with well-defined ion beams, on the other hand, are far away from such conditions. It will be shown that these experiments lead to a more fundamental understanding of the electron-emission processes. In Sect. 7.5, the conclusions will be given as well as an outlook for the future.

7.2 Role of Insulating Films in Plasma-Display Panels

Plasma Display Panels (PDPs) have been developed to achieve large, flat, and light-weight displays. Compared to other technologies, they still have lower brightness and they are expensive. A lot of research has been devoted to improve the performance of PDPs and to lower their cost. The principles of operation of PDPs has recently been described by Boeuf [2]. A PDP consists of a matrix of discharge cells, with groups of three cells corresponding to pixels. Each cell in a pixel contains a phosphor in one of the three primary colors, which is excited by UV light from the discharge. Most commercial PDPs are operated by an AC potential applied between the electrodes of the cell. The minimum potential where the discharge starts is called the firing voltage V_f (or breakdown voltage)[1]. The electrodes are decoupled from the plasma by a dielectric layer covered with a thin insulating film. This insulating film is of crucial importance for the performance of the cell: it provides electrons for the plasma and it protects the dielectric layer from ion bombardment. Since MgO is an excellent electron emitter and very stable under ion bombardment, it is the most common material used in present-day PDPs. A schematic drawing of a typical PDP cell is shown in Fig. 7.1.

[1] The value of the firing voltage V_f is important for the performance of a PDP, since it is directly related to the costs of the components and to the operating costs.

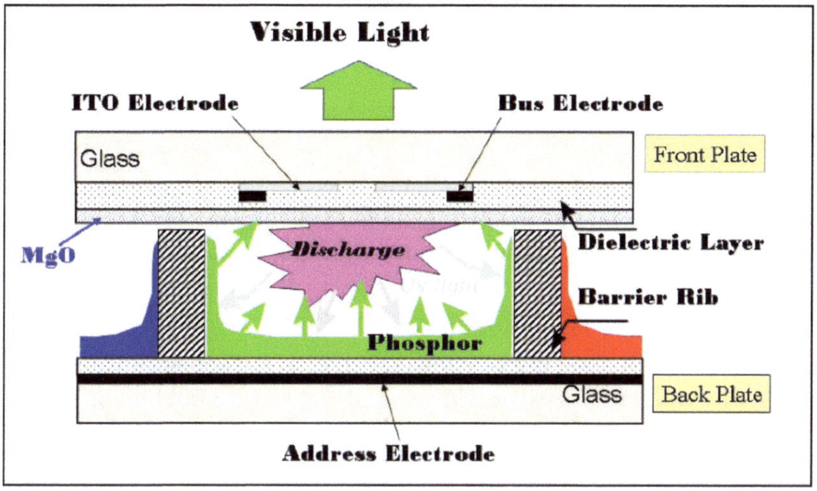

Fig. 7.1. Schematic drawing of a coplanar AC Plasma-Display-Panel cell. The cell is driven by an AC potential between the ITO electrodes. MgO provides electrons for the plasma and protects the dielectric layer from ion bombardment

In most commercial PDPs, the cells are filled with a rare-gas mixture of Ne and Xe (with a few up to 10% of Xe). Sometimes other noble-gas species are added [2, 3]. When a Ne/Xe cell is addressed, the discharge starts and Ne and Xe are excited (Ne^* resp. Xe^*) and/or ionized (Ne^+ resp. Xe^+) [2, 4]. Molecules are also formed in the discharge, in the ground state and in excited and ionized states [5, 6]. Excitation of the phosphor predominantly takes place by UV photons from de-excitation of Xe^* and Xe containing molecules. It is generally accepted that an important source of electrons for the plasma originates in the interaction of Ne^+ with the insulating film. The Ne^+ ions, accelerated by the potential of the cathode, hit the protecting film with average kinetic energies of tens of an eV[2]. Electron emission for these slow ions most likely originates in the neutralization process. The contribution of Xe^+ to the emission of electrons is small, because of its significantly lower potential energy (see also Sect. 7.4.2). It should be noted that long-lived (metastable) Ne^* and UV photons from radiative de-excitation of Ne^* may also contribute to the emission of electrons. For a cold-cathode discharge of Ar, Phelps and Petrović [7] have shown that the relative contributions of particles in different states strongly depend on the discharge parameters. Although in most cases the ions play a dominant role, contributions from particles in other states cannot be excluded.

The firing voltage V_f of a PDP is related to the secondary electron yield of particles hitting the cathode. Assuming, at the cathode, only ions produce

[2] The intensity of ions with kinetic energies of 50 eV and higher is many orders of magnitude smaller [2].

electrons and in the cell only electrons ionize atoms, a simple condition for breakdown is obtained [7]

$$1 - \gamma_i(e^{\alpha d} - 1) = 0 \tag{7.1}$$

with α the spatial ionization coefficient, d the distance between the two electrodes and γ_i the average number of emitted electrons per incoming ion on the cathode. α depends on the electric field E ($\simeq V_f/d$) in the discharge region and on the gas pressure p. Using a semi-empirical relation between α, E, and p,

$$\alpha/p = Ce^{-D\sqrt{p/E}} \tag{7.2}$$

with C and D parameters obtained from fitting α for a specific gas [8, 9], one arrives at the relation[3] between V_f, pd, and γ_i used by Bachmann et al. [11]

$$V_f \simeq \frac{D^2 pd}{\left(\ln \frac{Cpd}{\ln(1/\gamma_i+1)}\right)^2} \tag{7.3}$$

Equation (7.3) shows that an increase of γ_i leads to a decrease of V_f. Measurement of the functional dependence between V_f and pd, called the Paschen curve, can be used to determine a value for γ_i. This will be discussed further in Sect. 7.4.1.

7.3 Interaction of Ions with Insulators

The interaction of ions with insulators has recently received a lot of attention. A general overview can be found in two review papers by H. Winter [12, 13]. Interesting phenomena have been observed that are very distinct from observations at metals, such as large fractions of scattered ions in the negative charge state [14–17], population of exciton levels in the insulator [18], and large sputtering yields induced by multiply charged ions [19, 20]. Most of these observations, however, have been made for ions with kinetic energies above the threshold for kinetic processes. The discussion here will be limited to those processes that may also play a role in slow singly charged ion-surface interactions with ion kinetic energies of tens of an eV. For the phenomena exclusively occurring at higher kinetic energies and for higher charge states, the interested reader is referred to the cited references and to other contributions to this book (Chaps. 1, 3, and 4).

7.3.1 Atomic Energy Level Shifts

The atomic energy levels of a slow, positive, singly charged ion approaching a metal surface shift upwards due to the interaction between ion and its image

[3] Sometimes another choice for the exponential-like dependence of α is made. This leads to a different functional form of V_f (see e.g. [10]).

potential. For an elaborate discussion on image potentials we refer to Chap. 1 of this book. At sufficiently large distances, the static image potential $V_I(z)$ of a positive, singly charged ion reduces to the well known classical formula

$$V_I(z) = \frac{-1}{4z} \tag{7.4}$$

with z the ion-surface distance[4]. For a slow, positive, singly charged ion approaching an insulator surface, the situation is more complicated. We first consider a *resonant* process neutralizing the incoming ion. For the process to take place, one of the states in the valence band has to be resonant with an unfilled atomic state (see Fig. 7.2a). For the electron in the final state (on the atom), not only polarization of the insulator by ionic core and captured electron has to be considered, but also the presence of the screened hole (see Fig. 7.2b). As pointed out by Wirtz et al. [21, 22], considering the diabatic limit, i.e. assuming the hole stays in its original position, this leads to a strong reduction of the upward shift and possibly even to a downward shift. For sufficiently large distances, the shift $V(z,R)$ becomes

$$V(z,R) = V_P(z) + V_H(R) \simeq \frac{\epsilon - 1}{\epsilon + 1}\left(\frac{-1}{4z}\right) + \frac{2}{\epsilon + 1}\left(\frac{1}{R}\right) \tag{7.5}$$

with $V_P(z)$ the polarization potential induced by ionic core and captured electron, $V_H(R)$ the interaction between electron and screened hole, ϵ the static dielectric constant of the insulator, and R the ion-hole distance. Use of the static dielectric constant is justified since only very slow ions are considered. For LiF, with $\epsilon \simeq 9.1$ [21, 23], $V(z,R)$ is virtually 0 for a large range of distances[5]. Only very close to the surface, where the image potentials have to be cut off, equation (7.5) does not hold anymore. The Madelung potential will dominate and as a consequence $V(z,R)$ becomes negative. Similar arguments hold for MgO with $\epsilon \simeq 9.7$ [24]. In deriving equation (7.5), it was assumed that the hole stays in its original position. When this is not the case, in other words when hole dynamics is fast, $V_H(R)$ will be reduced. To our knowledge, so far no one has taken hole dynamics into account in a proper way. For the electron in the initial valence-band state, finally, it is important to note that the interaction with the screened ion needs to be taken into account. This leads to a downward shift of the valence-band level, a shift that is close to the second term in equation (7.5) (see [21]).

We now briefly discuss an *Auger* process neutralizing the incoming ion. For the time being, states in the band gap will not be considered. The condition for the Auger process to take place is that the excited electron has sufficient energy to escape to the conduction band (see Fig. 7.2c). Since the energy is

[4] In fact, for a metal surface, z denotes the distance between the ion and the surface image plane.

[5] This can easily be verified for $z = R$, i.e., when the ion is located on top of the hole.

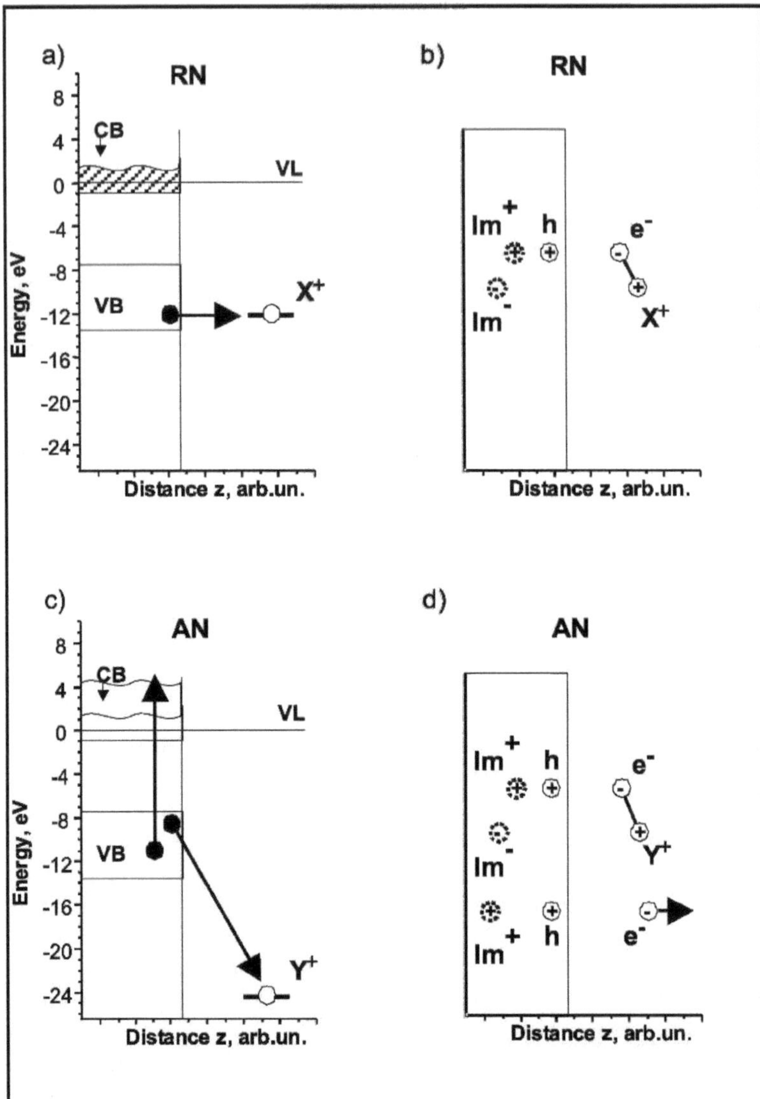

Fig. 7.2. (**a**) One-electron energy diagram for resonant neutralization (RN) of an ion X^+. One of the states in the valence band of the insulator has to be resonant with an unfilled atomic state. VB and CB denote valence and conduction band of the insulator respectively, VL the vacuum level. (**b**) Final state in the diabatic limit after resonant neutralization: not only polarization of the insulator by ionic core and captured electron has to be taken into account, but also the presence of the screened hole (h). (**c**) One-electron energy diagram for Auger neutralization (AN) of an ion Y^+. One electron is captured in an unfilled atomic state, the other electron picks up the released energy and is excited to the conduction band. (**d**) Final state in the diabatic limit after an Auger process: polarization of the insulator by ionic core and electrons as well as the presence of 2 screened holes (or a screened 'double' hole) has to be taken into account

released by capture of another electron in an unfilled atomic state, the exact position of its energy level is again important. For the level shift, similar arguments as in case of a resonant process hold, although the situation is more complicated since two electrons are removed from the insulator (see Fig. 7.2d). The electrons are removed either from two different (most likely adjacent) sites or from the same site. When two sites are involved, the shift, in the diabatic limit, of the final state is given by equation (7.5) with the second term replaced by the interaction between electron and two screened holes. In addition, an extra term appears, V_{hh}, which describes the hole-hole interaction. For LiF, V_{hh} has been determined experimentally for holes on adjacent sites to be about 3.1 eV [25, 26]. For the electrons in the initial valence band states, the interaction with the screened ion has to be taken into account also leading to a downward shift. When the electrons are removed from the same site, on the other hand, the second term in equation (7.5) has to be replaced by the interaction between electron and a 'double' screened hole. For this situation, an extra term appears in the initial state, which takes into account the extra energy needed to remove the second electron from the same site. For MgO, removal of two electrons from the same O center has been considered and the extra binding energy has been estimated to be about 1.6 eV [27]. Finally, for an Auger process, use of the static dielectric constant may not be correct, since the excited electron disappears rapidly from the insulator. It may therefore be better to use the optic dielectric constant in (the modified) equation (7.5).

7.3.2 Neutralization of Ions at Insulators

The neutralization of singly charged noble-gas ions at surfaces of three of the alkali halides will be discussed. The obtained results will be used to make a prediction on how we expect neutralization of slow noble-gas ions proceeds on a MgO surface.

Alkali-Halide Surfaces

The alkali halides are ionic crystals that have wide band gaps of typically 10 eV, with the valence-band electrons localized at the halide ions. At elevated temperatures, the ionic conductivity of these materials is known to be sufficiently good to perform ion-beam experiments [12]. Because of the wide band gap of the alkali halides, electrons emitted in the neutralization of singly charged ions have very low energies. In one of the few electron studies, this has been illustrated for 50-eV He^+ incident on a thin film of LiF [28]. Scattering experiments, on the other hand, in which surviving ion fractions are determined, are very useful to get information on the neutralization of the ions. Several experiments have been performed on single crystal surfaces.

In Fig. 7.3, a one-electron energy diagram is shown for He^+, Ne^+, and Ar^+ ions in front of KI, KCl, and LiF [12]. Hecht et al. [29] discuss scattering experiments on LiF(001) carried out at very grazing angles of incidence of about 1° with the surface. They show that for their lowest kinetic energies of about 1 keV, virtually no Ar^+ survives the collision with the surface as opposed to 75% of Ne^+ and slightly less than 40% of He^+. For Ar^+, the atomic energy level is close to in resonance with the valence band[6], so neutralization most likely proceeds via a resonant process. For Ne^+, resonant neutralization is not possible since its unfilled atomic energy level is not resonant with the valence band, whereas Auger neutralization does not seem likely since the excited electron ends up somewhere in the middle of the band gap. This explains the large fraction of surviving Ne^+ ions. Although a minority of the ions, it needs to be explained via which channel 25% of the Ne^+ ions is neutralized. Khemliche and Roncin et al. [25, 26, 30] carried out detailed experiments, also with keV noble-gas ions scattering off LiF(001) at very grazing angles. The energy loss of scattered surviving ions and neutralized particles was measured in coincidence with the emitted electrons. For the survival fractions they found similar results as Hecht et al. They have convincingly shown [25] that neutralization of Ne^+ proceeds via kinetically assisted Auger neutralization and via the formation of electron bihole complexes, so-called trions[7]. The authors found that the kinetically assisted Auger process is accompanied by an average inelastic energy loss of the projectile of about 7.5 eV, the trion excitation by an energy loss of about 4 eV. For the neutralization of He^+, similar processes can be invoked. The fact that for He^+ a larger fraction of the ions is neutralized, is probably related to the smaller energy deficit for the kinetically assisted processes. In [12], experimental results are also reported for the survival of keV Ar^+ from KCl and KI. For KCl, the survival probability amounts to more than 60%. Neutralization of Ar^+ can be understood in a similar way as for Ne^+ and He^+ on LiF: it can only proceed via kinetically assisted processes. Whether electrons can only be excited to the conduction band or also to trion levels has not been studied for KCl. For KI, finally, only a small fraction of ions survives the collision with the surface (about 10%). This is consistent with the fact that Auger neutralization of the ions can take place virtually without any kinetic energy of the ions involved. It may also be that trions are populated, without any involvement of the kinetic energy of the ions.

In summary, neutralization of noble-gas ions incident on wide-band gap alkali-halide surfaces is suppressed when neither resonant neutralization nor Auger neutralization is energetically allowed according to a one-electron energy diagram as shown in Fig. 7.3. For keV ions, however, kinetically assisted processes have been identified, leading to Auger neutralization of the ions, accompanied with excitation of electrons to the conduction band and to trion

[6] This also holds when the shift of the atomic level is downwards, since also the downward shift of the valence-band states needs to be considered.

[7] In a trion, two holes in the valence band accomodate an excited electron [31]. In the energy diagram shown in Fig. 7.3, the state lies within the band gap.

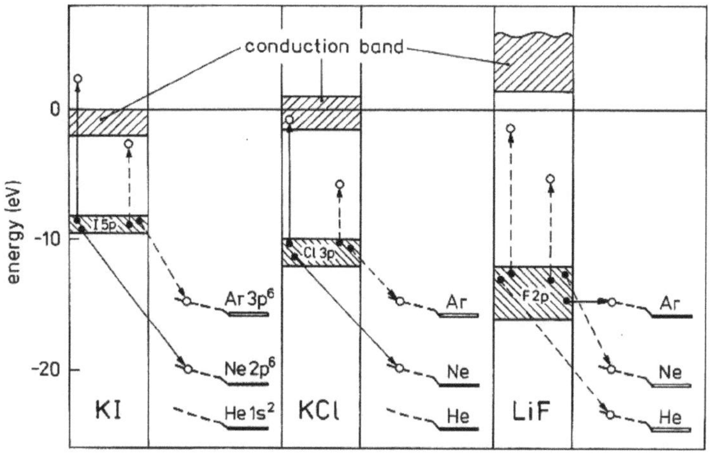

Fig. 7.3. Energy levels of He, Ne, and Ar with respect to the valence and conduction bands of KI, KCl, and LiF. Possible defect states in the band gap are not indicated, nor are exciton and trion levels. The atomic level shifts are drawn only taking into account polarization of the insulator by ionic core and captured electron. When the atomic energy level is resonant with the valence band, resonant neutralization is likely to take place. Auger neutralization takes place when the excited electron has sufficient energy to reach the conduction band. For keV ions, kinetically assisted processes have been identified, leading to considerable neutralization (see text). Reprinted from Progr. Surf. Sci., 63, H. Winter, 'Scattering of atoms and ions from insulator surfaces', pp. 177–247 (2000), Copyright 2000, with permission from Elsevier

levels. For very slow ions, with kinetic energies below 50 eV as in PDPs, on the other hand, kinetic processes are not likely to contribute. Such slow ions will be neutralized by Auger processes only when the excited electron has sufficient energy to reach either the conduction band or a trion level. It should finally be noted that in the discussion we have assumed that the amount of defect states in the band gap is so low that they do not play any role in the neutralization of the ions.

MgO Surfaces

A one-electron energy diagram of MgO is shown in Fig. 7.4. The bulk value of the band gap of MgO is about 7.8 eV [32], which at the surface is reduced to 6.7 eV [33]. For the affinity level of MgO several values can be found in the literature, we used a value of 0.8 eV [34]. The valence band of MgO consists of the O 2p electrons, with electrons in orbitals perpendicular to the surface having a slightly different binding energy than the ones in orbitals in the surface plane [35]. Defect states in MgO have been studied extensively, theoretically as well as experimentally. They are not shown in the energy diagram, nor are exciton and trion levels. For neutral and charged oxygen

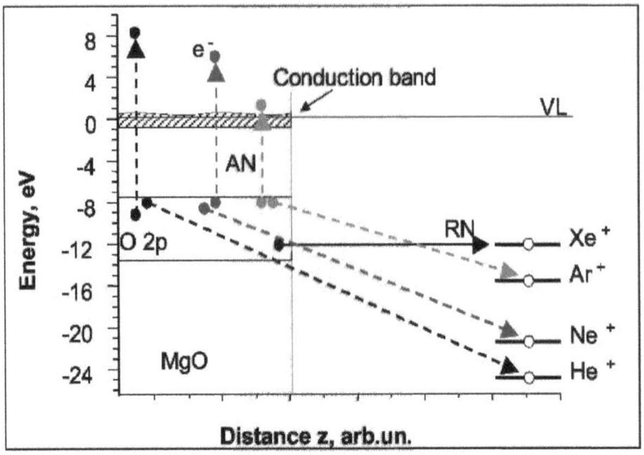

Fig. 7.4. One-electron energy diagram showing the energy levels of He^+, Ne^+, Ar^+, and Xe^+ with respect to the valence band of MgO. The band gap of MgO at the surface is taken to be 6.7 eV [33], the affinity level 0.8 eV [34]. VL denotes the vacuum level. Possible defect states in the band gap are not shown, nor are excitons and trions. The most likely neutralization processes, resonant neutralization (RN) for Xe^+ and Auger neutralization (AN) for Ar^+, Ne^+ and He^+ are also indicated

and magnesium vacancy defects (respectively F centers and V centers) we refer to [33, 36, 37], for the presence of peroxy ions (O_2^{2-}) and oxygen holes (O^-) to [38–44].

Assuming the amount of defect states is sufficiently small, neutralization of slow noble-gas ions incident on MgO is anticipated to proceed as follows: Xe^+ by a resonant process, since its energy level is resonant with the valence band; Ar^+, Ne^+, and He^+ by Auger processes, since the excited electrons can escape to the conduction band[8]. As has been well established for metal surfaces [45–47], for MgO it is also expected that higher ionization energy of the incoming projectiles results in emission of electrons with higher energies: so highest energies for He^+, somewhat lower for Ne^+, and lowest for Ar^+. The total electron yield is anticipated to decrease in the same manner (so highest yield for He^+). It should finally be noted that, if they exist in MgO, Auger processes may also lead to population of trions. If the position of the trion level is located somewhat below the bottom of the conduction band, for He^+, Ne^+ as well as for Ar^+ the trion can be populated without involvement of the kinetic energy of the projectile. This would reduce the total amount of emitted electrons. Depending on the relative position of the trion level with respect to the atomic level, slow Xe^+ may also be neutralized in this way.

[8] Note when the energy level of Ar^+ shifts upwards, Ar^+ may also be neutralized by a resonant process.

7.4 Experiments on Electron Emission

A brief overview will be given of the most important experimental results obtained with discharge cells on the electron emission from MgO thin films. A few results for other insulating materials will be addressed as well. Total electron yields and electron energy distributions obtained with well-defined ion beams will be presented afterwards. The focus will be on MgO films deposited on conductive substrates. It will be discussed how the experimental results fit in the neutralization picture outlined above.

7.4.1 Discharge Cells

In Sect. 7.2, the Paschen curve was discussed, the functional dependence between firing voltage V_f of a discharge cell and the product pd of gas pressure p and electrode distance d. Measurements of Paschen curves can be used to determine the total electron emission yield γ_{eff}. We write γ_{eff} and not γ_i, since in a discharge not only ions but also other particles may contribute to the emission of electrons, and the obtained values should be regarded as (effective) averages. This method has been used by several authors for insulating films as well as for metals (for the latter, see e.g. [48, 49]). Schematic drawings of two discharge cells as used by Philips Research Laboratories are shown in Fig. 7.5. Bachmann et al. [11] and van Elsbergen et al. [50] used the open-glass cell shown in Fig. 7.5a, pumped down to UHV conditions and subsequently filled with gas. The coatings to be tested were deposited on glass plates. The coated plates were kept at a well defined distance by a spacer and they were connected to two Au electrodes. The applied AC voltage V_a is not equal to the firing voltage V_f of the discharge equation (7.3), since the (dielectric) glass plates act as capacitors. The cell was therefore operated in series with a reference capacitor. Plotting the voltage across the reference capacitor as a function of V_a results, when the plasma ignites, in a Lissajous-like figure. V_f can be determined from this figure. For a more elaborate description of the method we refer to [8, 11, 51]). Vink et al. [52, 53] used the macroscopic discharge cell shown in Fig. 7.5b, to determine the firing voltage for a series of coatings. The dimensions of the cell are a factor of 50 larger than those of a real PDP cell. Since the properties of the discharge cell scale with pd, the pressure of the gas was chosen a factor of 50 smaller. The dimensions of the macroscopic cell allow for in-situ preparation of chemically reactive materials and for measurement of the breakdown voltage under realistic PDP conditions.

Measured Paschen curves, taken with the open-glass cell, are shown in Fig. 7.6a. Ne was used as discharge gas. For the upper curve, the glass plates were uncoated, for the lower curve they were coated with a MgO film. Prior to the measurements, the MgO film was baked at 400–450 °C. The surfaces were furthermore preconditioned by operating the cell with a 500-mbar Ne, 30-kHz sine-wave plasma discharge. Measurement of the Paschen curves was

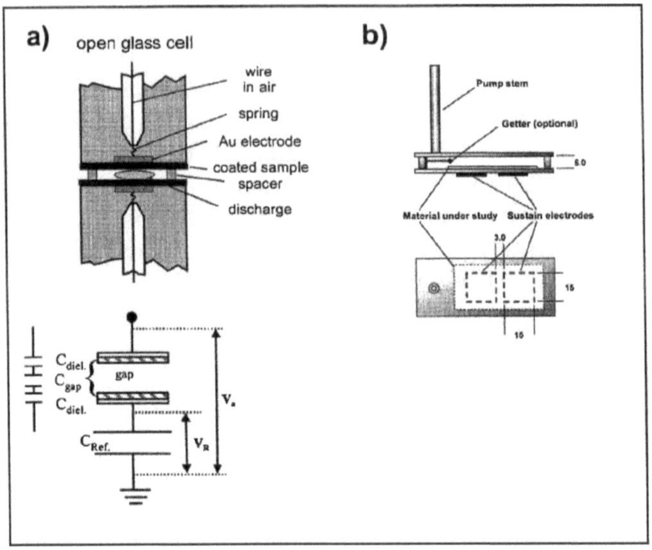

Fig. 7.5. Schematic drawings of two discharge cells used by Philips Research Laboratories. (**a**) Open-glass discharge cell, used to measure V_f as a function of *pd*, from which γ was determined (see Fig. 7.6). The cell was operated in series with a reference capacitor (see lower part of the figure). Reprinted from Diamond and Relat. Mater., 10, P.K. Bachmann, V. van Elsbergen, D.U. Wiechert, G. Zhong, J. Robertson, 'CVD diamond: a novel high γ-coating for plasma display panels?', pp. 809–817 (2001), Copyright 2001, with permission from Elsevier. (**b**) Macroscopic discharge cell, used to measure breakdown voltages for a series of chemically reactive materials (see Fig. 7.7). Reused with permission from T.J. Vink, A.R. Balkenende, R.G.F.A. Verbeek, H.A.M. van Hal, S.T. de Zwart, 'Materials with a high secondary-electron yield for use in plasma displays', Appl. Phys. Lett., 80, pp. 2216–2218 (2002). Copyright 2002, American Institute of Physics

carried out with a 2-kHz sine-wave discharge using the Lissajous method. To reproduce the shape of the theoretical Paschen curves, this frequency is about the minimum value to be used (see [8]). Figure 7.6 also shows fits of equation (7.3) to the experimental data, with γ_i (in fact $\gamma_{\rm eff}$) the only fit parameter. The fits reproduce the measurements quite well. It is clear that coating the glass plates with MgO strongly enhances electron emission and reduces the firing voltage. The Paschen curves can also be used to determine $\gamma_{\rm eff}$ as a function of $E/p \simeq V_f/pd$, with E the field strength in the plasma during breakdown: each data point in Fig. 7.6a corresponds to a specific value of E/p, which can be converted to α, the spatial ionization coefficient using tabulated first Townsend coefficients [54].[9] Substituting the such-obtained value for α in the breakdown equation (7.1) yields $\gamma_{\rm eff}(E/p)$. The results are shown in Fig. 7.6b.

[9] For the conversion, also equation (7.2) could have been used. The results should be similar.

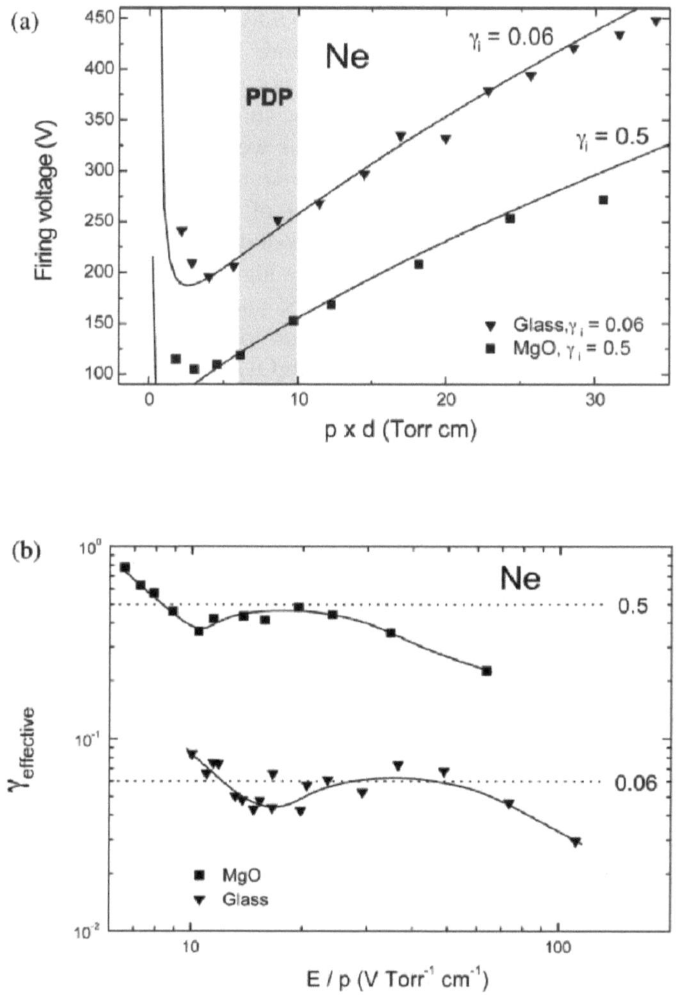

Fig. 7.6. Measurements taken with the open-glass cell shown in Fig. 7.5a. (**a**) Paschen curves for uncoated glass plates and for glass plates coated with a 500-nm thick MgO(111) film (by electron-beam evaporation). Also indicated is the regime where actual PDPs are operated. The experimental data are fitted to equation (7.3), with γ_i (in fact γ_{eff}) the only fit parameter. Coating the glass with MgO clearly increases γ_{eff} and decreases the firing voltage. (**b**) γ_{eff} as function of E/p as determined from the measured Paschen curves (see text). Reprinted from Diamond and Relat. Mater., 10, P.K. Bachmann, V. van Elsbergen, D.U. Wiechert, G. Zhong, J. Robertson, 'CVD diamond: a novel high γ-coating for plasma display panels?', pp. 809–817 (2001), Copyright 2001, with permission from Elsevier

The increase of γ_{eff} at low values for E/p may be related to photo-induced electron emission [7], although it seems more likely that in this regime the description of breakdown equation (7.1) is not sufficiently accurate.

The measurements described in [50] show that V_f and γ_{eff} critically depend on the quality of the deposited MgO film as well as on its crystal face. With preconditioning time, the authors report a decrease of the measured value of V_f of close to 50% and an increase of γ_{eff} of up to a factor of 6. In some cases, 20 hours of preconditioning was needed before constant values were obtained. The authors furthermore report a significant dependence of γ_{eff} on the face of the MgO crystal: $\gamma_{\text{eff}}(110) = 0.28$, $\gamma_{\text{eff}}(100) = 0.31$, and $\gamma_{\text{eff}}(111) = 0.5$. They give several possible explanations for these differences: according to scanning-electron-microscopy images, MgO(111) films have smaller grains with many small tips at the surface [55], which could possibly enhance electron emission and lower the firing voltage. An alternative explanation could be that the differences in electron emission originate in differences in the electronic structure of the three faces.

From the discharge experiments, as well as from the experience with actual PDPs, it is clear that MgO(111) is a very good electron emitter. Electron emission from many other materials has been studied as well; only very few of them seem to be as good as MgO(111). Bachmann et al. [11] report γ_{eff} measurements for CVD (Chemical Vapor Deposition) diamond. When the surface was H terminated, they found values of $\gamma_{\text{eff}} = 0.5$, similar as for MgO(111). In their paper, they discuss that CVD diamond may be an alternative for MgO for use in actual PDPs. They relate the excellent electron-emission properties to the negative electron affinity of the material: when replacing the H termination by O, with positive electron affinity, they found a strong decrease of γ_{eff} by more than a factor of 10. Vink et al. [53], on the other hand, measured V_f for a series of materials using the macroscopic cell shown in Fig. 7.5b. The materials were evaporated in-situ, either through evaporation using an electron beam or by using a getter. The cell was subsequently filled with a mixture of Ne with 3.5% Xe (pressure about 10 Torr). The measured firing voltages are shown in Fig. 7.7, plotted as a function of $E_k^{\max}$, the maximum kinetic energy of the emitted electrons. $E_k^{\max}$ was calculated from $E_i(\text{Ne}^+)$, the ionization energy of Ne (not taking into account the level shifts as discussed in Sect. 7.3.1) and tabulated values for E_g and χ, band gap and electron affinity of the material. The firing voltage of most materials follow a general decrease with increasing $E_k^{\max}$. The authors present a simple, qualitative model for Auger neutralization of the incoming ions, along the lines of the work of Hagstrum (see e.g. [56]) and Aboelfotoh and Lorenzen [57]. In their model, the decrease in firing voltage is simply related to a larger fraction of electrons excited to energies above the vacuum level (leading to an increase in γ_{eff}). The scatter at the smallest $E_k^{\max}$ can be related to the details of the neutralization process: the Auger rates critically depend on the detailed atomic energy-level shifts, density of final states etc. The firing voltages measured for TiO_2 and Si_3N_4 seem to be relatively high compared to the other

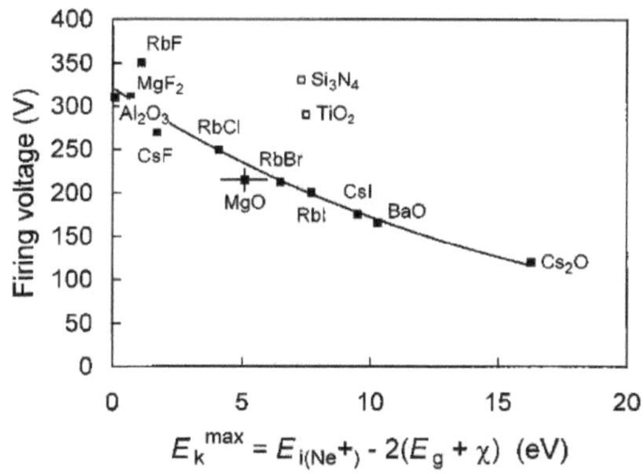

Fig. 7.7. Measured firing voltages for a series of materials plotted as a function of the maximum kinetic energy of the emitted electrons. The macroscopic cell shown in Fig. 7.5b was used with a NeXe(3.5%) mixture. The cell was operated with a 50 kHz square wave signal applied to the electrodes. The decrease of firing voltage with increasing $E_k^{\max}$ is related to a larger fraction of electrons excited to energies above the vacuum level (see text for details). Reused with permission from T.J. Vink, A.R. Balkenende, R.G.F.A. Verbeek, H.A.M. van Hal, S.T. de Zwart, 'Materials with a high secondary-electron yield for use in plasma displays', Appl. Phys. Lett., 80, pp. 2216–2218 (2002). Copyright 2002, American Institute of Physics

materials. The authors attribute this to their large electron affinities, making γ_{eff} relatively low, since a large fraction of excited electrons ends up in states below the vacuum level. Finally, MgO is the material still mostly used in actual PDPs. Although Fig. 7.7 may suggest that other materials could be even better candidates, one should realize that some of them are chemically very active whereas others are not as resistent against sputtering as MgO.

7.4.2 Ion-Beam Experiments

The advantage of using ion beams over discharge cells is that specific ions with a certain kinetic energy can be used. The results obtained are thus more detailed and may lead to a more fundamental understanding of the mechanisms underlying the emission of electrons. Most of the studies so far have concentrated on total electron emission yields γ, lesser on electron energy distributions. A major problem in all experiments is the charging of the insulating films, to which especially low-energy electrons are very sensitive. Moon et al. [58], e.g., discuss γ-measurements for low-energy noble-gas ions incident on thin (50–500 nm) polycrystalline MgO films. In as-deposited films, they found large instabilities in the measured currents, which they attribute to liberation of electrons trapped in defect states above the Fermi level: ion bombardment leads to charging and thus to an electric field across the film, possibly resulting in field emission of the trapped electrons. Prolonged ion

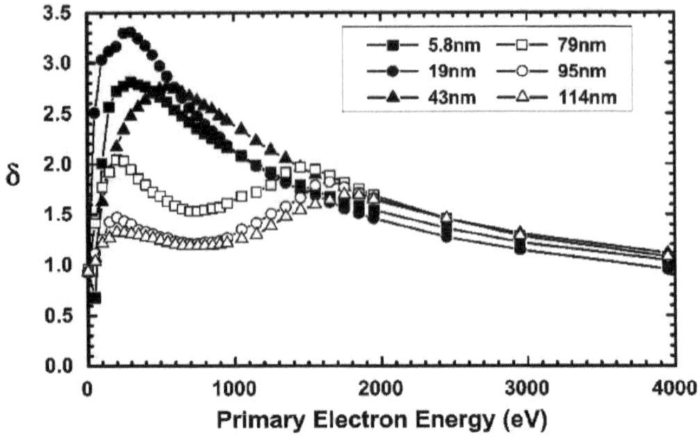

Fig. 7.8. Secondary electron emission yield δ, obtained with a primary electron beam normally incident on SiO_2 films of various thickness on a conductive Si substrate. The double-hump structure for the thicker films is indicative of charging and can be removed by applying a negative potential to the Si substrate. Reprinted from Thin Solid Films, 397, W. Yi, T. Jeong, SG. Yu, J. Lee, S. Jin, J. Heo, J.M. Kim, 'Study of the secondary-electron emission from thermally grown SiO_2 films on Si', pp. 170–175 (2001), Copyright 2001, with permission from Elsevier

bombardment[10] and in addition annealing of the film to 450°C was shown to remove the instabilities.

Beams of electrons have frequently been used to study charging phenomena in detail. In several studies (e.g. [59–62]), charging was shown to have a direct effect on the secondary-electron-emission yield δ[11]. Yi et al. [59] present results for an electron beam normally incident on thin SiO_2 films deposited on a conductive Si substrate. For the thicker films, they unambiguously show that charging occurs. An example of their results is shown in Fig. 7.8. For the thinnest films, the measurements show the characteristic functional dependence of δ on primary electron energy. For films with thickness larger than 43 nm, a double-hump structure is observed, which the authors attribute to charging. The authors propose that for primary electron energies of around 1000 eV (in between the two humps), the bulk of the secondary electrons and the remaining positive holes are created far away from the conductive Si substrate. The finite lifetime of the positive holes seems to lead to 'recapture' of secondary electrons, thereby reducing δ. The authors also showed that application of a large negative potential to the Si substrate, of the order of -1000 V, led to the removal of the double-hump structure. The authors propose that

[10] The authors do not report the fluence of the ion beam. They show that the instabilities disappear after about 30 minutes and mention that the fluence needed is roughly proportional to layer thickness.

[11] For primary electrons, the yield is normally indicated by $\delta = I_s/I_p$, with I_p the primary electron current and I_s the secondary electron current.

this negative bias potential increases the tunneling probability of electrons between conductive substrate and SiO_2 layer, thereby removing the positive holes. Similar results have been obtained for MgO films on a conductive Si substrate [60].

Total Electron Emission Yield

The remainder of this chapter will be devoted to ion-induced electron emission from MgO. The above experiments illustrate that special precautions have to be taken to prevent charging. Simply heating the sample to a few hundred °C, as used for the alkali halides [26], is not enough to sufficiently increase the ionic conductivity of MgO. In stead, researchers have used pulsed ion beams [63, 64], electron-flood guns [65], focussed ion beams in combination with scanning-area techniques [66, 67], and very thin insulating films deposited on good conductive substrates [27]. A schematic drawing of a typical set-up for γ measurements, as used by Moon et al. [58], is shown in Fig. 7.9a. The mass and energy selected ion beam enters the target region through a small opening in the collector. When biased at an appropriate positive potential with respect to the target, the collector detects most of the secondary electrons. The total electron yield directly follows from the measured currents: $\gamma = I_C/I_P$ with I_C the collector current and I_P the ion-beam current. Measured values for

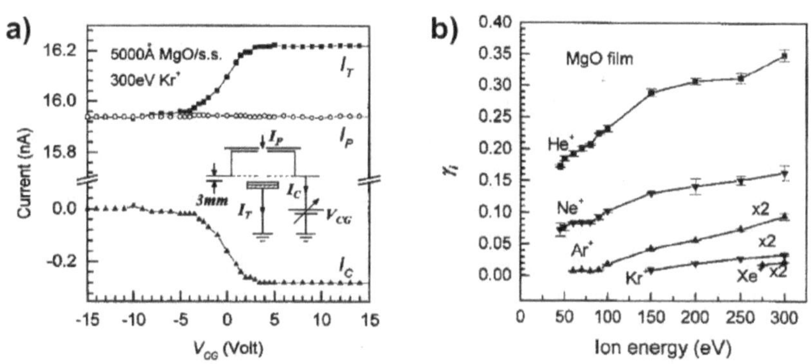

Fig. 7.9. Ion-induced total electron emission yield γ for a 500-nm polycrystalline MgO film. (**a**) The inset shows a schematic drawing of the experimental set-up used. The ion beam enters the target region through a small opening in the collector. The latter collects the secondary electrons when biased positively with respect to the target. (**b**) Measured γ for different noble-gas ions with kinetic energies ranging from 45 eV up to 300 eV. γ strongly increases with the ionization energy of the projectiles and with their kinetic energy. Reused with permission from K.S. Moon, J. Lee, K.-W. Whang, 'Electron ejection from MgO thin films by low energy noble gas ions: energy dependence and initial instability of the secondary electron emission coefficient', J. Appl. Phys., 86, 4049 (1999). Copyright 1999, American Institute of Physics

γ, for different noble-gas ions and for a range of kinetic energies, are shown in Fig. 7.9b. The most striking features are the strong increase of γ with increase of the ionization energy of the projectiles and the strong increase of γ with projectile kinetic energy. The strong dependence on ionization energy, in particular at the lowest kinetic energies, is a direct indication for the occurrence of Auger neutralization of the incoming ions (see discussion in Sect. 7.3.2, Fig. 7.4). The kinetic threshold for Ne^+ and Ar^+ seems to be at energies of about 50–100 eV, for He^+ at even lower energies, and for Kr^+ and Xe^+ at somewhat higher energies. The increase of γ with ion kinetic energy has been observed in many other experiments on MgO (see, e.g., [64, 66, 67]). The thresholds discussed above are close to what was found in [64]: for He^+, Ne^+, and Ar^+ kinetic thresholds at about 50 eV, for Kr^+ at about 100 eV and for Xe^+ at about 150 eV. It is obvious that above threshold the increase of γ is caused by some kind of kinetic process. For other insulating surfaces, similar observations have been made. For the alkali halides, e.g., kinetic processes were discussed in terms of promotion of electrons in ion-atom collisions at the surface [68–70] and in terms of the formation of intermediate negative ions and subsequent loss of electrons [18, 71]. A more elaborate discussion of these mechanisms can be found in Chap. 4 of this book. For metal oxides, kinetic processes have been related to collisional excitation and subsequent decay of molecular complexes [72–74]. The latter process may also lead to the emission of negative anions, which would contribute to the measured value of γ as well.

As discussed in Sect. 7.2, most present-day PDPs contain mixtures of Ne and Xe. Assuming singly charged ions contribute most to the emission of electrons, Fig. 7.9b clearly shows that the contribution from Ne^+ will dominate the one from Xe^+. According to Fig. 7.9b, for kinetic energies with tens of an eV, the γ-value for Ne^+ amounts to $\simeq 0.08$. For slow Ne^+ incident on MgO, a range of values has been reported in the literature: from $\gamma = 0.05$ to $\gamma = 0.45$ [27, 58, 66, 67, 75]. The origin of this variation of almost an order of magnitude is not clear. The different values cannot directly be related to the different physical parameters used in the experiments, such as film thickness and crystal face: although for single crystals only small values have been reported [66], for thin films the values range from $\gamma = 0.08$ (Fig. 7.9) to 0.45 [75]. Different crystal faces, on the other hand, show variations well within a factor of two [66, 75]. It should be noted that in most experiments the largest values were observed for the (111) face, consistent with the results reported in [50] for a gas discharge (see Sect. 7.5). Why then do the measured γ values vary with almost an order of magnitude? Differences in quality (especially cleanliness) of the MgO surfaces are likely to play a role. The surfaces in the different experiments were not identically treated: in some cases, the MgO surfaces were exposed to ambient conditions prior to the experiments and, in addition, not always UHV conditions were used in the experiments. For the experiments in a gas discharge, discussed in Sect. 7.5, it has been shown that quality indeed does have a large effect on γ: upon prolonged preconditioning γ_{eff} was shown to increase by a factor of 6. Another factor that is likely to contribute to the observed differences, in particular for the thicker films and single crystals, is charging of the sample. It is clear that slow (positive) ions produce positive holes far away from the substrate. It depends on the experimental conditions, such as ion-beam current, conductance of the insulating film, method used to prevent charging, etc. whether or not the holes are removed sufficiently fast. If charging occurs, one expects that this would lead to, as observed in the experiments with primary electrons (see Fig. 7.8), a decrease of γ.

Electron Energy Distributions

First electron energy distributions induced by ions incident on MgO have been measured with set-ups designed for γ measurements, utilizing the retarding-field method. Care needs to be taken with the such obtained distributions since, depending on the specific detection geometry, their shape may be strongly distorted. For MgO films with a thickness of about 100 nm, Ishimoto et al. [75] studied the maximum kinetic energy of emitted electrons for different noble-gas ions and observed an increase going from Xe^+ to Ar^+ to Ne^+. Although these observations are consistent with Auger neutralization of the ions at the MgO films, it should be noted that the primary ions had kinetic energies of 250 eV, i.e., above the threshold for kinetic emission. Only recently, set-ups equipped with energy and angle dispersive electrostatic analyzers have been used to carry out detailed experiments on MgO thin films and single crystals. An example of a typical experimental set-up is shown in Fig. 7.10.

Ion-induced electron emission from very thin films of MgO has been studied by Matulevich et al. [27, 76]. MgO(100) films were grown on a Mo(100) substrate by evaporating Mg in an oxygen atmosphere. Thickness and crystallinity of the samples were determined by low-energy ion scattering. UV photoelectron spectroscopy with the HeI line showed only minor intensities due to defect states in the MgO band gap. A bias potential of −7 V was applied to the sample to also detect the lowest-energy electrons. The results shown in Fig. 7.11 were obtained for a MgO film with a thickness of 1–5 nm, bombarded with 40 eV He^+, Ne^+, and Ar^+ ions incident at 40° with respect to the surface[12] [27]. The large number of electrons at energies slightly below 6 eV corresponds to electrons with virtually no kinetic energy at the MgO surface that are accelerated towards the analyzer. The peaks show up at somewhat lower energies than the applied acceleration voltage due to the work function difference between sample and analyzer. A potential difference of 2 V was observed between the MgO surface and the Mo substrate (with the latter at lower potential). From the sample current and the ion-beam current measured in a Faraday cup, the total electron emission per incoming ion was

[12] In fact, since the ions are accelerated to the sample by −7 V, they have a kinetic energy of 47 eV and are incident on the surface at an angle somewhat larger than 40°. For the present discussion this is of no importance.

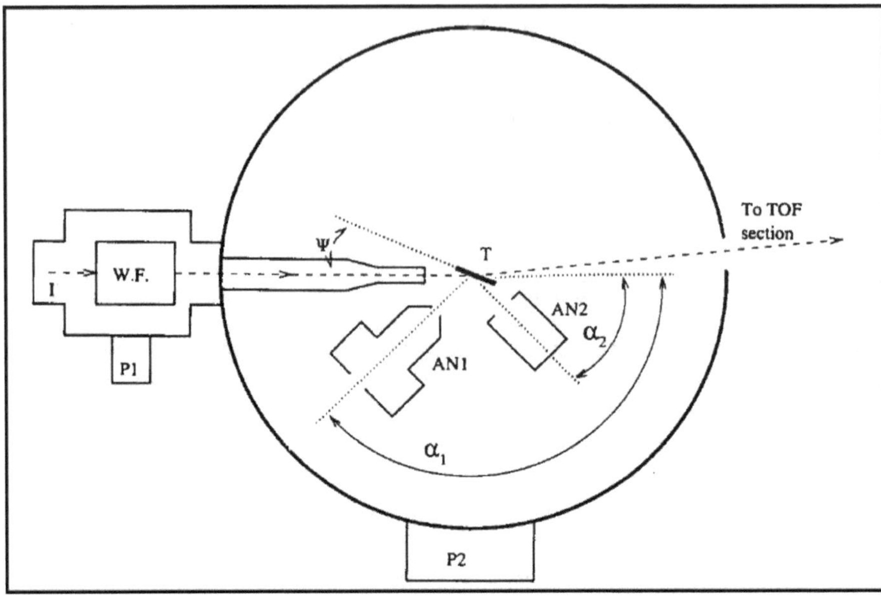

Fig. 7.10. Schematic drawing of one of the experimental set-ups used in Utrecht for studies on ion-induced electron emission. P1 and P2 denote the two pumping stages. The base pressure of the set-up is about 2×10^{-10} mbar. The electron-impact ion source (I) is indicated on the left and contains a Wien filter (WF) for velocity selection of the primary ions. The main chamber contains the target (T) and 2 electrostatic analyzers (AN1: hemispherical analyzer; AN2: double parallel plate analyzer). The set-up also contains a time-of-flight (TOF) tube for detection of scattered ions/neutrals. For these experiments the target has to be rotated in an upward position

found to be: $\gamma = 0.55 \pm 0.05$ (He^+), $\gamma = 0.41 \pm 0.04$ (Ne^+), and $\gamma = 0.10 \pm 0.01$ (Ar^+). These experimental observations are consistent with Auger neutralization of the incoming ions at the MgO film (see also Fig. 7.4 and discussion): highest electron energies and largest γ for He^+, both somewhat lower for Ne^+ and lowest for Ar^+. As discussed in [27], however, model simulations revealed that the electron spectra could not be explained by Auger neutralization of the incoming ions alone (especially the high-energy tail observed for Ar^+). An additional mechanism for electron emission was proposed: transport of holes through the MgO film to the Mo substrate, where they are neutralized by either a resonant or an Auger process with the latter leading to additional electron emission.

The above described model assumes that the holes, originating in the neutralization of noble-gas ions at the MgO surface and corresponding to O atoms and/or O^- ions in the O^{2-} sublattice, are sufficiently mobile to move through the MgO film to the Mo substrate. The experiments were carried out with ion beams with an intensity of ~1-nA, sufficiently low not to charge up the

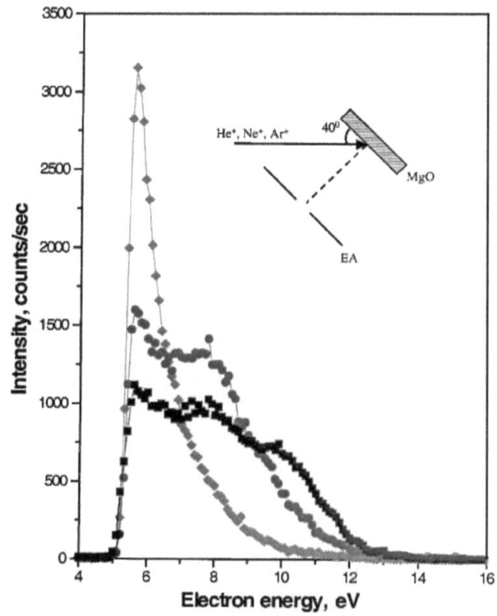

Fig. 7.11. Electron energy distributions obtained with ~1-nA beams of 40-eV Ar^+ (*solid diamonds*), Ne^+ (solid circles), and He^+ (*solid squares*) ions incident, in the ⟨100⟩ direction at 40° with the surface, on a thin MgO(100) layer (1–5 nm thick) deposited on a Mo(100) substrate. The size of the beam spot on the sample was ~ 5 mm^2. The inset shows the experimental situation. The sample was at room temperature and biased at −7 V. The energy of the electrons is given with respect to the vacuum level of the analyzer (EA). The large number of electrons at energies slightly below 6 eV corresponds to 0-eV electrons at the MgO surface that are accelerated towards the analyzer. Reprinted with permission from Y.T. Matulevich, T.J. Vink, P.A. Zeijlmans van Emmichoven, 'Low-energy ion-induced electron emission from a MgO(100) thin film: the role of the MgO-substrate interface', Phys. Rev. Lett., 89, pp. 167601 (2002). Copyright 2002 by the American Physical Society

MgO surface during the experiments[13]. This means that of the order of 10^{10} holes per second were produced in and removed from an area of about 5 mm^2 of a 1–5 nm thick MgO film. The presence of the potential difference of 2 V between MgO surface and Mo substrate, which establishes an electric field towards the substrate, may play a role in moving the holes to the substrate. Although localization of holes and conductivity related to holes has been addressed in several publications [38–44], these results cannot directly be applied to the present situation: the origin of the holes is different, i.e.,

[13] Although the potential difference between Mo substrate and MgO film indicates that the MgO may contain some charges, during the experiments no further charging took place: the low-energy peaks in Fig. 7.11, at energies slightly below 6 eV, remained in the same position during the experiments.

they mostly originate in spurious contaminants already present in MgO. The model proposed was further tested in an experiment with 40-eV Xe^+ ions, incident on slightly thicker films of MgO [76], where Auger neutralization of the incoming ions is virtually impossible (see Fig. 7.4). Electron emission was observed up to energies of 4–5 eV, with $\gamma = 0.03 \pm 0.02$, consistent with the above discussed model: resonant neutralization of Xe^+ at the MgO film and transport of the holes to the Mo substrate, where they are neutralized either by a resonant or by an Auger process, with the latter being responsible for the observed electron emission.

At very thin films of MgO as in the above experiments, Auger neutralization of noble-gas ions may in principle also proceed via population of trions. Since these states most likely lie below the vacuum level (as for LiF), this process will not contribute to the emission of electrons and it will only lead to a reduction of the total electron yield. Processes involving the kinetic energy of the incoming ions, on the other hand, are not likely to contribute since the kinetic energy of the incoming ions was very low. Other contributions to the emission of electrons as, e.g., field emission of electrons directly from the substrate, cannot completely be excluded with the present experiments. To underpin the proposed model, further studies are needed, e.g., with slow noble-gas ions incident on very thin films of MgO deposited on substrates with different band structure.

Ion-induced electron emission from thicker MgO films has been studied by Riccardi et al. [77, 78], from bulk MgO by Matulevich et al. [79]. In [79], first studies on bulk MgO(110) with 60 eV He^+ and Ne^+ ions surprisingly revealed the independence of measured electron energy distributions of ion type. Very low-intensity ion beams were used in the experiments to prevent charging. The results were soon confirmed by Riccardi et al. [77], who used He^+, Ne^+, Ar^+, and Na^+ ions incident on an approximately 100-nm thick polycrystalline MgO film deposited on a doped Si substrate. An example of their results is shown in Fig. 7.12. They also presented an explanation for the independence of ion type. In their experiments, the sample was biased at a potential of –4.9 V. Charging was minimized by flooding the sample with low-energy electrons prior to the experiments and by using low-intensity ion beams. The normalized measured spectra, obtained with 200 eV noble-gas ions and with 500 eV Na^+ ions are virtually identical. The kinetic energies of the ions are certainly above the kinetic threshold. The measured total electron yields amount to $\gamma \simeq 0.2$ for He^+ and Ne^+, $\gamma \simeq 0.1$ for Ar^+, and $\gamma \simeq 0.3$ for Na^+. The similarity of the measured values of γ for He^+ and Ne^+ is in clear contrast with other experiments (see, e.g., Fig. 7.9). To explain the independence of the measured spectra of ion type, the authors proposed that collisions between projectiles and O^{2-} ions of the MgO lead to promotion of O 2p electrons along quasi-molecular orbitals and to subsequent population of exciton levels (a kinetic process). Assuming the MgO film has a negative electron affinity, they argued that the exciton levels may lie above the vacuum level [80]. This implies that decay of excitons would lead to the emission of low-energy electrons, along

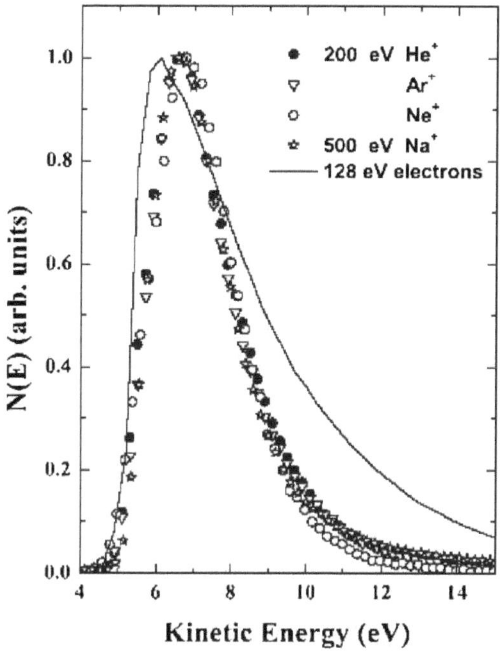

Fig. 7.12. Normalized electron energy distributions obtained with 200 eV noble-gas ions, 500 eV Na^+ ions and 128-eV electrons incident on a 100-nm thick MgO film on a doped Si substrate. The sample was biased at –4.9 eV. Surprisingly, the electron energy distributions obtained with the ions are virtually independent of ion type. Reprinted from Surf. Sci., 571, P. Riccardi, M. Ishimoto, P. Barone, R.A. Baragiola, 'Ion-induced electron emission from MgO by exciton decay into vacuum', L305-L310, Copyright 2004, with permission from Elsevier

the lines of observations at negative-electron-affinity diamond after excitation with photons [81] and electrons [82]. It should be noted that in the experiments on diamond, exciton decay only leads to the emission of electrons with very low energies just above the vacuum level. In [78], Riccardi et al. study the emission of electrons from higher energy ions incident on similar MgO films. With increasing ion kinetic energy, γ was shown to drastically increase and the electron spectra to broaden at the high-energy side. The authors proposed that electron emission takes place via promotion of O 2p electrons directly to the continuum as well as via population and decay of excitons. These promotion mechanisms are similar as to what has been proposed for collisions of protons at LiF surfaces [18, 68, 69].

In Sect. 7.3.2, it was argued that neutralization of a noble-gas ion at a MgO surface is expected to proceed via a resonant or via an Auger process (depending on the ionization energy of the ion). For very thin films of MgO, electron energy distributions indeed have shown that neutralization of noble-gas ions proceeds in this way. The experiments were carried out with ions with kinetic energies below 50 eV. For these low kinetic energies, neither excitation and decay of excitons nor direct promotion to the continuum is likely to give significant contributions to the overall electron emission. In electron energy distributions obtained with noble-gas ions incident on thicker MgO films and on bulk MgO, Auger electron emission has not yet been identified. The experiments on thicker MgO films were carried out with kinetic energies significantly above the kinetic threshold. The interpretation in terms of direct promotion of electrons to the continuum seems plausible. Population and decay of excitons may also contribute to the emission of electrons, although its quantitative contribution is not yet clear. Why there are no indications of Auger neutralization in the energy distributions remains for the time being an open question that needs clarification.

7.5 Concluding Remarks and Outlook

Slow ion-induced electron emission from thin insulating films has been discussed. Because of its use in most present-day PDPs, the focus has been on MgO. For slow Ne^+ incident on MgO, large differences in the total electron yield of up to an order of magnitude have been reported. Although the origin of these differences is not completely clear, it seems likely that the quality of the films plays an important role. Charging effects may also contribute. The results obtained with discharge cells largely agree with those obtained with ion beams. Significant variations in total electron yield have been observed for different crystal faces of MgO, with MgO(111) giving the highest yield. Measured electron energy distributions, obtained with noble-gas ions below the kinetic threshold and incident on very thin films of MgO, have revealed that Auger processes contribute to the neutralization of the incoming ions. This conclusion is supported by several total-yield experiments. Surprisingly, in electron energy distributions obtained with ions with somewhat higher kinetic energies incident on thicker MgO films and on bulk MgO, Auger neutralization of the ions has not yet been identified.

A systematic study of the electron emission from MgO films with a range of thicknesses, deposited on substrates with different band structure, is needed. Very slow ions should be used, with energies below 50 eV, to prevent, as much as possible, kinetic effects. The thickness should be varied from a few nanometer, where Auger neutralization has been shown to take place, up to a few hundred nanometer, relevant for actual PDPs. The total electron yield γ as well as electron energy distributions should both be measured. These measurements could reveal the detailed mechanisms by which neutralization of noble-gas ions incident on MgO proceeds, the role played by the substrate for the very thin films, and the role played by charging for the thicker films. As a comparison, it would help to have similar measurements for one of the alkali-halide surfaces, for which charging can easily be prevented by heating the sample (but preferably not LiF with its large band gap).

Many PDPs nowadays contain MgO which is intentionally doped. It is not clear how doping causes the firing voltage to be lowered. A systematic study of the influence of the purity of MgO and the effect of doping on the ion-induced electron emission would certainly be of interest.

Acknowledgments

Fruitful discussions with Dr. T.J. Vink and Prof. Dr. L.F. Feiner are gratefully acknowledged. This work was supported by the 'Stichting voor Fundamenteel Onderzoek der Materie (FOM),' which is financially supported by the 'Nederlandse Organisatie voor Wetenschappelijk Onderzoek (NWO).'

References

1. G.W. Fraser: Int. J. Mass Spectrom. **215**, 13 (2002)
2. J.P. Boeuf: J. Phys. D: Appl. Phys. **36**, R53 (2003)
3. M.F. Gillies, G. Oversluizen: J. Appl. Phys. **31**, 6315 (2002)
4. S. Uchida, H. Sugawara, Y. Sakai, T. Watanabe, Byoung-Hee Hong: J. Phys. D: Appl. Phys. **33**, 62 (2000)
5. O.B. Postel, M.A. Cappelli: Appl. Phys. Lett. **76**, 544 (2000)
6. O.B. Postel, M.A. Cappelli: J. Appl. Phys. **89**, 4719 (2001)
7. A.V. Phelps, Z.Lj. Petrović: Plasma Sources Sci. Technol. **8**, R21 (1999)
8. V. van Elsbergen, P.K. Bachmann, T. Juestel: SID 00 Digest **220**, (2000)
9. Yu. P. Raizer: *Gas Discharge Physics* (Springer, Berlin, 1997) p. 56
10. K. Wasa, S. Hayakawa: *Handbook of sputter deposition technology*, reprint edn (Noyes Publications, New Jersey 1992) p. 84
11. P.K. Bachmann, V. van Elsbergen, D.U. Wiechert, G. Zhong, J. Robertson: Diamond and Relat. Mater. **10**, 809 (2001)
12. H. Winter: Progr. Surf. Sci. **63**, 177 (2000)
13. H. Winter: Phys. Rep. **367**, 387 (2002)
14. R. Souda, K. Yamamoto, W. Hayami, B. Tilley, T. Aizawa, Y. Ishizawa: Surf. Sci. **324**, L349 (1995)
15. A.G. Borisov, V. Sidis, H. Winter: Phys. Rev. Lett. **77**, 1893 (1996)
16. P. Roncin, A.G. Borisov, H. Khemliche, A. Momeni, A. Mertens, H. Winter: Phys. Rev. Lett. **89**, 043201 (2002)
17. M. Maazouz, L. Guillemot, S. Lacombe, V.A. Esaulov: Phys. Rev. Lett. **77**, 4265 (1996)
18. P. Roncin, J. Villette, J.P. Atanas, H. Khemliche: Phys. Rev. Lett. **83**, 864 (1999)
19. G. Hayderer, S. Cernusca, M. Schmid, P. Varga, HP. Winter, F. Aumayr, D. Niemann, V. Hoffmann, N. Stolterfoht, C. Lemell, L. Wirtz, J. Burgdörfer: Phys. Rev. Lett. **86**, 3530 (2001)
20. J. Stöckl, T. Suta, F. Ditroi, HP. Winter, F. Aumayr: Phys. Rev. Lett. **93**, 263201 (2004)
21. L. Wirtz, G. Hayderer, C. Lemell, J. Burgdörfer, L. Hägg, C.O. Reinhold, P. Varga, HP. Winter, F. Aumayr: Surf. Sci. **451**, 197 (2000)

22. L. Wirtz, J. Burgdörfer, M. Dallos, T. Müller, H. Lischka: Phys. Rev. A **68**, 032902 (2003)
23. *Handbook of Optical Constants*, ed. by E.D. Palik (Academic Press, New York 1985) p. 675
24. *Handbook of Chemistry and Physics*, ed. by D.R. Lide, 81st edn (CRC Press, New York 2000) p. 12.56
25. H. Khemliche, J. Villette, A.G. Borisov, A. Momeni, P. Rhoncin: Phys. Rev. Lett. **86**, 5699 (2001)
26. H. Khemliche, A.G. Borisov, A. Momeni, P. Rhoncin: Nucl. Instr. and Meth. in Phys. Res. B **191**, 221 (2002)
27. Y.T. Matulevich, T.J. Vink, P.A. Zeijlmans van Emmichoven: Phys. Rev. Lett. **89**, 167601 (2002)
28. F. Wiegershaus, S. Krischok, D. Ochs, W. Maus-Friedrichs, V. Kempter: Surf. Sci. **345**, 91 (1996)
29. T. Hecht, C. Auth, A.G. Borisov, H. Winter: Phys. Lett. A **220**, 102 (1996)
30. P. Rousseau, M. Gugiu, H. Khemliche, P. Roncin: Nucl. Instr. and Meth. in Phys. Res. B **230**, 361 (2005)
31. A. Thilagam: Phys. Rev. B **55**, 7804 (1997)
32. V.E. Henrich, P.A. Cox: *The Surface Science of Metal Oxides*, (Cambridge University Press, Cambridge 1994) pp. 129–135
33. P.V. Sushko, A.L. Shluger, C.R.A. Catlow: Surf. Sci. **450**, 153 (2000)
34. R.C. Alig, S. Bloom: J. Appl. Phys. **49**, 3476 (1975)
35. D. Ochs, W. Maus-Friedrichs, M. Brause, J. Günster, V. Kempter, V. Puchin, A. Shluger, L. Kantorovich: Surf. Sci. **365**, 557 (1996)
36. A. Gibson, R. Haydock, J.P. LaFemina: Phys. Rev. B **50**, 2582 (1994)
37. C. Sousa, G. Pacchioni, F. Illas: Surf. Sci. **429**, 217 (1999)
38. H. Kathrein, F. Freund: J. Phys. Chem. Solids **44**, 177 (1983)
39. B.V. King, F. Freund: Phys. Rev. B **29**, 5814 (1984)
40. A.S. Kuznetsov, É.É. Vil't, V.P. Nagirnyĭ: Sov. Phys. Solid State **26**, 1248 (1984)
41. M.M. Freund, F. Freund, F. Batllo: Phys. Rev. Lett. **63**, 2096 (1989)
42. F. Batllo, R.C. LeRoy, K. Parvin, F. Freund, M.M. Freund: J. Appl. Phys. **69**, 6031 (1991)
43. A.L. Shluger, E.N. Heifets, J.D. Gale, C.R.A. Catlow: J. Phys.: Condens. Matter **4**, 5711 (1992)
44. J.T. Devreese, V.M. Fomin, E.P. Pokatilov, E.A. Kotomin, R. Eglitis, Yu.F. Zhukovskii: Phys. Rev. B **63**, 184304 (2001)
45. H.D. Hagstrum: Phys. Rev. **96**, 325 (1954)
46. H.D. Hagstrum: Phys. Rev. **104**, 672 (1956)
47. H.D. Hagstrum: Studies of Adsorbate Electronic Structure Using Ion Neutralization and Photoemission Spectroscopies. In: *Electron and Ion Spectroscopy of Solids*, ed. by L. Fiermans (Plenum Press, New York 1978) pp. 273–323
48. Ph. Guillot, Ph. Belenguer, L. Therese, V. Lavoine, H. Chollet: Surf. Interface Anal. **35**, 590 (2003)
49. Y. Sosov, C.E. Theodosiou: J. Appl. Phys. **95**, 4385 (2004)
50. V. van Elsbergen, P.K. Bachmann, G. Zhong: IDW'00 **687**, (2000)
51. T. Tamida, A. Iwata, M. Tanaka: Tech. Meeting on El. Discharge (IEEJ), ED-96-274 **113**, (1996)
52. T.J. Vink, R.G.F.A. Verbeek, S.T. de Zwart: IDRC'00 **163**, (2000)
53. T.J. Vink, A.R. Balkenende, R.G.F.A. Verbeek, H.A.M. van Hal, S.T. de Zwart: Appl. Phys. Lett. **80**, 2216 (2002)

54. A.A. Kruithof: Physica **7**, 519 (1940)
55. V. Hoffmann, A. Klöppel, J. Trube: Euro Display 99 **289**, (1999)
56. H.D. Hagstrum: Phys. Rev. **96**, 336 (1954)
57. M.O. Aboelfotoh, J.A. Lorenzen: J. Appl. Phys. **48**, 4754 (1977)
58. K.S. Moon, J. Lee, K-W. Whang: J. Appl. Phys. **86**, 4049 (1999)
59. W. Yi, T. Jeong, S.G. Yu., J. Lee, S. Jin, J. Heo, J.M. Kim: Thin Solid Films **397**, 170 (2001)
60. J. Lee, T. Jeong, S.G. Yu., S. Jin, J. Heo, W. Yi, D. Jeon, J.M. Kim: Appl. Surf. Sci. **174**, 62 (2001)
61. J.J. Scholtz, R.W.A. Schmitz, B.H.W. Hendriks, S.T. de Zwart: Appl. Surf. Sci. **111**, 259 (1997)
62. J. Cazaux, K.H. Kim, O. Jbara, G. Salace: J. Appl. Phys. **70**, 960 (1991)
63. N.J. Chou: J. Vac. Sci. Technol. **14**, 307 (1977)
64. S.K. Lee, J.K. Kim, J.H. Ryu, J. Lee: IDW'03 **981**, (2003)
65. M. Ishimoto, R.A. Baragiola, T. Shinoda: IDW'00 **683**, (2000)
66. E.H. Choi, J.Y. Lim, Y.G. Kim, J.J. Ko, D.I. Kim, C.W. Lee, G.S. Cho: J. Appl. Phys. **86**, 6525 (1999)
67. D.I. Kim, J.Y. Lim, Y.G. Kim, J.J. Ko, C.W. Lee, G.S. Cho, E.H. Choi: Jpn. J. Appl. Phys. **39**, 1890 (2000)
68. P. Stracke, F. Wiegershaus, S. Krischok, H. Müller, V. Kempter, P.A. Zeijlmans van Emmichoven, A. Niehaus, F.J. García de Abajo: Nucl. Instrum. and Methods Phys. Res. B **125**, 67 (1997)
69. P.A. Zeijlmans van Emmichoven, A. Niehaus, P. Stracke, F. Wiegershaus, S. Krischok, V. Kempter, A. Arnau, F.J. García de Abajo, M. Peñalba: Phys. Rev. B **59**, 10950 (1999)
70. H. Eder, A. Mertens, K. Maass, H. Winter, HP. Winter, F. Aumayr: Phys. Rev. A **62**, 052901 (2000)
71. C. Auth, A. Mertens, H. Winter, A. Borisov: Phys. Rev. Lett. **81**, 4831 (1998)
72. J.C. Tucek, R.L. Champion: Surf. Sci. **382**, 137 (1997)
73. J.C. Tucek, S.G. Walton, R.L. Champion: Surf. Sci. **410**, 258 (1998)
74. W.S. Vogan, R.L. Champion, V.A. Esaulov: Surf. Sci. **538**, 211 (2003)
75. M. Ishimoto, S. Hidaka, K. Betsui, T. Shinoda: SID 99 **552**, (1999)
76. Y.T. Matulevich, P.A. Zeijlmans van Emmichoven: Phys. Rev. B **69**, 245414 (2004)
77. P. Riccardi, M. Ishimoto, P. Barone, R.A. Baragiola: Surf. Sci. **571**, L305 (2004)
78. P. Riccardi, P. Barone, A. Bonanno, A. Oliva, P. Vetrò, M. Ishimoto, R.A. Baragiola: Nucl. Instrum. and Methods Phys. Res. B **230**, 455 (2005)
79. Y.T. Matulevich, T.J. Vink, L.F. Feiner, P.A. Zeijlmans van Emmichoven: Nucl. Instrum. and Methods Phys. Res. B **193**, 632 (2002)
80. P.A. Cox, A.A. Williams: Surf. Sci. **175**, L782 (1986)
81. C. Bandis, B.B. Pate: Phys. Rev. Lett. **74**, 777 (1995)
82. V.M. Asnin, I.L. Krainsky: Appl. Phys. Lett. **73**, 3727 (1998)

Index

Springer Tracts in Modern Physics

Springer Tracts in Modern Physics

203 **Effective Field Theories in Flavour Physics**
By T. Mannel 2004. 29 figs., VIII, 175 pages

204 **Stopping of Heavy Ions**
By P. Sigmund 2004. 43 figs., XIV, 157 pages

205 **Three-Dimensional X-Ray Diffraction Microscopy**
Mapping Polycrystals and Their Dynamics
By H. Poulsen 2004. 49 figs., XI, 154 pages

206 **Ultrathin Metal Films**
Magnetic and Structural Properties
By M. Wuttig and X. Liu 2004. 234 figs., XII, 375 pages

207 **Dynamics of Spatio-Temporal Cellular Structures**
Henri Benard Centenary Review
Edited by I. Mutabazi, J.E. Wesfreid, and E. Guyon 2005. approx. 50 figs., 150 pages

208 **Nuclear Condensed Matter Physics with Synchrotron Radiation**
Basic Principles, Methodology and Applications
By R. Röhlsberger 2004. 152 figs., XVI, 318 pages

209 **Infrared Ellipsometry on Semiconductor Layer Structures**
Phonons, Plasmons, and Polaritons
By M. Schubert 2004. 77 figs., XI, 193 pages

210 **Cosmology**
By D.-E. Liebscher 2005. Approx. 100 figs., 300 pages

211 **Evaluating Feynman Integrals**
By V.A. Smirnov 2004. 48 figs., IX, 247 pages

213 **Parametric X-ray Radiation in Crystals**
By V.G. Baryshevsky, I.D. Feranchuk, and A.P. Ulyanenkov 2006. 63 figs., IX, 172 pages

214 **Unconventional Superconductors**
Experimental Investigation of the Order-Parameter Symmetry
By G. Goll 2006. 67 figs., XII, 172 pages

215 **Control Theory in Physics and other Fields of Science**
Concepts, Tools, and Applications
By M. Schulz 2006. 46 figs., X, 294 pages

216 **Theory of the Muon Anomalous Magnetic Moment**
By K. Melnikov, A. Vainshtein 2006. 33 figs., XII, 176 pages

217 **The Flow Equation Approach to Many-Particle Systems**
By S. Kehrein 2006. 24 figs., XII, 170 pages

219 **Inelastic Light Scattering of Semiconductor Nanostructures**
By C. Schüller 2007. 105 figs., XII, 178 pages

220 **Precision Electroweak Physics at Electron-Positron Colliders**
By S. Roth 2007. 107 figs., X, 174 pages

221 **Free Surface Flows under Compensated Gravity Conditions**
By M. Dreyer 2007. 128 figs., X, 272 pages

222 **Theory of Light Hydrogenic Bound States**
By M.I. Eides, H. Grotch, and V.A. Shelyuto 2007. 108 figs., XVI, 260 pages

223 **Electrical Resistivity of Thin Metal Films**
By P. Wißmann, H.-U. Finzel 2007. 110 figs., VII, 150 pages

224 **Quantum Tunneling in Complex Systems**
By J. Ankerhold 2007. 62 figs., XII, 210 pages

225 **Slow Heavy-Particle Induced Electron Emission from Solid Surfaces**
By H. Winter and J. Burgdörfer 2007. 119 figs., XIV, 244 pages

MIX
Papier aus verantwortungsvollen Quellen
Paper from responsible sources
FSC® C105338

If you have any concerns about our products,
you can contact us on
ProductSafety@springernature.com

In case Publisher is established outside the EU,
the EU authorized representative is:
Springer Nature Customer Service Center GmbH
Europaplatz 3, 69115 Heidelberg, Germany

Printed by Libri Plureos GmbH
in Hamburg, Germany